AF797046

The Nature of Our Existence

HUMANKIND, THE COSMOS AND OUR REALITY

Written by

RICHARD LAWSON SINGLEY

PROMETHEUS

Front Cover: Hubble Deep Field

Copyright © 2023 by Richard Lawson Singley

ISBN: 978-1-7321844-9-7

All rights reserved. This book or any portion thereof may not be reproduced or used in any manner whatsoever without the prior written permission of Richard Lawson Singley except in the case of brief quotations embodied in critical reviews and certain other noncommercial uses permitted by copyright law.

The eBook assumes no credit or liabilities for the video content linked within. The management of videos is the responsibility of YouTube and videos may not be present due to external restrictions or control. They are linked here to support the narratives contained within. Selection of the link is at the discretion of the user and may not be available on some devices.

Cover design by Adrian Christopher Singley and Richard Lawson Singley

Published by Prometheus Educational and Business Consultants, Germantown, MD, 20876

Amazon: https://www.amazon.com/Richard-Singley/e/B07FC71R87?ref=sr_ntt_srch_lnk_fkmr0_1&qid=1563812520&sr=8-1-fkmr0

Medium.com: https://medium.com/@richardsingley

Email: rich.singley5234@gmail.com

LinkedIn: https://www.linkedin.com/in/richard-singley-a7218332

Also, by Richard Lawson Singley:

In a Nutshell, A New Perspective, Renaissance Man, Mystic Chords of Memory, Unholy Crusades, The Pursuit of Happiness, Triumvirate, Negritude and Interesting Times

DEDICATION: This book is dedicated to my mother who gave me my first telescope and opened my eyes to the heavens.

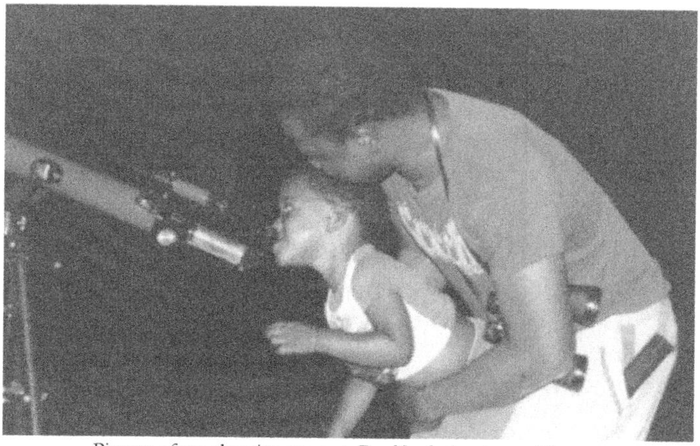

Picture of me showing my son, Bradford, the rings of Saturn

Special thanks to Glenn Yanis for inspiring me to write this book.

Author's note:

Many have asked why I include my middle name, Lawson, in everything that I write. It is to honor my father Lawson Singley. He told me once that he did not want to name any of his kids after him, but since I was born just before Father's Day, he decided to give me his name as a middle name.

Fundamental constants are finely tuned for life. A remarkable fact about our universe is that physical constants have just the right values needed to allow for complex structures, including living things. Steven Weinberg, Martin Rees, Leonard Susskind and others contend that an exotic multiverse provides a tidy explanation for this apparent coincidence: if all possible values occur in a large enough collection of universes, then viable ones for life will surely be found somewhere[1]

1 Does the Multiverse Really Exist? - Scientific American

Table of Contents

Contents

LIST OF ILLUSTRATIONS

All illustrations are in the Public Domain unless otherwise specified. Illustrations from Wikipedia Commons are in accordance with usage guidelines.

[2] By Giant Magellan Telescope - GMTO Corporation - http://www.gmto.org/Resources/Still-GMT-S21-hi-res.jpg, CC BY-SA 3.0, https://commons.wikimedia.org/w/index.php?curid=28291818

3 How Does Our Sun Compare With Other Stars? | NASA Space Place – NASA Science for Kids

4 By CactiStaccingCrane - Own work, CC BY-SA 4.0, https://commons.wikimedia.org/w/index.php?curid=117012949

5 By Luc Viatour, CC BY-SA 3.0, https://commons.wikimedia.org/w/index.php?curid=60793

6 https://exoplanets.nasa.gov/search-for-life/habitable-zone/

---

[7] By Durbed - http://durbed.deviantart.com/art/Happy-New-Year-from-Hell-Creek-345906630, CC BY-SA 3.0, https://commons.wikimedia.org/w/index.php?curid=37131266

[8] By Fama Clamosa - Own work, CC BY-SA 4.0, https://commons.wikimedia.org/w/index.php?curid=85466018

[9] ,Charles Dawson, Arthur Smith Woodward. Front row: A. S. Underwood, Arthur Keith, W. P. Pycraft, and Ray Lankester. Note the portrait of Charles Darwin on the wall. Painting by John Cooke, 1915.

[10] by W. Schnaubelt & N. Kieser (Atelier WILD LIFE ART)Homo erectus\.JPG: photographed by User:Lillyundfreya—Homo_erectus.JPG, CC BY-SA 3.0, https://commons.wikimedia.org/w/index.php?curid=9503611

[11] By Nilenbert, Nicolas Perrault III - File:Range_of_Homo_neanderthalensis.png, CC BY-SA 3.0, https://commons.wikimedia.org/w/index.php?curid=60634820

[12] By Bacon Cph - Cro-Magnons Conquered Europe but Left Neanderthals Alone" From PLOS ONE, an open-access article distributed under the terms of the Creative Commons Attribution License, which permits unrestricted use, distribution, and reproduction in any medium, provided the original author and source are credited."© 2004 Public Library of Science. https://commons.wikimedia.org/w/index.php?curid=78873692

[13] By Mysid - Own work. Self -made in Blender & Inkscape., CC BY-SA 3.0, https://commons.wikimedia.org/w/index.php?curid=45121761

[14] https://www.discovermagazine.com/the-sciences/einsteins-lost-theory-describes-a-universe-without-a-big-bang

[15] CC BY-SA 3.0, https://commons.wikimedia.org/w/index.php?curid=2974242

[16] By Original: Penubag Vector: Victor Blacus - Own work based on: Electromagnetic-Spectrum.png,

CC BY-SA 3.0, https://commons.wikimedia.org/w/index.php?curid=22428451

17 EHT Collaboration By Event Horizon Telescope, uploader cropped and converted TIF to JPG - https://www.eso.org/public/images/eso1907a/ (image link) The highest-quality image (7416x4320 pixels, TIF, 16-bit, 180 Mb), ESO Article, ESO TIF, CC BY 4.0, https://commons.wikimedia.org/w/index.php?curid=77925953

18 23.3 The Unification of Forces - Physics | OpenStax

19 By Arpad Horvath - Own work, CC BY-SA 2.5, https://commons.wikimedia.org/w/index.php?curid=637353

20 By Swiggity.Swag.YOLO.Bro - Own work, CC BY-SA 4.0, https://commons.wikimedia.org/w/index.php?curid=40660207

21 By Eduemoni - Own work, CC BY-SA 4.0, https://commons.wikimedia.org/w/index.php?curid=86585553

22 By Doctor C - Own work, CC BY-SA 4.0, https://commons.wikimedia.org/w/index.php?curid=48960438

23 By Richard Powell - The Hertzsprung Russell Diagram, CC BY-SA 2.5, https://commons.wikimedia.org/w/index.php?curid=1736396

24 By R.N. Bailey - Own work, CC BY 4.0, https://commons.wikimedia.org/w/index.php?curid=59672008

25 By User:Rursus - R. J. Hall, CC BY 2.5, https://commons.wikimedia.org/w/index.php?curid=2565862

26 By Andrew J. Hanson - Ticket#2014010910010981, CC BY-SA 3.0, https://commons.wikimedia.org/w/index.php?curid=30579072

27 Quarks (gsu.edu)

28 https://chandra.harvard.edu/photo/2000/0065/more.html

Preface

Imagination is more important than knowledge. For knowledge is limited to all we now know and understand, while imagination embraces the entire world, and all there ever will be to know and understand

Albert Einstein

FROM THE BEGINNING, HUMANS HAVE ALWAYS LOOKED TO THE STARS. It is as if they intuitively realized that stars held the key to their existence. Each star appearing as a diamond against the dark blue sky, were in fact, precious stones of our beginning calling out to us from a distant past. Some believed that by observing their movements they could tell the future. Others believed that they were there to guide us at night. Although the spiritual connection to the stars may have been the first step taken toward spiritual enlightenment, the physical link is a more recent one. To quote the renowned astronomer, Carl Sagan, "we are all made of stardust". Stardust... yes stardust – for each and every one of us contains a little piece of a star that sacrificed its existence so that we may live. Thus, the cosmic connection converges on a fallen star and science has finally found its diamond in the rough.

The history of our universe is divided into epochs in which a particular phase of development succeeds a previous stage and becomes dominant during that phase.[29] This process was repeated during the evolution stages of the earth, the age of heating and cooling of the earth, cell development, the Age of Dinosaurs, etc., all paving the way for humans to inhabit the earth. The current scientific theory of the creation of the cosmos is called: The Big Bang theory. In essence, it says that in the beginning there was this large ball of cosmic matter that exploded, triggering the start of the cosmos. From this "Cosmic Egg" the stars, planets and lifeforms were developed. Each star was originally composed of, Hydrogen, the simplest element. The source of energy of this simple star was nuclear fusion which is the conversion of four Hydrogen atoms into the next element Helium.[30] It is through such means that a star is born and that a star lives. The force of gravity that gave birth to the star is also the force that eventually destroys it. And in between its birth and death, the star must burn to stay alive by maintaining a

29 [Lederman, Schramm: From Quarks to the Cosmos, p.152]

30 Actually, some nuclear fusion took place before the formation of stars. When the Cosmos was three minutes old, 30 percent of its hydrogen was converted to Helium. [Jastrow, Robert: God and the Astronomers, p.126]

state of equilibrium between itself and gravity. Perhaps the most interesting discovery was that this process was cumulative. Based on the initial size of a star, it will continue to create more complex elements until it effectively "burns out" and the process of nuclear expansion ends.

At the start of the 20th century, much progress was made in understanding both the macro and the micro worlds that constitute our existence. We now know that we live on a planet that is just a microcosm of a much greater cosmos. We live in a galaxy of over 200 billion stars in a cosmos with as many galaxies as there are stars. We now know that the atom is not the smallest entity as the Greeks once thought, yet another microcosm with denizens unimaginable a century before. Like the angels, many of them could dance upon a pinhead, luckily for us, unlike the angels dancing on a pinhead; the dancing of atoms can be measured and moreover understood. We now know that dwellers of the microscopic world float like our planets in an infinite yet bounded universe of their own.

In this regard, it becomes a matter not of existence but of scale. Most atoms are indeed composed of nothingness in terms of mass and space. Yet together, like points on a line, they manage to form something giving new credence to the saying *Creatio ex nihilo* of nothing comes something. But what is the source of this nothingness? Are we composed of energy moving through various dimensions of which we are just beginning to scratch the surface of our understanding of their existence? If you looked down upon the earth from the heavens, with all its majestic beauty, you would think that there was peace on earth and goodwill toward men. Yet throughout the ages, on this blue dot spinning around in the heavens, it has been consumed by the wickedness and war that its inhabitants have waged against each other. The simple fact remains that like the star that we revolve around, death is part of the nature of our existence and without it, life could not exist. Yet we have the capability to conceive a timeless non-entropic world among the stars that we call paradise in which we live blissfully forever.

It has become standard to think of our three-dimensional universe as the sum of all things, when in fact the evidence suggests that what we see in terms of baryonic matter (matter composed of protons and neutrons) is just the tip of the iceberg. Discoveries during the 20th century opened a new window into our cosmic understanding hitherto unseen. What mattered in the past no longer holds true, and we now know that matter may not be the crucible that we once thought it to be; that energy and mass are two sides of the same coin, and that space and time are bonded in an unshakable marriage. Such concepts would not only have been considered bizarre in the preceding centuries, but ludicrous and even heretical. I realize that many of the ideas expressed in the essays of this book lie on the periphery of what is considered contemporary science. Perhaps, we are on the verge of breaking new barriers in science that will answer many of the questions raised by scientists a century ago. We stand on their shoulders, and from there we gaze into new worlds both large and small and observe such phenomena as black holes and quantum particles acting in ways that seem supernatural. Throughout the 20th century, the cosmos revealed itself in all its majesty and beauty to creatures that are infinitesimal within its grand scheme.

We marvel at the vastness and splendor of the cosmos and its ability to create sentient beings that seek to understand it. But do stars speak to us in a special language that until recently was unintelligible? The answer is yes. Buried deep within the twinkle of each little star are encoded messages that reveal its composition, characteristics, and position in the cosmic family. Using modern tools such as Infrared, X-ray, Gamma ray, Radio, and Optical telescopes, we are translating the cosmic messages of the stars into a universal *Consciousness.*

Although science has opened many doors for us, many doors remain locked awaiting the key to our understanding. In this book, I have emphasized the *Conservation of Energy* and the Laws of Thermodynamics and the role they play in defining our universe. Yet according to *Gödel's Incompleteness Theorem,* our laws remain incomplete. We have defined the relationships between energy and entropy, but we have not defined the process from which order is achieved within the context of the two. For example, at the local level, our solar system is a very ordered system, the earth is a very ordered system, and life is a very ordered even down to the cell level. However, what we do not know is by what process such ordered systems reached their ordered state.

To this end, there must be another *law* one perhaps based solely on information or *Consciousness.* Such a law would explain "how" and perhaps "why" we came into existence. It would be independent yet coeval with the *First and the Second Laws of Thermodynamics* and may even be the causation of those laws. For all we know, it could exist outside of the boundaries of science and beyond our dimension. Yet we must continue to search for the source of something that is eluding us that we as human beings through our *Consciousness* intuitively know of its existence but not the nature of its existence.

The annals of science are full of serendipitous discoveries and fortuitous events that stem from the zeal and curiosity of sapient beings that revolve around a star in a galaxy of billions, in a cosmos of billions of galaxies. Yet humans, in their isolation, seek to know if there are other sapient beings revolving likewise around other stars. Recently, we have discovered thousands of exoplanets (planets that orbit their star outside the solar system in habitable zones) and nature seldom makes one of anything. Despite the improbable nature of our existence, the universe appears to be fine-tuned or programmed for life. Based on the *Anthropic Principle*, the process that produced us could be repeated *ad infinitum.*

I believe that each element in this vast universe has its purpose and I seek to know our purpose within the context of the whole. I not only believe that there are other worlds within our cosmos, but that we dwell in a universe among universes within a dimension of dimensions. Nikola Tesla once said: "It is paradoxical, yet true, to say, that the more we know, the more ignorant we become in the absolute sense, for it is only through enlightenment that we become conscious of our limitations. Precisely one of the most gratifying results of intellectual evolution is the continuous opening of new and greater prospects."

I also believe that we are connected to a higher dimension through a *Consciousness* yet discovered or understood. Although the gears of the clock have since been revealed, the watchmaker remains obscure. I contend that we are limited by the laws of our cosmos and more importantly, that humankind is not the measure of all things, but only of all things that could be measured within the limits of our intellect. However, science has shown, at least at the quantum level, when we try to measure something, we change its behavior.

I have tried to use my imagination and my limited knowledge of science to propose scenarios that I believe are plausible. In it, I concede that there are many gaps and contradictions. However, truth often speaks to us in riddles or paradoxes. Like others, I seek to solve the riddle of our reality by proposing ideas that will stimulate thought and engage dialogue on issues that may lie outside of the reach of humans but within the grasp of the watchmaker. Because I believe in the words of the late Joseph Campbell: "The goal of life is to make your heartbeat match the beat of the universe, to match your nature with Nature."[31]

THE NATURE OF OUR EXISTENCE

Sir Isaac Newton once said: *If I have seen further than others it is by standing on the shoulders of giants.* At the turn of the 20th century, when the old paradigms of science gave birth to the concepts of modern realities, we did not envision two world wars, landing a person on the moon or advanced computers in the hands of Earth's denizens. Yet through time and circumstance, those events did unfold in less than one century. We are the titans of our children and their progenies who someday may live among the heavens that we so much adore. Looking down from the heavens upon this blue marble floating in the sky, they will continue to ponder our existence within a universe that seems boundless as we progress one small step for a human, one giant leap for humankind. I spent most of my adult life as an engineer, and I have never lost my love and appreciation for science. I wrote this book to help future generations understand what has brought them to a juncture that my generation, like the ones before, will never see. After all, "Space is the final frontier."

This book is comprised of three major parts. The first part consists of a general background of what we know about our place in the cosmos. The second contains a series of essays using science as a tool to propose ideas about the numerous possibilities that confront us. The last is a glossary of scientific terms and timelines that can be extremely helpful to STEM students.

Richard Lawson Singley
December 30, 2022

[31] http://www.brainyquote.com/quotes/authors/j/joseph_campbell.html#WhRcR67PtvIov2AT.99

Part I
A SCIENTIFIC OVERVIEW

Einstein's official portrait after receiving the 1921 Nobel Prize in Physics

ON THE SHOULDERS OF GIANTS

I do not feel obliged to believe that the same God who has endowed us with sense, reason, and intellect has intended us to forgo their use

Galileo Galilei, Letter to the Grand Duchess Christina

Since the early humans gazed into the heavens and imagined the stars as distant campfires, humankind has wondered if we are alone in the Universe. Are we orphans revolving around a distant sun? Thus, began our search for the nature of our existence. For thousands of years before the birth of Christ, humans have been watching the sky. In fact, at least by 3,000 BCE, the calendar that we use today was developed by the Egyptians. They developed it by using the heliacal rising of the star Sirius to determine that there were three hundred and sixty-five and one-quarter days in a year. However, the system that dominated the world and was embraced by the Catholic Church was developed by Claudius Ptolemy.

Claudius Ptolemäus, Picture from a16th century book

The Early Astronomers

Claudius Ptolemy was born in Egypt around 100 CE, and he lived much of his life in Alexandria. After spending twenty-five years studying stars in the night sky, in 150 CE, he published *The Almagest* which contained accounts of his observations. *The Almagest* included charts for over a thousand stars and forty-eight constellations. [32]. Ptolemy's astronomy assumed that the motions of the heavenly bodies could be explained with mathematics. His book established his views on cosmology, and the Ptolemaic model, which erroneously placed Earth at the center of the universe, would persist as an accepted standard for more than a thousand years.

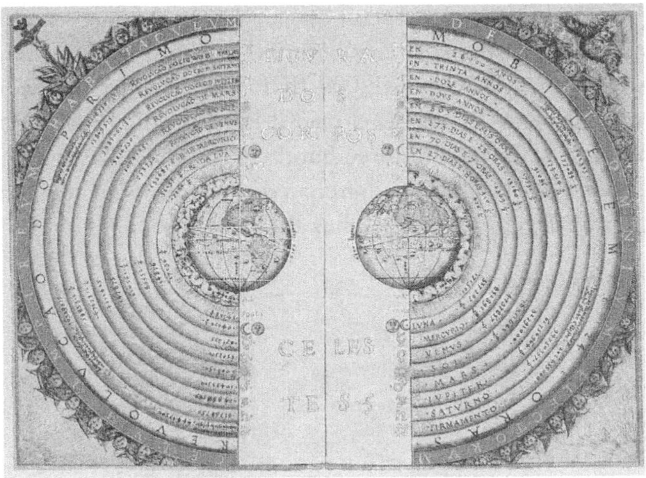

Ptolemaic Model of the Universe

At the turn of the first millennium, many Islamic astronomers began to question the Ptolemaic Model. The observations and cataloging of stars by Islamic scholars laid the foundation for European astronomers. Building on the work of medieval Islamic astronomers, at the beginning of the 16th century, Nicolaus Copernicus questioned the geocentric model of Ptolemy.[33] Copernicus suggested that

32 https://study.com/academy/lesson/ptolemy-and-the-ptolemaic-system.html

33 Muslim astronomers began questioning the framework of the Ptolemaic system of astronomy. These criticisms, however, remained within the geocentric framework and followed Ptolemy's astronomical paradigm; one historian described their work as "a reformist project intended to consolidate Ptolemaic astronomy by bringing it into line with its own principles."[15] In 1070, Abu Ubayd al-Juzjani published the Tarik al-Aflak. In his work, he indicated the so-called "equant" problem of the Ptolemaic model. Al-

his sun-centered heliocentric model could explain the apparent movement of the stars and planets better than Ptolemy's geocentric model. Although a step in the right direction, Copernicus' heliocentric model failed to fully explain the motion of the planets and was based on the planets moving in perfect circles. Because of its controversial nature, Copernicus waited to publish his book *De revolutionibus orbium coelestium* (On the Revolutions of the Celestial Spheres) until 1543, the year of his death.

Nicolaus Copernicus portrait from Town Hall in Toruń - 1580)

Johannes Kepler, an assistant of Tycho Brahe, picked up the mantle from Copernicus. An avid wealthy astronomer, Brahe had collected a lifetime of astronomical observations, which, on his death, passed into Kepler's hands.[34] Using these observations, Kepler found that the orbits of the planets followed three laws. Kepler refined Copernicus' model by stating that the planets moved in elliptical orbits rather than circular. He used simple mathematics, also developed by Islamic scholars,

Juzjani even proposed a solution for the problem. In al-Andalus, the anonymous work al-Istidrak alaBatlamyus (meaning "Recapitulation regarding Ptolemy"), included a list of objections to the Ptolemic astronomy. The most important work, however, was Al-Shuku ala Batlamyus (meaning "Doubts on Ptolemy"). In this, the author summed up the inconsistencies of the Ptolemic models. Many astronomers took up the challenge posed in this work, namely, to develop alternate models that evaded such errors. https://1library.net/article/astronomy-in-medieval-islam-islam-lifestyle-modern-world.qvg835rq

[34] Planetary Motion: The History of an Idea That Launched the Scientific Revolution (nasa.gov)

to formulate three laws of planetary motion. *Kepler's First Law* stated that planets move in elliptical paths around the Sun. He also discovered that planets move proportionally faster in their orbits when they are closer to the Sun; this became *Kepler's Second Law*. Finally, *Kepler's Third Law* explained the relationship between the distance of a planet from the Sun and the amount of time it took to orbit the Sun. Kepler's model then matched observations perfectly. These laws of celestial mechanics revolutionized astronomy.[35]

Portrait of Johannes Kepler

Although the groundwork was laid by Copernicus and Kepler, the main turning point for heliocentric acceptance was in 1610 when Galileo Galilei looked up at the sky with a telescope.[36] From the seventeenth century onward, Galileo has been seen by many as the "hero" of modern science. He made numerous discoveries but two confirmed

35 https://www.jpl.nasa.gov/news/the-history-of-johannes-kepler

36 Galileo worked on various versions of telescopes; the first version was a three-power instrument that consisted of a tube containing two lenses. The other one magnified the objects nine times. As mentioned in "Galileo at Work: His Scientific Biography", Venetian Senate was impressed by his works and offered him a position as a lecturer with a handsome salary. https://www.businessapac.com/invention-of-telescope/

the heliocentric paradigm of the solar system. He is renowned for his breakthroughs, and he was the first to report major telescopic observations of celestial bodies and phenomena. First, he discovered moons orbiting around Jupiter which proved objects could orbit objects other than Earth. Secondly, he discovered that like our Moon, Venus, being an interior planet had phases, which further confirmed the theory that Venus (and the other planets in the solar system) orbit the Sun.[37] The discovery that Venus undergoes a full sequence of phases (like the Moon), implied that Venus revolved around the Sun, and suggests that the Earth is likewise a celestial body moving around the Sun. The phases of Venus contradicted the Ptolemaic ordering of the planets but clearly supported the heliocentric model. Along with the craters and mountains on the Moon, he could identify the rings of Saturn and four moons of Jupiter.

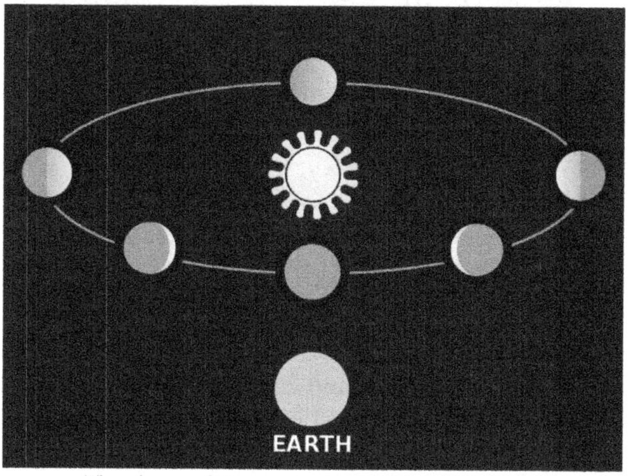

The phases of Venus, observed by Galileo in 1610

Galileo's scholarship can be divided into three parts: First, his physics, second, his astronomy, and third his methodology, which might also include his method of Biblical exegesis or his thoughts about the nature of proof or demonstration. However, when Galileo once proposed that the tides were caused by the movement of the Earth around the Sun, he was wrong. Yet he dared to question the status quo and from his courage and observations, Newton and others were able to stand on his shoulders and usher in a new era of science that continues to this day.

37 https://www.space.com/geocentric-model

Justus Sustermans - Portrait of Galileo Galilei, 1636

The Inquisition

The Ptolemaic model did not go down without a fight. Notwithstanding the overwhelming empirical evidence, on June 22, 1633, Galileo was taken to the church of Santa Maria sopra Minerva and ordered to kneel while his condemnation was read. He was declared guilty of "vehement suspicion of heresy," and made to recite and sign a formal abjuration:[38]

> I have been judged vehemently suspect of heresy, that is, of having held and believed that the sun in the center of the universe and immoveable, and that the Earth is not at the center of same, and that it does move. Wishing however, to remove from the minds of your Eminences and all faithful Christians this vehement suspicion reasonably conceived against me, I abjure with a sincere heart and unfeigned faith, I curse and detest

[38] (Quoted in Shea and Artigas 2003, 194)

the said errors and heresies, and generally all and every error, heresy, and sect contrary to the Holy Catholic Church. [39]

Cristiano Banti's 1857 painting Galileo facing the Roman Inquisition

Although Galileo's inquisition was by far the most famous example of the ignorance, hubris and inerrant authority of the Church, it was not the only one. Before Galileo, Giordano Bruno was an Italian scientist, philosopher, and former Catholic priest who espoused the Copernican idea of a heliocentric universe as opposed to the church's teachings of an Earth-centered universe. He also believed in an infinite universe with numerous inhabited worlds. Asked by the Inquisition to recant his beliefs, Bruno refused. He replied: "You may be more afraid to bring that sentence against me than I am to accept it." On February 12, 1600, the inquisitors stripped Bruno naked, bound his tongue, and burned him alive.[40]

Bruno was a luminary far beyond his time, who chose truth over fiction. In *The Ash Wednesday Supper*, he was one of the first to propose the existence of an infinite universe, which contained an infinite number of worlds similar to the Earth. In this

39 (Quoted in Shea and Artigas 2003, 194 https://plato.stanford.edu/entries/galileo/

40 In 1992, after 12 years of deliberations, the Roman Catholic Church grudgingly admitted that Galileo Galilei had been right in supporting the theories of Copernicus. The Holy Inquisition had forced an aged Galileo to recant his ideas under threat of torture in 1633. But no such admission has been made in the case of Bruno. His writings are still on the Vatican's list of forbidden texts. Giordano Bruno, philosopher and scientist, burnt at the stake 400 years ago - World Socialist Web Site (wsws.org)

regard, he was light years ahead of his contemporaries in his comprehension of the universe. In doing so, he rejected the limits of the Copernican system, which suggested a finite universe limited by a fixed sphere of stars just beyond the solar system. He argued that the Sun was not the center of the universe, saying that if the Sun were observed from any of the other stars it would appear no different from them. Bruno even speculated that the other worlds would be inhabited.[41] The Catholic Church had to defend its "monopoly on God" at a time when it was under siege from multiple directions, e.g., the Protestant Reformation. Thus, Bruno was a casualty of their ignorance and thirst for power.

The earliest depiction of Bruno is an engraving published in 1715 in Germany based on a lost contemporary portrait.

41 Giordano Bruno, philosopher and scientist, burnt at the stake 400 years ago - World Socialist Web Site (wsws.org)

The Inquisition delivered its verdict on January 20, 1600, stating:

> We hereby, in these documents ... pronounce sentence and declare the aforesaid Brother Giordano Bruno to be an impenitent and pertinacious heretic, and therefore to have incurred all the ecclesiastical censures and pains of the Holy Canon.... We ordain and command that thou must be delivered to the Secular Court ... that thou mayest be punished with the punishment deserved, though we earnestly pray that he (the Roman Governor) will mitigate the rigor of the laws concerning the pains of thy person, that thou mayest not be in danger of death or of mutilation of thy members.[42]

The Bible says: *The truth shall set you free* (John 8:31-32) and in the coming years, the inerrancy of the Church yielded to the truth.

Newton's Clockwork Universe

It was, however, Sir Isaac Newton that opened the eyes to the mechanics of the universe. Newton was able to solve not only the riddle of the tides; but how the earth and the planets move amid the heavens in a Clockwork Universe. Born on Christmas day in the year that Galileo died (old calendar, a few days short of one year after Galileo's death), he was able to put the pieces of the astronomical puzzle together. Building on the empirical observations of others, he was able to scientifically define the motion of the heavens and to explain events that were only hitherto defined by religion.

Newton's clockwork universe, as it was henceforth called, shepherded a new Scientific Revolution and a period of Enlightenment in which the existence or role of God was questioned by men such as Laplace and others. Unlike the days of Galileo, they would not be silenced by the Church as the lines between the religious and secular worlds were clearly drawn and therefore, science could operate independently of the Church in its search for the truth. In order to explain his universe, Newton invented Calculus, which became an indispensable mathematical tool to explain celestial and terrestrial phenomena.[43] The success of his research in celestial mechanics based on the *Principia*

---

42 Giordano Bruno, philosopher and scientist, burnt at the stake 400 years ago - World Socialist Web Site (wsws.org)

43 In mathematics, Newton was the first to develop a full range of algorithms for symbolically

was extraordinary. Nothing of comparable scope and accuracy had ever occurred before.

Portrait of Isaac Newton (1642-1727)

Still, Newton was not one who removed God from the equation of the universe. He believed that the force of gravity on its own would not have been sufficient to create the motions of the planets. He stated: Gravity may put the planets into motion, but without the divine power it could never put them into such a circulating motion as they have about the Sun; and therefore, for this, as well as other reasons, I am compelled to ascribe the frame of this system to an intelligent agent.[44] In other words, the clock could not exist without a clockmaker. The following illustrates Newton's influence on modern science:

determining what we now call integrals and derivatives, but he subsequently became fundamentally opposed to the idea, championed by Leibniz, of transforming mathematics into a discipline grounded in symbol manipulation.

44 (Quoted in Shea and Artigas 2003, 194) SnobelenMythofClockworkuniverse2012.pdf

A giant even among the brilliant minds that drove the Scientific Revolution, Newton is remembered as a transformative scholar, inventor and writer. He eradicated any doubts about the heliocentric model of the universe by establishing celestial mechanics, his precise methodology giving birth to what is known as the scientific method. Although his theories of space-time and gravity eventually gave way to those of Albert Einstein, his work remains the bedrock on which modern physics was built.[45]

His most famous discoveries are delineated below:

1. Newton's three laws of motion set the foundation for modern classical mechanics.
   (a) His first law: An object at rest remains at rest, and an object in motion remains in motion at a constant speed and in a straight line unless acted on by an unbalanced force.
   (b) His second law: The acceleration of an object depends on the mass of the object and the amount of force applied. `F=ma
   (c) His third law: Whenever one object exerts a force on another object, the second object exerts an equal and opposite on the first.
2. He also formulated his law of Universal Gravitation in the *Principia*, which states that every point mass attracts every single other point mass by a force pointing along the line intersecting both points. His discovery of the gravitational force validated the Copernican heliocentric model of the solar system, and Kepler's laws of motion.
3. His co-invention of calculus provided a strong mathematical tool, aiding the precise analytical treatment of the physical world. Newton invented differential calculus. Whereas Leibniz is best-known for the creation of integral calculus.[46] He also discovered the generalized binomial theorem.
4. He did groundbreaking work in optics identifying light as a particle and invented the reflecting telescope.

---

45 Isaac Newton - Facts, Biography & Laws - HISTORY

46 10 Discoveries By Newton That Changed The World | Wonders of Physics: A Blog About Physics, Astronomy and Science History

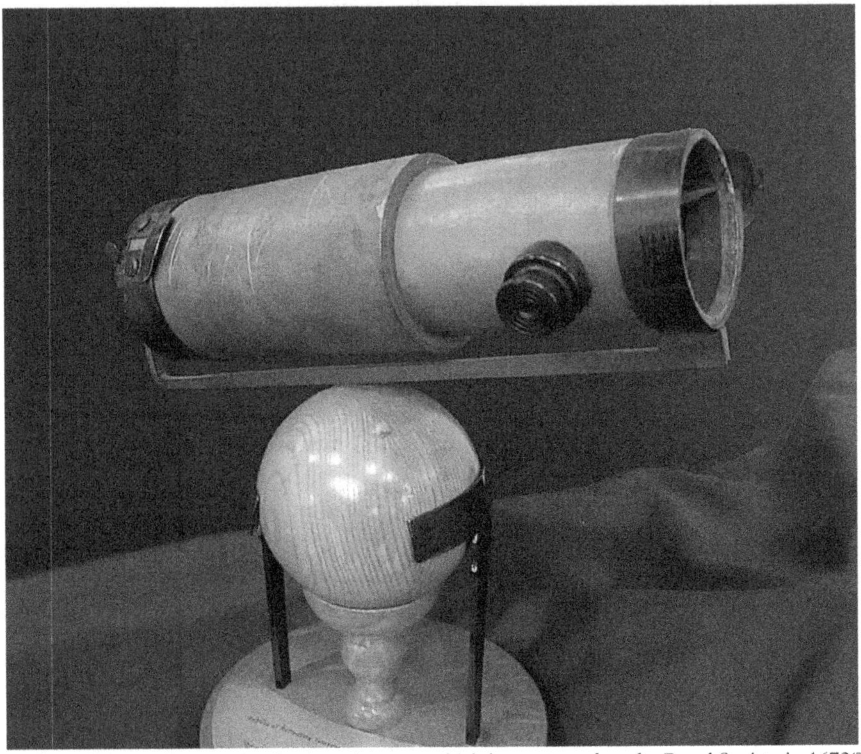

Replica of Newton's second reflecting telescope, which he presented to the Royal Society in 1672[47]

47 By User:Solipsist (Andrew Dunn) - www.andrewdunnphoto.com, CC BY-SA 2.0, https://commons.wikimedia.org/w/index.php?curid=513483

WHAT'S A HEAVEN FOR?

Beyond the range of natural vision, the telescope enables us to descry numerous objects of amazing magnitude; and, in proportion to the excellence of the instrument and the powers applied, objects still more remote in the spaces of immensity are unfolded to our view, leaving us no room to doubt that countless globes and masses of matter lie concealed in the still remoter regions of infinity, far beyond the utmost stretch of mortal vision

William Herschel

Although humans had been watching the sky for thousands of years, it was Galileo who established the telescope as the quintessential instrument to make celestial observations. Over the next several centuries, the study of the sky moved from astrology to astronomy, from religious skepticism to scientific interpretation as more advanced terrestrial and celestial telescopes were created. Time and circumstance had lifted the shroud of ignorance so that mortals could peep deep into the beauty and vastness of the heavenly abode once reserved for gods.

History of the Telescope

While the telescope may have been serendipitously developed, the history of the telescope begins with the Islamic world. During the 10th century, Ibn Sahl, a Muslim mathematician of the Islamic Golden Age, in his treatise *On Burning Mirrors and Lenses* describes how curved mirrors and lenses bend and focus light. He is credited with first discovering the law of refraction, usually called Snell's law. He used the law of refraction to derive lens shapes that focus light with no geometric aberrations.[48] Roger Bacon (1214–1292) was instrumental in introducing the knowledge gained by Islamic scholars to Europe. Other European scholars would follow in Bacon's footsteps. European educational institutions such as the University of Paris and Oxford began to emulate Islamic institutions that were in existence for hundreds of years.

In 1608, when Kepler was developing his laws of celestial motions, the first telescope was built by Hans Lippershey a German-born lens grinder and spectacle maker.

[48] Galileo Galilei: The Invention of the Telescope and Modern Astronomy (juliantrubin.com)

According to legend, Lippershey, got the idea when he saw two young children holding two lenses together, which he noticed made a weathervane in the distance appear closer. his design was able to achieve a very respectable magnification of three times the image but most important was that he had the right idea at the right time in the history of science. Hearing of Hans Lippershey's design, Galileo was able to construct a telescope with a magnification ten times greater than Lippershey's.[49] With the strength of his telescope, he could see the craters and other close-up details on the Moon.

Following in Galileo's footsteps, Sir Isaac Newton built the first reflecting telescope. His design was to use a series of mirrors instead of glass lenses that were currently in use in the older telescopes at the time.[50] Newton's design was emulated by other astronomers and as time progressed, larger and more sophisticated telescopes were constructed. William Herschel was as renowned as a maker of telescopes as he was as an astronomer, and he oversaw the construction of several hundred instruments, which were sold in England and the European continent. On March 13, 1781, he discovered Uranus with one of his telescopes. Herschel was the first to discover another planet using a telescope. On 28 August 1789, Herschel began exploring the cosmos with his forty-foot telescope, which would remain the largest in the world for the next half-century.[51] His telescope was the gateway of large terrestrial telescopes and others like Percival Lowell and George Hale would construct more sophisticated telescopes during the bourgeoning days of America.

---

49 http://www.historyoftelescope.com/

50 Newton did not do this on his own but using almost two thousand years of slowly growing experience that was gathered by Indian, Egyptian, Chinese, Greece, Roman and Arab scientists. All those ancient accounts were gathered and examined in the famous "Book of Optics" by Arab scholar Alhazen (965 – 1040 AD), which became the foundation of knowledge of optics in Europe. In the beginning, Venice and Florence started creating their own lenses that were used for spectacles in 13th century, with Germany and Netherlands coming to prominence century later. http://www.historyoftelescope.com/
Reflecting telescopes are used not only to examine the visible region of the electromagnetic spectrum but also to explore both the shorter- and longer-wavelength regions adjacent to it (i.e., the ultraviolet and the infrared). The name of this type of instrument is derived from the fact that the primary mirror reflects the light back to a focus instead of refracting it. telescope - Reflecting telescopes | Britannica

51 Kathleen Lundeen, "On Herschel's Forty-Foot Telescope, 1789" | BRANCH (branchcollective.org)
The fame of the 40-foot telescope made it a visual rallying point for discussions about religion and science. According to Richard Holmes, many prominent atheists, including Pierre-Simon Laplace and Jerome Lalande, increasingly used Herschel's papers about nebulae and the origins of stars to justify an alternative genesis story that did not require God. A Giant of Astronomy | Science History Institute

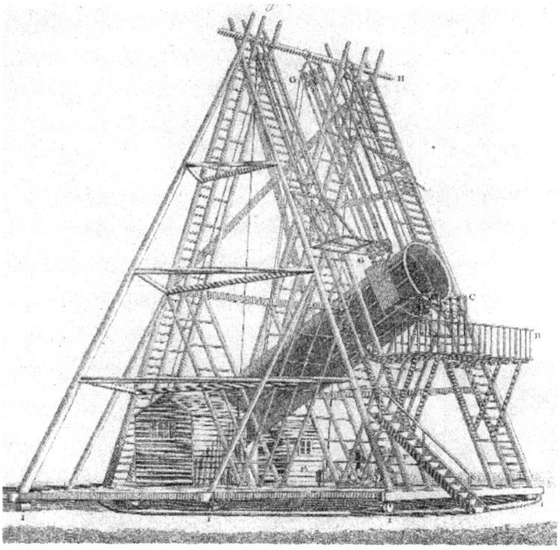

William Herschel's 49-inch (1,200 mm) "40-foot" telescope of 1789

The Mount Wilson Observatory

In 1904, the Mount Wilson Observatory was founded by George Hale under the auspices of the Carnegie Institution of Washington. Like Herschel before him, Hale always had a fascination with telescopes. Hale wanted to understand the physical processes that took place in the Sun and other more distant stars. He built what would become the world's leading astronomical research facility, and in 1919, he built the largest telescope in the world. Hale and his colleagues developed groundbreaking technologies to obtain information encoded in the light from distant astronomical objects. This small group of innovative scientists began the prolonged process of deciphering the light from distant objects with telescopes and instruments that could witness the birth, life and death of stars. Each new answer brought new questions, and each triumph brought new challenges as the universe unveiled.[52]

Hale's 100-inch telescope photographed stars and nebulae hitherto unseen[53]. But the best was yet to come. In 1919, Edwin Hubble joined the staff of Mount Wilson. Hubble was a Rhodes scholar, and it was as if Mount Wilson was built for him. He was the right man, at the right place at the right time. Hubble had two major

52 Mount Wilson Observatory | Our Story (mtwilson.edu)

53 The 100-inch telescope's reign of three decades as the world's largest ended when the 200-inch Hale telescope at Palomar Observatory saw "first light" in January 1949.

discoveries that changed the scientific view of the cosmos forever. Using a Cepheid variable star in Andromeda, he empirically proved that the universe extended beyond the Milky Way Galaxy and that the myriad spiral nebulae were galaxies outside the Milky Way. He also demonstrated that the universe is expanding and discovered the velocity-distance relationship that is fundamental to that expansion. The overwhelming success of Hubble and Mount Wilson was the key that unlocked the door to a new era in astronomy that would eventually move a telescope(named in his honor) from the terrestrial sphere to the celestial realm.

George Ellery Hale (1868 – 1938) and Edwin Hubble (1889-1953)

Space Probes and Telescopes

It has been nearly a century since Edwin Hubble discovered the vastness of our universe by peering vicariously through the terrestrial telescope at the Mount Wilson Observatory. Such a view changed not only our perception of the cosmos but set in motion a series of events that have led to a greater understanding of our beginning and the heavens that we so much adore. Over the last several decades, and through a celestial telescope named in Hubble's honor, we have unveiled some of the secrets and beauty of our cosmos. Thanks to geniuses like Albert Einstein and his scions, we now know that space and time are relativistic, that mass and energy are two sides of the same coin and that we all are the progeny of stars. Yet there is so much more to be understood. Throughout the 20th century, we learned more about the cosmos than in all the centuries hitherto as the mask of the cosmos was slowly lifted by the hands of science.

Regardless of how powerful terrestrial telescopes were, they were limited by the earth's atmosphere and clouds that obstructed them. In a space age world of rockets and satellites, scientists remove this obstruction by placing telescopes and probes above the earth. In 1977, Voyager 1 and 2 probes were launched from earth. Their mission was to explore Jupiter and Saturn—and beyond to the outer planets of our solar system. The reason the Voyager missions in the 1970s and 80s were so successful is that they took advantage of the fact that all the outer solar system planets were roughly aligned, so the two Voyager spacecraft could visit them all using gravity assists. Voyager 1 targeted Jupiter and Saturn before continuing to chart the far edges of our solar system. Voyager 2 targeted Jupiter, Saturn, Uranus and Neptune before joining its sister probe on their interstellar mission. The two spacecraft took tens of thousands of pictures of Jupiter and Saturn and their moons.

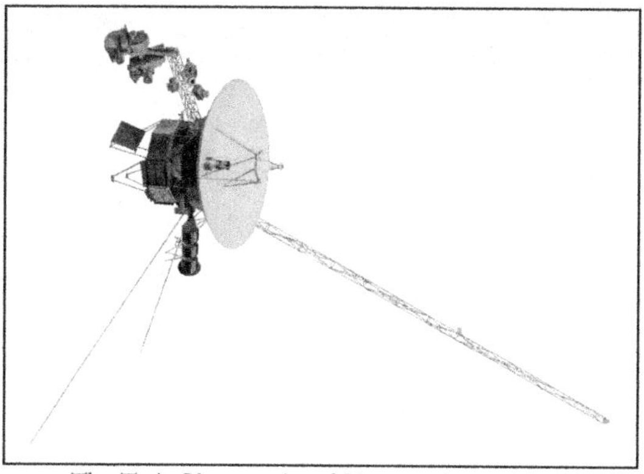

The Twin Voyager 1 and Voyager 2 spacecraft

The pictures from Voyager 1 and 2 allowed us to view many things for the first time. For example, they captured detailed photos of Jupiter's clouds and storms, and the structure of Saturn's rings. The Voyagers have enough fuel and power to operate at least until 2025 and beyond. Some time after this, they will not be able to communicate with Earth any longer. Unless something stops them, they will continue to travel on and on, passing other stars after many thousands of years.[54]

54 Each Voyager spacecraft also carries a message. Both spacecraft carry a golden record with scenes and

The Hubble Space Telescope (HST) was launched from the Space Shuttle Discovery and deployed in April 1990. The first major optical telescope to be placed in space, it commemorated the most significant advancement in astronomy since Galileo's telescope. Although Hubble was initially hampered by a defectively shaped main mirror, however, during a 1993 repair mission, Shuttle astronauts installed COSTAR, a device that resolved the problem, and scientists were finally able to make full use of the telescope's incredible capabilities.[55] Hubble orbits about 350 miles (560 kilometers) above Earth's surface. Unlike terrestrial telescopes, it operates far above the distortion of Earth's atmosphere, clouds and light pollution, and therefore has an unobstructed view of the universe. Hubble completes 15 orbits per day, approximately one every 95 minutes.[56]

Hubble Space Telescope

In addition to the aforementioned space telescope and probes, NASA's Great Observatory Program planned to explore the universe with large space telescopes, the Compton Gamma-Ray Observatory, the Chandra X-Ray Observatory,

sounds from Earth. The records also contain music and greetings in different languages. So, if intelligent life ever finds these spacecraft, they may learn something about Earth and us as well! Voyager 1 and 2: The Interstellar Mission | NASA Space Place – NASA Science for Kids

55 I was one of the engineers that worked on COSTAR. COSTAR's long box holds a clever device: tiny nickel-size mirrors that extended on short arms into the light path inside the telescope. These mirrors were precisely shaped to cancel the flaw in the large main mirror, and precisely positioned to send corrected, focused light into three instruments. A new fourth instrument, a wide-field camera, came with its own built-in corrective optics. COSTAR was an ingenious solution to the unexpected challenge of correcting the Hubble's distorted vision, https://airandspace.si.edu/hubble-space-telescope

56 In Depth | Hubble Space Telescope – NASA Solar System Exploration

and Spitzer Space Telescope, each specialized in viewing the universe in different wavelengths of light. By combining data from different kinds of telescopes, scientists can paint a fuller picture of our universe. [57] On April 5, 1991, the Compton Gamma-ray Observatory (CGRO) was launched aboard the Space Shuttle Atlantis. The purpose of the CGRO mission was to use its collection of four primary instruments to detect the broad range of high energy radiation known as gamma-rays.[58] The first goal of CGRO was to map for the first time the entire sky over a broad range of gamma-ray energies. As the most energetic form of light, gamma rays reveal information about some of the most energetic phenomena in the universe, including gamma-ray bursts, supernovae, galactic structure and dynamics, pulsars, quasars, and black holes.[59]

NASA's Compton Gamma Ray Observatory drifts away from the space shuttle Atlantis on April 7, 1991, following its deployment during the STS-37 mission.

The Chandra X-ray Observatory was launched on July 23, 1999, into low Earth orbit by the Space Shuttle Columbia. The observatory was deployed from the shuttle's cargo bay at 155 miles above the Earth. It allowed scientists to obtain unprecedented X-ray images of these and other exotic environments to help understand the structure and evolution of the universe. The observatory not only helped to probe these mysteries but also served as a unique tool to study detailed physics in a laboratory that cannot be replicated here on earth[60]

---

57 10 Things: Spitzer Space Telescope – NASA Solar System Exploration

58 Launch of the Compton Gamma-Ray Observatory | NASA

59 Looking Back: The Legacy of the Compton Gamma Ray Observatory | NASA Blueshift

60 https://www.nasa.gov/centers/marshall/news/background/facts/axafnew.html

Artist illustration of the Chandra X-ray Observatory.

The Spitzer Space Telescope was launched on August 25, 2003. It was designed to view the universe in Infrared. Infrared wavelengths of light, which primarily come from heat radiation, are too long to be seen with human eyes, but are important for exploring space specifically when it comes to getting information about something extremely far away.

NASA's Spitzer Space Telescope. Artist's conception of the Spitzer Space Telescope.

In February 2017, NASA announced that Spitzer had revealed the first known system of seven Earth-size planets around a single star. Three of the planets are firmly located in the habitable zone, the area around the parent star where a rocky planet is most likely to have liquid water. At the time, Spitzer set a record for the greatest number of habitable-zone planets found around a single star outside our solar system.[61]

Cassini along with its Huygens probe was launched on October 15, 1997. The Cassini/Huygens mission to Saturn was a collaboration between NASA and the European Space Agency (ESA). Cassini was designed by NASA and the Huygens probe by the ESA. Although the goal was to study Saturn and its moons for six months, in 2000, the spacecraft studied Jupiter before reaching Saturn in 2004. It conducted a string of flybys of Saturn's moons and on Christmas day, Cassini deployed the Huygens probe to Saturn's largest moon, Titan, which landed on January 4, 2005. Huygens performed a study of Titan's atmosphere and surface composition while Cassini made the first observations of a complete seasonal period of Saturn and its moons, flew between Saturn's rings, and descended into the planet's atmosphere.[62]

Cassini telescope.: Original Caption Released with Image: This is an artist's concept of Cassini during the Saturn Orbit

The Kepler Space Telescope was launched on March 7, 2009, into an initial Earth orbit at 115 × 115 miles (185 × 185 kilometers). Kepler's mission was to search for Earth-like planets exoplanets, orbiting other stars. It was equipped to look for planets with size spans from one-half to twice the size of Earth (terrestrial planets) in the habitable zone of their stars where liquid water might be present in the natural state on the surface of the planet.[63]. After nine years in deep space collecting data that

61 In Depth | Spitzer Space Telescope – NASA Solar System Exploration

62 Titan - The atmosphere | Britannica

uncovered our night sky to be packed with billions of unknown planets (more planets even than stars), Kepler was retired. It leaves a legacy of more than 2,600 planet discoveries from outside our solar system, many of which could be promising places for life.[64]

Illustration of the Kepler Space Telescope

The Transiting Exoplanet Survey Satellite (TESS) is an MIT-led NASA mission. It continues Kepler's exploration of planets outside of our solar system that could support life. It is an all-sky survey for transiting exoplanets. Transiting planets are those that go in front of the star as seen from the telescope and, to date, is the most successful discovery technique for finding small exoplanets. The stars that TESS studied are 30 to 100 times brighter than those of Kepler. TESS will create an inventory of thousands of exoplanet candidates using this transit photometry method. This will reveal whether the planets are rocky (like Earth), gas giants (like Jupiter) or

63 https://solarsystem.nasa.gov/missions/kepler/in-depth/

64 Kepler's legacy: discoveries and more – Exoplanet Exploration: Planets Beyond our Solar System (nasa.gov)

something even more unusual. TESS is just one more tool in the arsenal of a new generation of telescopes ushering in a new era of space exploration capable of helping us understand our role in the cosmos. The James Webb and Giant Magellan telescopes will work with TESS and allow scientists to study the atmospheres of many of these planets.[65]

The fully integrated Transiting Exoplanet Survey Satellite (TESS), which was launched in 2018 to find thousands of new planets orbiting other stars

The James Webb telescope, also called JWST or Webb, was launched on December 25, 2021, and builds on the scientific achievements and lessons learned from the Hubble Space Telescope (HST). Webb is the largest and most powerful space telescope ever built. It will allow scientists to look at what our universe was like about 200 million years after the Big Bang. The HST forced us to look to longer wavelengths and to "go beyond" what it had achieved. HST's spectacular deep- field photograph helped to explain the vastness of our universe. From it, we learned that more distant objects are more highly redshifted, and their light is pushed from the UV and optical into the near infrared. Thus, observations of very distant objects (like the first galaxies formed in the universe) require an infrared telescope. Webb, therefore, will primarily look at the universe in the infrared, while Hubble studied it primarily at optical and ultraviolet wavelengths (although it has some infrared capability).

65 https://tess.mit.edu/
https://www.nasa.gov/content/about-tess
https://exoplanets.nasa.gov/resources/2190/animation-kepler-passes-the-torch-to-tess/

Webb also has a much bigger mirror and light gathering capability than Hubble which means that it can look farther back in time. Even though Hubble's view exceeds that of any terrestrial telescope, its orbit is around the earth. However, Webb will be 1.5 million kilometers (km) away at the second Lagrange (L2) point.[66] Webb's orbit is more than four times farther away from Earth than the Moon. Furthermore, with significantly more infrared capability, Webb can document the light of older galaxies and stars that the Hubble might not even register with its technology. Therefore, the James Webb telescope allows it to search further into deep space, theoretically back to galaxies that formed close to the Big Bang.

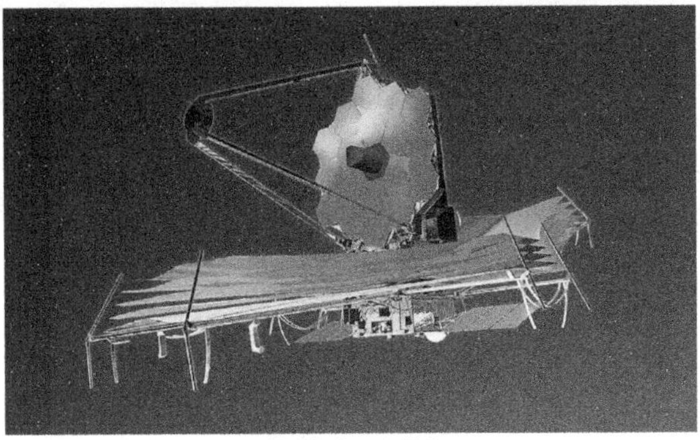

James Webb Space Telescope

66 Comparison: Webb vs Hubble Telescope - Webb/NASA
L2 is one of the so-called Lagrangian points, discovered by mathematician Joseph Louis Lagrange. Lagrangian points are locations in space where gravitational forces and the orbital motion of a body balance each other. Therefore, they can be used by spacecraft to 'hover'. L2 is located 1.5 million kilometres directly 'behind' the Earth as viewed from the Sun. It is about four times further away from the Earth than the Moon ever gets and orbits the Sun at the same rate as the Earth. ESA - L2, the second Lagrangian Point

The James Webb Space Telescope Deep Field each dot is another galaxy with billions and billions of stars. Webb's image is approximately the size of a grain of sand held at arm's length, a tiny sliver of the vast universe.[67]Each dot is not another star but another galaxy with billions of stars like our sun

The Giant Magellan Telescope (GMT) is a terrestrial telescope more advanced than traditional terrestrial telescopes. Furthermore, the ground-based follow-up will be able to determine the planets' compositions. The GMT can accommodate up to ten instruments at once, more than any other telescope. Each one has a unique ability to explore space: from analyzing a distant planet's atmosphere in search of life to looking back in time to when the Universe first formed. The GMT will employ seven of the world's largest mirrors to see farther into deep space than ever before. Its unique design will produce the highest possible resolution of the universe over the widest field of view.

67 https://webbtelescope.org/contents/news-releases/2022/news-2022-035?Collection=First%20Images Hubble vs. James Webb telescope images: See the difference - ABC News (go.com)

The astonishing image clarity of the GMT will enable scientists around the world to obtain new clues to the fundamental nature and evolution of the universe. The GMT will overlap many of the activities currently in process, e.g., from searching for signs of life on distant exoplanets, to investigating the cosmic origins of chemical elements providing new pieces to the cosmic puzzle.[68] It has 50 million times the light gathering power of the human eye and will be up to 200 times more powerful than existing research telescopes and 10 times more powerful than Hubble. The GMT is being built in the mountains of Chile's Atacama Desert, one of the best locations on Earth to explore the heavens.[69] First light is scheduled for 2029.

Artist's concept of the completed Giant Magellan Telescope which will be situated in the Atacama Desert some 115 km (71 mi) north-northeast of La Serena, Chile.[70]

In the coming years, sentient denizens of this blue sphere will continue to build more powerful telescopes in search of our beginning. A search, at times, that appears boundless, a journey that seems endless and as remote as the stars that we gaze upon. Nonetheless, in less than 500 years, we have come a long way from Galileo's telescope to the James Webb, to beyond the terrestrial limits and imagination of those visionaries of yesteryear.

68 https://giantmagellan.org/explore-the-design/

69 Giant Magellan Telescope

70 By Giant Magellan Telescope - GMTO Corporation - http://www.gmto.org/Resources/Still-GMT-S21-hi-res.jpg, CC BY-SA 3.0, https://commons.wikimedia.org/w/index.php?curid=28291818

THE COSMIC NEIGHBORHOOD

Our solar system consists of our star, the Sun, and everything bound to it by gravity – the planets Mercury, Venus, Earth, Mars, Jupiter, Saturn, Uranus, and Neptune; dwarf planets such as Pluto; dozens of moons; and millions of asteroids, comets, and meteoroids. Beyond our own solar system, we have discovered thousands of planetary systems orbiting other stars in the Milky Way[71]

Humankind has indeed come a long way since Galileo first turned his telescope to the sky to observe the celestial bodies in our cosmic neighborhood. During the 20th century, our knowledge of our place in the universe increased exponentially. Our universe is so vast that it must be measured in light years, A light-year is a measurement of distance and not time (as the name might imply). It is the distance that a beam of light travels in a single Earth year, which equates to approximately 6 trillion miles, that is a 6 with 12 zeros behind it, or 9.7 trillion kilometers. The galaxy GN-z11 is thought to be the farthest detectable galaxy from Earth at 13.4 billion light-years away.[72] Therefore, when we use powerful telescopes to look at distant objects in space, we are actually looking back in time because light travels at a constant speed of 186,000 miles (or 300,000 km) per second. Although this may seem very fast, objects in space are so far away that it takes a long time for their light to reach us. Therefore, the farther an object is, the farther in the past we see it.[73]

The Milky Way Galaxy

We -live on a planet that revolves around an ordinary star in an average galaxy called the Milky Way. Our galaxy is 100 light-years wide, and our closest galactic neighbor is Andromeda, also known as M31, is 2.5 million light-years away. Andromeda can be thought of as the big sister to our Milky Way galaxy.[74]. Our Sun is

71 Overview | Our Solar System – NASA Solar System Exploration

72 What Is a light-year? | Space

73 What Is a Light-Year? | NASA Space Place – NASA Science for Kids

74 Andromeda is the nearest large galaxy to the Milky Way, and a new study shows that our neighbor has consumed other galaxies in two distinct epochs

one of the two hundred billion stars in the Milky Way galaxy. Moreover, there are as many galaxies in our universe as there are stars in the Milky Way

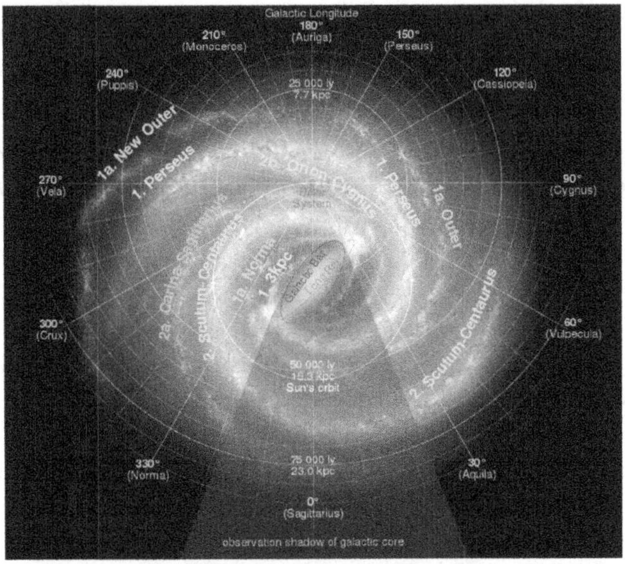

The Milky Way Galaxy and our Sun

A Galaxy Evolution Explorer image of the Andromeda Galaxy.

Our universe contains trillions upon trillions of stars. In fact, there are more stars than there are grains of sand on the Earth and more planets than there are stars. Moreover, we are just beginning to scratch the surface of the enormity of the universe. When gazing upon the heavens, in the larger sense, there is nothing special about our Sun or the rocks and gases that swirl around it. At best, we can only be used as a yardstick to measure others.

However, unlike ours, many stars, revolve around a binary or multiple star system. A binary star is a star system that is made up of two stars that orbit a common center of mass because of gravitational attraction. The brighter and larger star is usually called the primary and the other one the companion star.[75] A binary star system is different than a double star, which occurs when two stars appear very close together when observed from Earth but are actually very distant. Additionally, a binary star system occurs when two stars are trapped in a barycenter in which mass is most concentrated. Binary and multiple star systems produce about four-fifths of the light seen in the sky.[76]

The nearest *binary* star system is Sirius, which is also our brightest star, it consists of a Blue Giant Star (Sirius A) also known as the Dog Star and a White Dwarf (Sirius B) also known as the Pup. Sirius A is a blue-white star 25.4 times as luminous as the Sun. It has a radius 1.71 times that of the Sun and a surface temperature of 9,940 kelvins (K), which is more than 4,000 K higher than that of the Sun. Its name comes from a Greek word meaning "sparkling" or "scorching. Its distance from our solar system is 8.6 light-years, only twice the distance of the Alpha Centauri system the nearest known star system.

Around 120 million years ago, Sirius B was a large white star, larger and brighter than Sirius A and five times more massive than our Sun. However, it has since passed through the red giant phase and can no longer conduct nuclear fusion. Today, its luminosity is about 10,000 times less than Sirius A's.[77] All of the stars that we see in the sky belong to the Milky Way. The Milky Way and Andromeda galaxies are members of the Local Group, a cluster of more than 50 galaxies spread across a spherical volume with a diameter of 10 million light years.[78]

---

75 What Is The Binary Star System – Describing The Binary Star System (planetfacts.org)

76 What Is A Binary Star? - WorldAtlas

77 EarthSky | Sirius B: How to see Sirius' companion

78 The Milky Way and Andromeda galaxies are members of the Local Group, a cluster of more than 50 - Brainly.com

An artist's impression of Sirius A and Sirius B, A – bigger, B – smaller white dwarf (Credit:NASA)

Stars with more than two stars, such as Alpha Centauri, are not binary star systems. Alpha Centauri, the closest star system to the Earth, has three stars in the system, *Rigil Kentaurus, Toliman and Proxima Centauri.* The dimmest, *Proxima Centauri*, is the nearest star to the solar system. Alpha Centauri is 4.35 light years away from us, two are close and could be classed as a binary star system with the third a little further out as shown below.

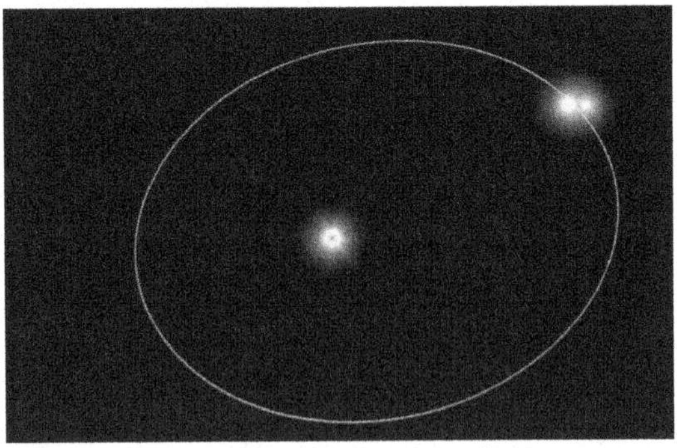

The orbit of HD 188753 BC around A, trinary stat system

Our Sun is an average size star. It is 864,000 miles in diameter and 10,000 degrees Fahrenheit on the surface. There are bigger stars, and there are smaller stars within the Milky Way. Some stars are one hundred times bigger in diameter than our Sun and others are just one-tenth the size of our Sun.[79] Notwithstanding, one million earths can fit in the Sun.

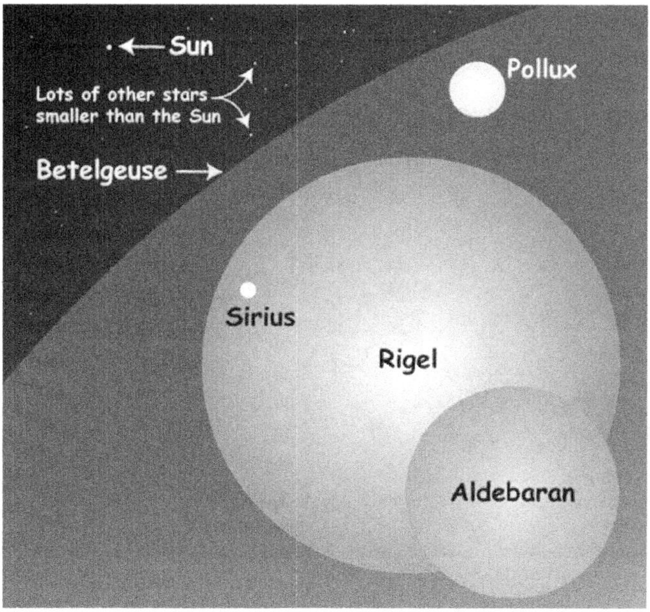

Relative sizes of stars[80]

Our Solar System

Our solar system began to form about 4.6 billion years ago within a concentration of interstellar dust and hydrogen gas called a molecular cloud. Like all stars, the cloud shrunk under its own gravity and our protostar formed in the hot dense center. The Sun, like other stars, was born from a cloud of gas and dust called the "solar nebula" and its remnants made planets. As the cloud collapsed, it began to spin around. The remainder of the cloud formed a swirling disk called the solar nebula.[81] Our solar system consists of our star, the Sun, and everything bound to it by the Sun's gravity.

79 How Does Our Sun Compare With Other Stars? | NASA Space Place – NASA Science for Kids

80 How Does Our Sun Compare With Other Stars? | NASA Space Place – NASA Science for Kids

81 Formation of the Solar System: Birth of Worlds (nasa.gov)

The planets Mercury, Venus, Earth, Mars, Jupiter, Saturn, Uranus, and Neptune; dwarf planets such as Pluto; dozens of moons; and millions of asteroids, comets, and meteoroids, revolves around this mediocre star at the center of our solar system.[82] The Sun revolves clockwise around the center of the galaxy, but the planets of the solar system rotate counterclockwise around the Sun.

The solar system, with sizes of the Sun and planets and moons to scale. The small terrestrial planets are on the right, the gas and ice giants are on the left.[83]

In ancient times, only six planets were known: Mercury, Venus, Earth, Mars, Jupiter and Saturn. In 1781, Uranus was discovered by William Herschel and Neptune was discovered by John Couch Adams in 1846.[84] In recent times, however, we have discovered thousands of planetary systems orbiting other stars in the Milky Way galaxy and beyond.[85] Planets, asteroids, and comets have elliptical orbits around

82 Pluto had been officially listed among the planets since it was discovered in 1930 orbiting beyond Neptune, but in 1992 an icy object was discovered still farther from the Sun than Pluto. Many other such discoveries followed, including an object named Eris that appears to be at least as large as Pluto. It became apparent that Pluto was simply one of the larger members of this new group of objects, collectively known as the Kuiper belt. Accordingly, in August 2006 the International Astronomical Union (IAU), the organization charged by the scientific community with classifying astronomical objects, voted to revoke Pluto's planetary status and place it under a new classification called dwarf planet. solar system | Definition, Planets, Diagram, Videos, & Facts | Britannica

83 By CactiStaccingCrane - Own work, CC BY-SA 4.0, https://commons.wikimedia.org/w/index.php?curid=117012949

84 https://starchild.gsfc.nasa.gov/docs/StarChild/questions/question16.html

our Sun. It takes the Earth one year to go around the Sun, however, Mercury revolves around the Sun in only 88 days. All the planets and dwarf planets, the rocky asteroids, and the icy bodies in the Kuiper belt move around the Sun in the same direction that the Sun rotates. This motion is termed prograde, or direct, motion. Most of the planets rotate on their axes in a counterclockwise motion similar to their orbital revolution. Most of the moons and asteroids also rotate in a counterclockwise direction, with six moons moving in a retrograde or east-to-west movement. Venus is one exception to the counterclockwise rotation of the other planets. This odd clockwise rotation makes the Venus day longer than its year. Uranus is the other exception to the counterclockwise rotation of planets. However, Uranus lies at a 90-degree angle to the north/south poles, causing it to revolve on its side instead of its axis. Furthermore, Uranus is the only planet named after a Greek and not a Roman god.

Uranus in 2005. Rings, southern collar and a bright cloud in the northern hemisphere are visible (HST ACS image).

The rocky, terrestrial planets: Mercury, Venus, Earth, and Mars all formed in the inner, hotter part of our solar system. It was so hot that volatile materials (materials that evaporate easily at normal temperatures) and pressures could not condense.

---

85 While our own inner solar system has four planets, Kepler found systems with considerably more planets — up to eight — orbiting close to their parent stars. The existence of these compact systems raises questions about how solar systems form: Are these planets "born" close to their parent star, or do they form farther out and migrate in? Kepler's legacy: discoveries and more – Exoplanet Exploration: Planets Beyond our Solar System (nasa.gov)

The terrestrial planets of the Solar System: Mercury, Venus, Earth and Mars, sized to scale

In our solar system, our outer four gas giants are called "Jovian planets," named after Jupiter, the Roman king of all gods. They are Jupiter, Saturn, Uranus, and Neptune. To qualify as a gas giant planet, it must be larger and made up of mostly gases, such as helium and hydrogen, and have a small core. The cores of all four Jovian planets are made of some combination of rock, metal and hydrogen compounds. The difference between the Jovian interiors remains in the layers that surround the core.

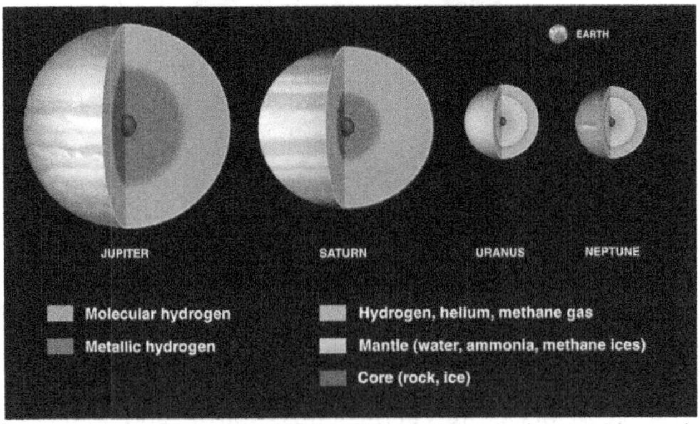

These cut-aways illustrate interior models of the giant planets. Jupiter is shown with a rocky core overlaid by a deep layer of metallic hydrogen.

Planet – Rotation Period – Revolution Period – Rotation speed at the equator – Mean orbital velocity around the Sun[86]

Mercury – 58.6 days – 87.97 days – 10.83 km/h (6.73 mph) – 47.36 km/s (29.43 mi/s)	**Venus** – 243 days – 224.7 days – 6.52 km/h (4.5 mph) – 35.02 km/s (21.76 mi/s) retrograde	**Earth** – 1 day – 365.26 days – 1,674 km/h (1,040 mph) – 29.78 km/s (18.5 mi/s)	**Mars** – 1.03 days – 1.88 years – 866 km/h (538.1 mph) – 24 km/s (14.91 mi/s)	**Jupiter** – 0.41 days – 11.86 years – 45,583 km/h (28,324 mph) – 13.07 km/s (8.51 mi/s)
Saturn – 0.45 days – 29.46 years – 36,840 km/h (22,891.3 mph) – 9.68 km/s (6 mi/s)	**Uranus** – 0.72 days – 84.01 years – 14,794 km/h (9,193 mph) – 6.8 km/s (4.225 mi/s) retrograde	**Neptune** – 0.67 days – 164.79 years – 9,719 km/h (6,039 mph) – 5.43 km/s (3.374 mi/s)	**Pluto** (dwarf planet) – 6.39 days – 248.59 years – 123.21 km/h (76.56 mph) – 4.743 km/s (2.947 mi/s)	

Jupiter is the largest planet in the solar system and is more than 300 times the mass of Earth. For instance, you could fit almost 1,300 Earth's inside Jupiter. Like our Sun, Jupiter is a giant sphere of mostly hydrogen and helium. It has a red spot, which is a powerful 400mph swirling storm larger than Earth itself. Furthermore, Jupiter is 2.5 times more massive than all the other planets in our solar system. Jupiter's rotation is the fastest of all the solar system's planets, completing a single rotation on its axis in slightly less than ten hours (9 hours, 55 minutes and 30 seconds). Therefore, a single Jovian year lasts 10,475.8 Jovian solar days. Jupiter has 53 named moons and another 26 awaiting official names, according to NASA. Among them are the four Galilean moons: Io, Europa, Ganymede, and Callisto.[87] Ganymede is the largest moon in the solar system. Mercury and Venus do not have any moons and Saturn has more moons than Jupiter.

86 https://ourplnt.com/relative-rotation-speeds-of-the-planets/

87 https://www.newsweek.com/facts-jupiter-moons-io-europa-ganymede-callisto-nasa-1659174

Number of Potential Moons Per Planet

Saturn	83
Jupiter	80
Uranus	27
Neptune	14
Mars	2
Earth	1

Jupiter with its red spot its moon Europa on the left. Earth's diameter is 11 times smaller than Jupiter's, and 4 times larger than Europa's.

The Galilean moons. From left to right, in order of increasing distance from Jupiter: Io; Europa; Ganymede; Callisto. Ganymede is bigger than the planet Mercury, Pluto and Saturn's largest moon, Titan It is only called a moon because it orbits Jupiter. If it weren't orbiting Jupiter, it would be a planet.

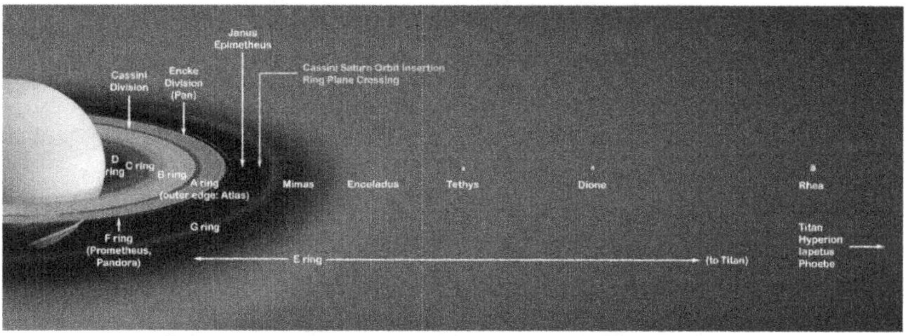

Saturn, its rings and major icy moons—from Mimas to Rhea

Although Saturn's rings are the best known, all the gaseous planets in our solar system have a ring system. Thus, four planets in our system that have rings: Jupiter, Saturn, Uranus and Neptune. From Earth, the rings look like a solid sheet of material, but they are actually made up of billions of particles of rock, ice and dust.

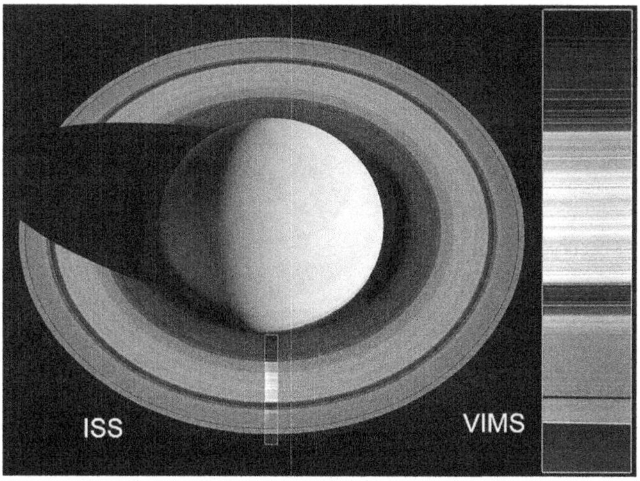

Saturn and its A, B and C rings in visible and (inset) infrared light. In the false-color IR view, greater water ice content and larger grain size lead to blue-green color, while greater non-ice content and smaller grain size yield a reddish hue.

The particles range in size from minuscule and microscopic to the size of houses and cars. Recent research and measurements taken by NASA's Cassini spacecraft have revealed that the ring material, which is mostly water ice, is being gradually pulled into Saturn.[88] Eventually, there will be almost nothing left. [89]

Uranus is also one of the planets in the solar system that has its own ring system. In fact, it is the most important ring system after Saturn. However, it is a ring system that goes much more unnoticed due to both the greater distance from the Sun and the composition of the rings, which makes them extremely narrow. Currently, a total of 13 rings are known on Uranus, which are named after different Greek letters and, of all of them, the most important is the epsilon (ε) ring.[90]

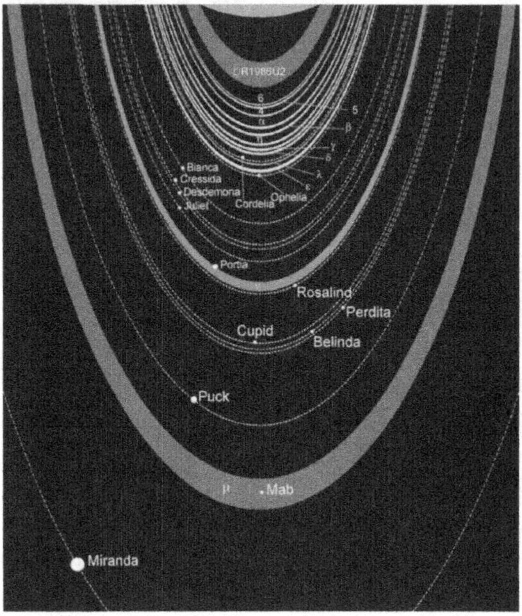

The scheme of Uranus's ring-moon system. Solid lines denote rings; dashed lines denote orbits of moons.

88 It's possible the Saturnian rings formed during or around the time dinosaurs ruled planet Earth. For a long time, scientists thought Saturn's rings were simply leftover material from the planet's birth 4.6 billion years ago but recent measurements from Cassini have challenged this assumption. How do planets get rings? | The Planetary Society

89 How do planets get rings? | The Planetary Society

90 https://agrocorrn.com/planets-with-rings-in-the-solar-system/

Neptune is slightly smaller than Uranus in diameter, which makes it the smallest of the Jovian planets, but it is the third most massive after Jupiter and Saturn. Its similar size and blueish color make it and Uranus appear as cosmic twins at the end of our solar system. The eighth and outermost planet from the Sun, Neptune's orbit is almost perfectly circular and one day on Neptune takes about 16 hours. However, a year takes about 165 Earth years (60,190 Earth days)![91] Like all Jovian planets, Neptune has several moons and rings (at least five main rings and four prominent ring arcs). Its biggest moon, Triton, is the only large moon in the solar system that circles its planet in a direction opposite to the planet's rotation (a retrograde orbit). Before Pluto was downgraded to a dwarf planet, Neptune was considered the penultimate planet in our solar system. Although fascinating, none of our cosmic neighbors can support life as we know it on earth.

Voyager 2 Narrow Angle Camera image of Neptune taken on August 20, 1989

91 Neptune | Planet, Moons, Rings, Temperature, Mass, Diameter, & Facts | Britannica

OUR BLUE PLANET

Earth hasn't always looked like the blue orb we know so well. The variety of contending creatures that have come and gone over billions of years, in a sense, paints a picture of the many planets [92]

It has been said that God speaks to us through our imagination and intuition. According to scripture, *in the beginning, God created the heavens and the earth. Now the earth was formless and empty, darkness was over the surface of the deep, and the Spirit of God was hovering over the waters.* (Genesis 1,2)." It was indeed our imagination that moved us away from superstition and the Ptolemaic view into our current understanding. Humankind and the planet we dwell on are not absolute, but rather infinitesimal specs of a much larger whole, and perhaps it is only our hubris and lack of understanding that makes us think that we are the cynosure of God's eye of Providence.

In the Beginnings

The history of Earth and the life that it gave birth to is an arduous and complicated one. Notwithstanding, there are some definite milestones that have shaped our planet. It all starts with the Hadean Eon. The Hadean Eon lasted between.4.6 billion and 4.0 billion years ago. 4.6 billion years ago, our blue planet was almost unrecognizable. Earth's surface was incredibly unstable during the early part of the Hadean Eon. Convection currents brought molten rock to the surface and caused the cooling rock to descend into a sea of magma. The temperature was hot with lava flowing. It suffered from higher incidences of impacts from asteroids and comets that repeatedly bombarded Earth. Earth's magnetic field was just developing, and it started deflecting solar winds making our planet more habitable.

Heavier elements, such as iron, fell to become the core, whereas lighter elements, such as silicon, rose to become part of the growing crust. Although no one knows exactly when the first outer crust of the planet formed, some scientists believe that the existence of a few grains of zircon dated to about 4.4 billion years ago confirms the presence of land, liquid water, and surface temperatures that were probably less than

100 °C (212 °F). As for the chemical makeup of the earth's crust, there are around 12 elements that make up almost 99% of the earth's crust. Oxygen is the most abundant (47%) of these elements, with silicon (27%) taking second place. Some of the other major constituents of the earth's crust are aluminum, iron, calcium, sodium, potassium, and magnesium.[93] The most common minerals found in soil that support plant growth are Phosphorus, Potassium and, Nitrogen. Other, less common minerals include Calcium, Magnesium, and Sulfur.

Scientists think that an object the size of Mars, Theia, hit Earth at a tremendous speed. It grazed Earth thus creating the Moon, and Earth's gravity was able to put the Moon into its current orbit. The present Pacific Ocean basin is the most prevalent spot for the part of the Earth from which the Moon came.[94] This happened around the start of the Hadean Eon. However, during the Archean Eon, Earth finally started to cool down with a more stable climate. Because it was cool enough, water could finally condense to form its first oceans and continents.

Earth's hydrosphere. Water covers 71% of Earth's surface, with the global ocean accounting for 97.3% of the water distribution on Earth.

Our blue planet is just one of eight planets in our solar system. It is the third terrestrial planet from the Sun. Like most planets, when viewed from the celestial north pole, the motion of Earth and its axial rotation appear counterclockwise. From the vantage point above the north poles of both the Sun and Earth, Earth orbits the

93 What is the earth's crust made of? (euresisjournal.org)

94 Some specular that when this object hit the earth that it brought life with it.

Sun in a counterclockwise direction. Its axis is tilted also 23.4° towards the ecliptic of the Sun, which is responsible for producing seasonal variations on the planet's surface.

In addition to producing temperature variations, this also resulted in variations in the amount of sunlight a hemisphere receives during the year. In early January, the Earth is closest to the Sun, and the northern hemisphere is tilted away from the Sun and the southern hemisphere is tilted towards the Sun. This is called the perihelion and the earth is 91,399,453 miles or 147,093,162 million kilometers from the Sun. Likewise, there is a point during Earth's annual orbit around the Sun when it reaches its farthest point and is called the aphelion. This happens around July 5, and the Earth is 94,510,889 miles or 152,100,533 kilometers from the Sun.[95]

Even though Earth has only one moon, it is unique. Although the Sun's diameter is about four hundred times larger than our Moon, however, the Sun is also about 400 times farther away. So, the Sun and the Moon *appear* nearly the same size as seen from Earth. This phenomenon does not occur with any other planets and moons in our immediate neighborhood of space. This accounts for solar eclipses as the Moon comes between the Earth and the Sun. There are four types of solar eclipses depending on how the Sun, Moon and Earth are aligned at the time of the event. They are as follows: total solar eclipse, partial solar eclipse, annular solar eclipse, and hybrid solar eclipse. For a solar eclipse to take place, the Sun, the Moon, and Earth must be aligned in a perfect or near perfect straight line. A rough alignment of the three bodies happens every lunar month, at the New Moon. A solar eclipse always occurs about two weeks before or after a lunar eclipse.[96] The maximum duration for a total solar eclipse is only 7 ½ minutes.

95 Due to variations in the eccentricity of the Earth's orbit, the dates when the Earth reaches its perihelion or aphelion are not fixed. In 1246, the December Solstice was on the same day as the Earth reached its perihelion. Since then, the perihelion and aphelion dates have drifted by a day every 58 years. In the short-term, the dates can vary up to two days from one year to another. The words perihelion and aphelion come from ancient Greek, where peri means close, apo means far, and helios means the Sun?https://www.timeanddate.com/astronomy/perihelion-aphelion-solstice.html

96 Solar eclipses 2023: When, where & how to see them | Space

From Earth, the Moon and the Sun appear the same size, as seen in the 1999 solar eclipse. [97]

Because the Moon rotates on its own axis at the same rate that it orbits around the Earth, we always see the same side of the Moon from our position on Earth. For example, during a Crescent Moon, the part of the Moon that faces Earth is mostly in shadow, and the far side of the Moon is mostly sunlit. The observable crescent is the only part of the lunar nearside that is experiencing daytime. The Moon has almost no atmosphere and only about one-sixth of Earth's gravity. It also causes tides, creating a rhythm that has guided humans for thousands of years.[98]

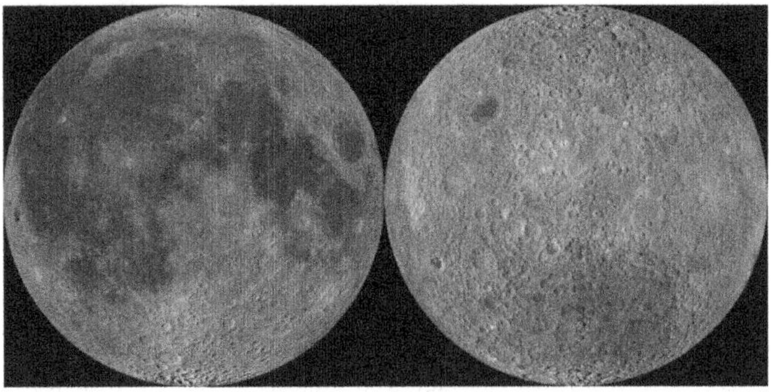

Near and far side of the Moon

97 By Luc Viatour, CC BY-SA 3.0, https://commons.wikimedia.org/w/index.php?curid=60793

98 Overview | Earth's Moon – NASA Solar System Exploration

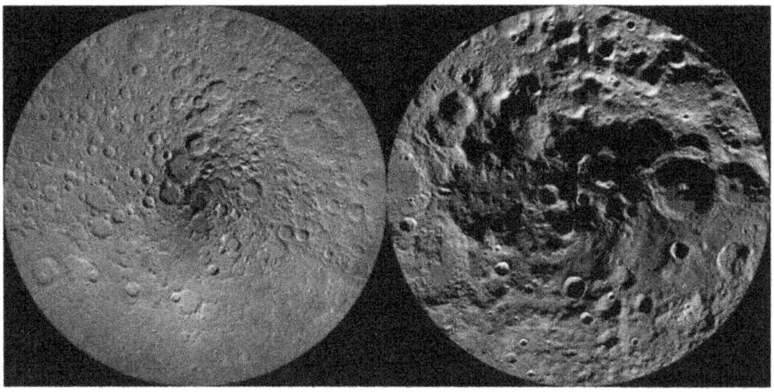

North and south poles of the Moon

Twice each month, the Moon lines up with the Earth and Sun and are called new and full moons. During a New Moon, when the Moon is between the Earth and the Sun, it is in the Sun's shadow and appears dark. Likewise, a Full Moon reflects sunlight and is when the Earth is between the Sun and the Moon. When the Sun, Moon and Earth are aligned, the Sun's tidal force works with the moon's tidal force. The combined pull can cause the highest and lowest tides, called spring tides. Spring tides happen whenever there is a new moon or a full moon and have nothing to do with the season of spring. In the period between the two spring tides, the Moon faces the Earth at a right angle to the Sun. When this happens, the pull of the Sun and the Moon are weak. This causes tides that are lower than usual. These tides are known as neap tides.[99]

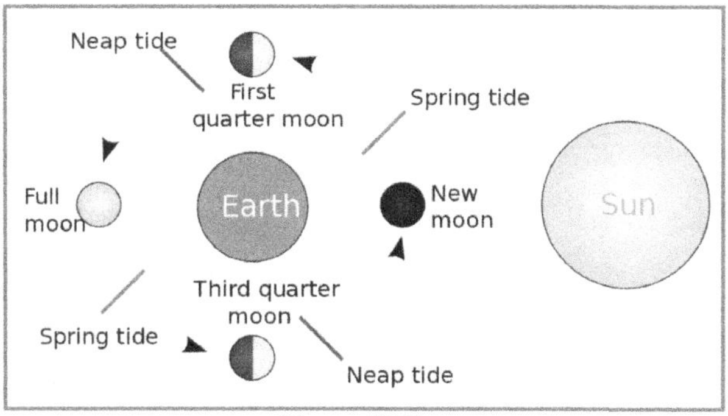

Types of Tides

99 Cause and Effect: Tides | National Geographic Society

The Earth also lies within the "habitable zone" or the distance from a star at which liquid water could exist on orbiting planets' surfaces. Habitable zones are also known as Goldilocks' zones, where conditions might be just right – neither too hot nor too cold to support life. First speculated in 1953, in recent times, numerous exoplanets have been discovered within habitable zones. Not all biologists and astronomers agree that a planet must exist in a habitable zone to support life or even intelligent life. However, it appears that one in five stars has an Earth-sized planet in a circumstellar habitable zone. Therefore, it could be as many as 40 billion habitable zone planets in the Milky Way Galaxy.[100] Eleven billion of these estimated exoplanets may be orbiting sun-like stars. According to scientists, the nearest such planet may be 12 light-years away.[101]

Habitable zone Image credit: NASA[102]

A perfect example of an exoplanet solar system is the TRAPPIST-1 system located only 40 light-years away. Scientists think that TRAPPIST-1 is between 5.4 and 9.8

100 https://fscj.pressbooks.pub/introductionastronomy/chapter/the-circumstellar-habitable-zone/
Given that the Milky Way has about 100 billion stars, this would suggest at least 17 billion Earth-sized worlds in our galaxy. (In November 2013, this number was revised up to 40 billion) In Depth | Kepler – NASA Solar System Exploration

101 Habitable Zones – Planetary Sciences, Inc. (planetary-science.org)
Kepler has shown us our galaxy is teeming with terrestrial-size worlds; the most recent analysis of Kepler's discoveries concludes that 20 to 50 percent of the stars in the sky are likely to have small, possibly rocky planets similar in size to Earth within the habitable zone of their parent stars, where water could pool on the planet surface. We still have much to learn about whether any of them could host life Kepler's

102 https://exoplanets.nasa.gov/search-for-life/habitable-zone/

billion years old. This is up to twice as old as our own solar system, which formed some 4.5 billion years ago.[103] It contains a total of seven planets revolving a red star around the size of Earth. Three of them, TRAPPIST-1 e, f and g, dwell in their star's habitable zone.[104] This system of seven rocky worlds, some with the potential for water on their surface, is a breathtaking discovery in the search for life on other worlds. There is the likelihood that future findings could uncover environments suitable for life. As previously suggested, the discovery proved that we are not the only planet or solar system in the universe.

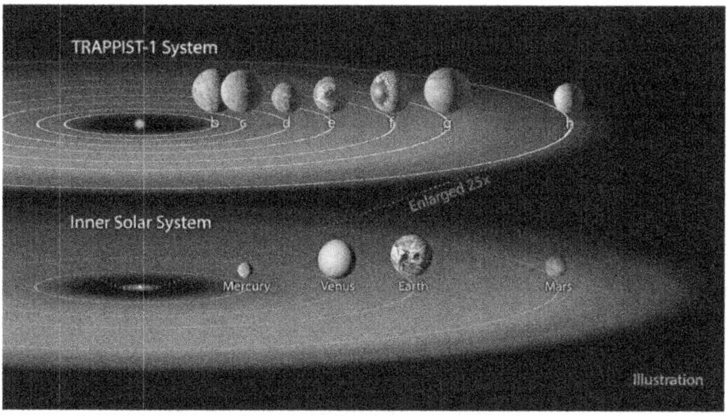

The illustrations shown for the various planets depict a range of possible scenarios of what they could look like when compared to our solar system

Early Life on Earth

The Precambrian period occurred 4.5 billion to 542 million years ago. During this time, the first areas of continental crust appeared and began coalescing into larger landmasses. Collisions with objects from space and volcanic activity decreased. During this period, continents formed, and our modern atmosphere developed, while early life evolved and flourished. Life first appeared on Earth during the Archean Eon.[105] The earliest fossil evidence of life on earth dates to 3.7 billion years ago. They were found preserved in ancient rock samples from Greenland. The fossils were once

103 10 Things: All About TRAPPIST-1 – NASA Solar System Exploration

104 For more information about the Spitzer mission, visit http://www.nasa.gov/spitzer and http://spitzer.caltech.edu

105 The Archean Eon began about 4 billion years ago with the formation of Earth's crust and extended to the start of the Proterozoic Eon 2.5 billion years ago; the latter is the second formal division of Precambrian time https://www.britannica.com/science/Archean-Eon

part of a now-extinct seabed and thought to be the remains of ancient microbes. For more than a billion-years, life on earth existed in primal unicell forms. All life was single-celled organisms, made up mainly of bacteria and algae.[106]. Common Precambrian fossils include stromatolites and similar structures, which are traces of algae-like microorganisms, and microfossils of other microorganisms.

In the absence of Oxygen, microscopic cyanobacteria (named after their blue/green color) converted sunlight to energy by conducting photosynthesis in the oceans and metabolizing their own food. Cyanobacteria (aerobic organisms) were the first to start exhaling Oxygen into the atmosphere. They are responsible for the oxygenation of the atmosphere and oceans since the Great Oxidation Event (GOE), a geological episode occurring around 2.35 billion years ago. The first complex form of multicellular life formed around 600 million years ago. Nevertheless, there are still many unicellular species on the planet, including bacteria, plankton and amoeba, as well as all protists (which are eukaryotes), and some fungi.[107] The first steps into multicellularity can be explained by the tendency towards the colonization of single-celled organisms in colonies of bacteria, mold and fungi forming group clusters. Over time, these clusters became accustomed to "group living", and were no longer purely autonomous, instead relying on other parts of the group for survival and function. From this point onwards, evolution could only move in one direction—towards greater cell complexity.[108]

A prime example of this complexity is the evolution of eukaryotic cells that contain membrane-bound structures. They form the basis for every multicellular organism, including animals, plants, and humans as well as some unicellular organisms. All living organisms can be broadly divided into two groups — prokaryotes (before nuclei). and eukaryotes (good or true nuclei) and are distinguished by the relative complexity of their cells. Prokaryotes are basic single-celled organisms, such as bacteria and archaea.[109]

106 The Evolution and Complete Timeline of Life on Earth (humanoriginproject.com)

107 Eukaryotic cells are the cells that contain a membrane bound nucleus and organelles. Eukaryotes evolved during the Proterozoic eon approximately 1.6 BYA. Prior to the origin of eukaryotes, all life on Earth was prokaryotic (lacking nucleus or other membrane-bound organelles). The leading hypothesis, called the endosymbiotic theory, is that eukaryotes arose as a result of a fusion of Archaean cells with bacteria, where an ancient Archaean engulfed (but did not eat) an ancient, aerobic bacterial cell. The engulfed (endosymbiosed) bacterial cell remained within the Archaean cell in what may have been a mutualistic relationship.

108 First Multicellular Organisms: When Did Multicellular Organisms Appear? (scienceabc.com)

109 What is an amoeba? | Live Science

Eukaryotic cells have many chromosomes which undergo meiosis and mitosis during cell division, while most prokaryotic cells consist of just one circular chromosome.[110] Eukaryotic cells evolved from prokaryotic cells about 2 billion years ago. The primary difference between prokaryotic and eukaryotic cells is that a nucleus and other membrane-bound organelles are only present in eukaryotic cells. Furthermore, eukaryotic cells are more than 100 to 10,000 times larger than their prokaryotic companions. [111]

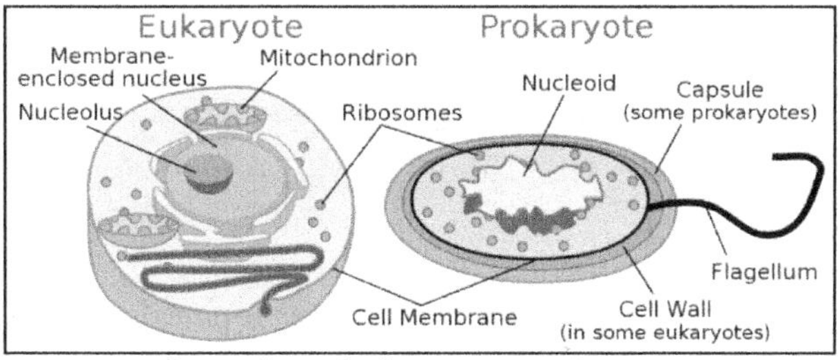

Comparison of eukaryotes vs. prokaryotes

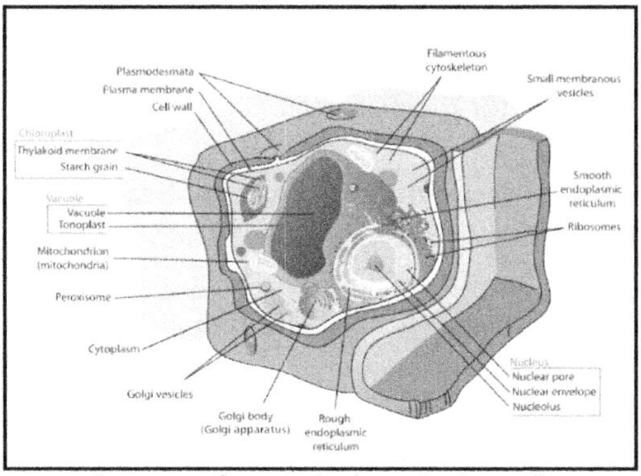

Structure of a typical eukaryotic plant cell

110 The difference between prokaryotic and eukaryotic cells? | Live Science Meiosis is a type of cell division that results in the formation of four daughter cells each with half the number of chromosomes as the parent cell. Mitosis is the type of cell division that results in the formation of two daughter cells each with the same number and kind of chromosomes as the parent cell. Mitosis is how new body cells are produced, whereas meiosis is used to produce gametes (i.e., sperm and egg cells). https://byjus.com/biology/mitosis-and-meiosis/

111 Eukaryotic Cell: What Is It, Difference from Prokaryotic Cells, and More | Osmosis

Amoebas are eukaryotes. They are amorphous and appear as jelly-like blobs as they move about. Amoebas move by changing their shape, exhibiting a unique type of crawling motion that has come to be known as amoeboid movement. Like other eukaryotes, their single cells possess certain characteristic features: For example, their cellular contents are enclosed within a cell membrane, and their DNA is packaged in the nucleus. Structurally, amoebas closely resemble the cells of higher organisms. They are like our cells, and when they are moving, they look very much like white blood cells.[112] Amoebas live in salt water and freshwater aquatic environments, wet soils, and some parasitic amoebas may inhabit animals and humans. Amoebas commonly reproduce by binary fission, a process in which the cell divides into two identical cells.[113] Like animals, amoeba take in Oxygen and during metabolic reactions, Oxygen is converted to Carbon Dioxide. The Carbon Dioxide is also released through the same process of diffusion.[114]

The growth of eukaryotic multicellular organisms opened two separate paths of evolutions: plants that could make their own food and animals that had to extract their food from other living organisms. Thus, amoebas were the first organisms that preyed on other forms of life, to take in Oxygen and release Carbon Dioxide and to form a symbiotic relationship, although primitive, between two distinct forms of life both dependent on the other for their survival.

The increase in Oxygen levels resulting from cyanobacteria helped to stimulate the development of multicellular organisms.[115] The irony of cyanobacteria is that the Oxygen they released as a waste product was toxic to them. Cyanobacteria did not completely vanish in the Proterozoic Eon. By hiding in low Oxygen environments, cyanobacteria avoided their mass extinction.[116]

112 The male and the female gametes i.e., the sperm and the egg or ova are also single celled. Is the sperm or an egg is a single cell or multicellular? (toppr.com)

113 Amoeba Anatomy, Digestion, and Reproduction (thoughtco.com)

114 Respiration in Amoeba | Fun Science

115 The appearance of free oxygen in the Earth's atmosphere led to the Great Oxidation Event. This was triggered by cyanobacteria producing the oxygen which developed into multicellular forms as early as 2.3 billion years ago. As evolutionary biologists have shown, this multicellularity was linked to the rise in oxygen and thus played a significant role for life on Earth as it is today. https://www.sciencedaily.com/releases/2013/01/130117084856.htm

116 Proterozoic Eon: Eukaryotes to Multicellular Life (2500 to 541 million years ago) - Earth How

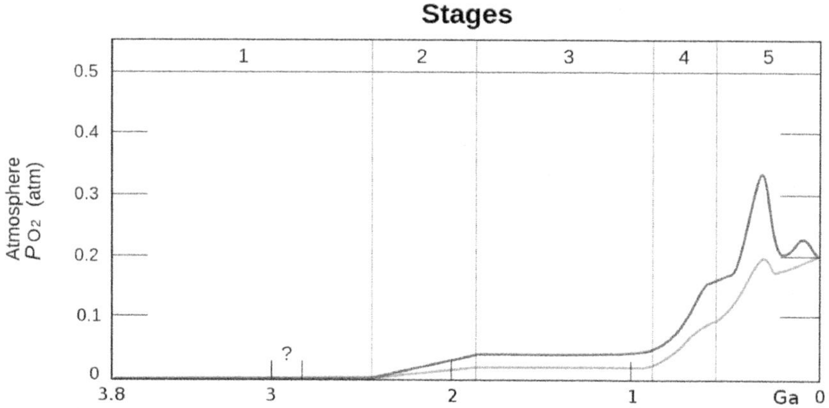

O2 build-up in the Earth's atmosphere. Red and green lines represent the range of the estimates while time is measured in billions of years ago (Ga). Stage 1 (3.85–2.45 Ga): Practically no O2 in the atmosphere. The oceans were also largely anoxic - except for O2 in the shallow oceans. Stage 2 (2.45–1.85 Ga): O2 produced, rising to values of 0.02 and 0.04 atm, but absorbed in oceans and seabed rock. Stage 3 (1.85–0.85 Ga): O2 starts to gas out of the oceans but is absorbed by land surfaces. No notable change in oxygen level. Stages 4 and 5 (0.85 Ga – present): Other O2 reservoirs filled; gas accumulates in the atmosphere.[117]

The following explains the impact of an Oxygen rich atmosphere:

> The rise in oxygen concentration allowed for the flourishing of aerobic (those using oxygen to make energy through cellular metabolism), organisms, which also supported multicellular life and greater complexity. Earlier in Earth's history, despite widespread unicellular life forms, these were largely anaerobic, including all the Archaea domain, the earliest form of life. When the atmosphere shifted to an oxygen-rich one, anaerobic organisms were forced to retreat to anaerobic environments, while aerobic organisms and thus multicellular organisms, were free to spread across the planet. With the exception of three microscopic marine animals, all multicellular life is aerobic in nature.[118]

[117] By Oxygenation-atm.svg: Heinrich D. Hollandderivative work: Loudubewe (talk) - Oxygenation-atm.svg, CC BY-SA 3.0, https://commons.wikimedia.org/w/index.php?curid=12776502

[118] First Multicellular Organisms: When Did Multicellular Organisms Appear? (scienceabc.com) An anaerobic organism or anaerobe is any organism that does not require molecular oxygen for growth. It may react negatively or even die if free oxygen is present.

Photosynthesis, first conducted by aerobic organisms, is the most important process of life and without it all forms of life could not exists. It is the chemical process by which plants, algae, and some bacteria use the energy from sunlight to transform Carbon Dioxide from the atmosphere, and water, into organic compounds such as sugars. Photosynthesis occurs in eukaryotic and prokaryotic cell structures called chloroplasts. A chloroplast is a type of plant cell organelle known as a plastid. Plastids assist in storing and harvesting needed substances for energy production. A chloroplast contains a green pigment called chlorophyll, which absorbs light energy for photosynthesis. Hence, the name chloroplast indicates that these structures are chlorophyll-containing plastids.[119]

Plants use sunlight, water, and the gases in the air to make glucose, which is a form of sugar that plants need to survive. Plants are the only form of life that can make its own food or better yet, trap and store the energy from the Sun. Scientists believe that "Algae borrowed this photosynthetic system from cyanobacteria, and plants are just a group of algae that took photosynthesis on land, so we think with this finding we're looking at the inception of the molecular machinery that would give rise to oxygen." [120] Energy flows through the biosphere as herbivores eat plants to retrieve the energy from the Sun. Chlorophyll.[121] Plants absorb CO_2 from the surrounding air and release water and Oxygen via microscopic pores on their leaves. The photosynthesis equation is: [122]

$$\mathbf{6CO_2 + 12H_2O + Light\ Energy \rightarrow C_6H_{12}O_6 + 6O_2 + 6H_2O}$$

[119] Chloroplast Function in Photosynthesis (thoughtco.com)

[120] Life on Earth before the rise of oxygen - Futurity

[121] Chloroplast, structure within the cells of plants and green algae that is the site of photosynthesis, the process by which light energy is converted to chemical energy, resulting in the production of oxygen and energy-rich organic compounds. Photosynthetic cyanobacteria are free-living close relatives of chloroplasts; endosymbiotic theory posits that chloroplasts and mitochondria (energy-producing organelles in eukaryotic cells) are descended from such organisms. Chloroplast | Definition, Function, Structure, Location, & Diagram | Britannica

[122] What is photosynthesis? | Live Science

The leaf is the primary site of photosynthesis in plants. Plant chloroplasts are commonly found in guard cells located in plant leaves. Guard cells surround tiny pores called stomata, opening and closing them to allow for gas exchange required for photosynthesis.[123]

The first animals, including the common ancestor of all animals today, evolved in the sea over half a billion years ago during the early Cambrian period. The creatures had bodies built from multiple cells with specialized roles the natural evolution of unicellular groups coalescing and creating symbiotic relationships. Earth's environment, as it has always been, was in flux during the Cambrian period, and the Ediacaran period that came before it.

A major rise in sea level washed calcium, phosphate and other life-essential chemicals into the ocean.[124] This served as a catalyst for new forms of life hitherto unseen. Animals were able to build skeletons and shells for the first time. As a result, evolution in the underwater world, expanded rapidly. Life became much more complex, and the oceans were filled with a stunning level of diversity and precursors of what would become denizens of land. Hidden in the genetic code of modern animals are signs of when the major animal groups first arose. Vertebrates (animals with a backbone) also developed during the Cambrian Explosion, a significant change from the largely jellylike, invertebrates (animals lacking a backbone).

Before the Cambrian Explosion, very few animals were mobile. During this period, however, animals evolved such essential parts as shells, skeletons, gills, eyes, jaws, teeth, legs, and even central nervous systems and brains.[125] This gave rise to an

123 Chloroplast Function in Photosynthesis (thoughtco.com)
124 First Animals (ox.ac.uk)

enhanced system of predators and prey and the associated offensive and defensive adaptations required to survive which is a fundamental part of all animal life (survival of the fittest). It was also at this time that a greater symbiotic relationship between plants and animals developed. Some animals(herbivores) ate plants and other animals(carnivores) ate the animals that ate the plants and others (omnivores) ate both. Moreover, animals also began to exhibit complex and even social behaviors such as building early reefs, cohabitation, and competition.

During the Early Cambrian period, most of the biosphere was confined to the sea.[126] However toward the end, the stage was set for the migration from sea to land, and the Earth would never be the same. The massive step in evolutionary terms began with environmental, both on land and in the sea, for plants and trees to develop on land and for animal life to follow. From a geological perspective, the Cambrian explosion lasted roughly 40 million years, which is an extremely short time given the exponential growth and diversity of life. For example, unicellular life appeared on Earth approximately 3.5 billion years ago and it took more than a billion years before multicellular life, the catalyst to such growth, to emerge.

The Silurian period (443 to 419 million years ago) followed the Cambrian Explosion. During this and the following Devonian Period (419 to 359 million years ago) known as the "Age of the Fishes," marine organisms flourished and diversified. It also witnessed the first clear evidence of life on land. The oldest fossilized tree `remains date to 385 million years ago releasing more Oxygen into the atmosphere.[127] Primitive plant life began to flourish on land, and insects were the first animals to come out of the water.[128] Around 400 million years ago, the first four-legged animals started to develop in the sea. These animals, known as tetrapods (meaning having four limbs), ` are the ancestors of all land dwellers including birds, mammals, reptiles and amphibians. The first amphibians appeared soon after, living on sea surfaces and in shallow marine ecologies. They were the first species to branch off from sea tetrapods.[129]

---

125 The Cambrian Explosion Timeline & Cause | What Is the Cambrian Explosion? | Study.com

126 The Cambrian Explosion Timeline & Cause | What Is the Cambrian Explosion? | Study.com

127 https://humanoriginproject.com/how-did-earth-evolve-the-story-of-earth/

128 The earliest plants were algae living in the oceans more than 700 million years ago. Modern-day plants evolved from these aquatic algae that did not have stems or roots. Plant evolution on land has continued throughout the geologic eras of time.). A Timeline of Plant Evolution (kremp.com)

129 The Evolution and Complete Timeline of Life on Earth (humanoriginproject.com)

Eryops is an extinct amphibian that lived approximately 295-270 million years ago.

The Oxygen released by plants and trees as waste products was yet another catalyst for animals to emerge and thrive on land. When an animal breathes, it takes in Oxygen from the atmosphere and releases Carbon Dioxide back into the atmosphere which is used by plants. Thus, the symbiotic relationship between plants and animals was duplicated on land as animals continued to evolve in the sea. By the beginning of the Devonian Period, animals with jaws and real bones were swimming the in water and during the Carboniferous period that followed, (359 to 299 million years ago), animals and plants, both on land and in the sea adapted to changes in their environment. For example, plants adapted to complex forest environments, and vertebrates underwent extensive evolution. Amphibians, land and sea dwellers, became widespread and diverse, and reptiles appeared for the first time and rapidly adapted to many land habitats. Such collective and extensive changes helped to prepare for "the Age of the Dinosaurs" that would last for 165 million years!

Time Periods and Events

	TIME PERIOD	EVENT/ACTIVITY
1	**The Hadean Eon** 4.6 to 4.0 billion years ago	The Hadean is the first geologic eon of Earth and lies before the Archean. It began with the formation of the Earth about 4600 million years ago and ended as defined by the ICS 4,000 million years ago. During the Hadean Eon, we obtain a heat source, magnetic field, and formation of a moon. At ~4.5 bya, young earth collided with planetoid, ejecting large mass of debris, which was held in orbit by earth's gravity and eventually formed the Moon. Because of the hot temperatures, any water on the planet would have evaporated into space.
2	**The Archean Eon** 4 to 2.5 billion years ago	Oldest known fossil came from this eon -Started out as simple single celled organisms, more advanced forms later evolved, including cyanobacteria. -Cyanobacteria were photosynthetic and produce enormous amounts of Oxygen gases. -These bacteria formed structures called stromatolites, which look like lumpy stones, but when you cut them in half you can see the layers that were created as secreted sticky films that trapped particles of sediment. Because Earth cooled down, it was able to support oceans and continents. -Oxygen filled the oceans from cyanobacteria. Interestingly, most of the world's Iron ore deposits were produced in this eon from banded Iron formation. Eventually, the ozone layer forms. This layer of protection was essential for life to form. In this eon, Earth's atmosphere was mostly Methane and Nitrogen. The only life forms that could exist were anaerobic cyanobacteria

		(blue-green algae).
3	**The Proterozoic Eon** 2.5 billion to 541 million years ago	Without oxygen, only anaerobic life existed in the Proterozoic Eon. -At approx. 1.7 billion years ago, the first multi-cellular organisms evolved. evidence that eukaryotes and multicellular organisms appeared on Earth. Another key event in the Proterozoic Eon was that Earth completely froze over. Again, Oxygen played a key role in forming a "Snowball Earth" or "Slushball Earth".
4	The Ediacaran Period is an interval of geological time ranging 635 to 541 million years ago	Within the Proterozoic Eon, the time span from 630 to 542 million years ago known as the Ediacaran period. Within this period large forms of life with some harder parts evolved, including the first animal life. It was a time of immense geological and biological change and records the transition from a planet largely dominated by microscopic organisms, to a Cambrian world swarming with animals.
5	**The Phanerozoic Eon** - 541 to 0 million years Ago	The Phanerozoic Eon is the current eon in the geologic time scale. It began around 541 million years ago (mya) and encompasses Earth's history from then to the present day. It represents around 12% of Earth's total history. Preceding the Phanerozoic Eon was the Proterozoic Eon.
6	The Paleozoic Era 541 to 252 million years ago (part of the Phanerozoic Eon)	-Animal life was restricted to primitive invertebrates living in the oceans, but, by its close, great forests covered the land and teamed with replies, amphibians, and insects. -Contains the Cambrian Period 541 - 485 million years ago
7	Cambrian Period 541 to 485 million years ago (within the Paleozoic Era)	The Cambrian Period is the first period of the Paleozoic era, during which all animals and plants lived in the Earth's oceans. Many organisms that we recognize as members of

		modern animal groups (including the arthropods, sponges, chordates, and molluscs) made their first unmistakable appearance in the fossil record during the Cambrian. Beginning of this period marked dramatic diversification of qualitative animal life. Often referred to as the "Cambrian Explosion" -Sponges, mollusks, worms, and many kinds of arthropods -A close relative of vertebrates called Pikaia, didn't have vertebrae, but had several other features found in vertebrates. Cambrian explosion witnessed the unparalleled emergence of organisms between 541 million and approximately 530 million years ago that included representatives of many major phyla still extant today.
8	<u>Ordovician Period</u> 485 to 443 million years ago (within the Paleozoic Era)	-Life in the oceans continued to diversify, with fish increasingly becoming the dominant large aquatic animals -End of Ordovician mass extinction affected only marine organisms, but at that time terrestrial organisms had only just begun to evolve.
9	<u>The Silurian Period</u> 443 to 419 million years ago (within the Paleozoic Era)	The Silurian lasted about 28 million years. Fish had not yet evolved jaws. With the evolution of jaws came the evolution of large predatory fish. It is also in the Silurian that we find the first clear evidence of life on land. Primitive plant life began to flourish on land
10	<u>Devonian Period</u> 419 to 359 million years ago (within the Paleozoic Era)	Along with the Silurian, the Devonian Period is called the "Age of Fishes." First forests appeared on land -huge jaw fishes, like Dunkleosteus, evolved in the seas, and the first true sharks appeared. -Late Devonian mass extinction was also largely limited to marine organisms, including some early vertebrate clades.

		Insects appeared in the fossil record not too long (geologically speaking) after the early land plants developed (Ordovician) but before amphibians began to populate the continents. By the end of the Devonian Period, the proliferation of plants increased the Oxygen content of the atmosphere considerably, which was important for development of terrestrial animals.
11	<u>The Carboniferous Period</u> 359 to 299 million years ago (within the Paleozoic Era) (within the Paleozoic Era)	This period began with a high concentration of Carbon Dioxide, leading to a high global average temperature during the early years. However, the rapid growth of the forests led to the Carbon Dioxide levels dropping by eight times what it was. Amphibians were widespread in the abundant swamps, and reptiles, the first amniotes, evolved. -Much of the coal that is mined today is formed from the rotting plants of Carboniferous swamps -Contains the Permian Period Plants made great advances in adapting to complex forest environments, and vertebrates underwent extensive evolution. Amphibians became widespread and diverse, and reptiles appeared for the first time and rapidly adapted to many habitats.
12	<u>The Permian Period</u> 299 to 252 million years ago (within the Carboniferous Period)	Continents collided together and formed a single supercontinent called Pangaea -Reptiles evolved into three main lineages 1. Anapsids (evolve into turtles 2. Synapsids (evolve into mammals) 3. Diapsids (evolve into lizards, snakes, crocodilians, and dinosaurs) -During this time represented dry, desert, environments -The single greatest mass extinction in our

		planet's history occurred at the end of the Permian. -Most notable was the trilobite's extinction -Palaeontologists agree that about 70% of all terrestrial vertebrate species, and 90-95% of all marine species, went extinct in a short span of time -Largest mass extinction of insects -The cause remains uncertain -Lava deposits, known as Siberian traps, formed at this time. -The volcanic eruptions that formed these deposits may have been ongoing for 200,000 years or more -Large amounts of volcanic gases were leaked into the air, which lead to a greenhouse effect and increased global temperatures. -May have also resulted in the melting of frozen chemicals called methane hydrates deep in the ocean, which in turn would have contributed to more global warming and even more melting of methane hydrates, and so on.
		What Happened in the Archean Eon? [4.0 to 2.5 billion years ago] - Earth How Geologic Time Scale Flashcards | Quizlet Ediacaran Period | Definition, Biota, and Facts | Britannica

THE AGE OF THE DINOSAURS

After the Permian Extinction wiped out over 95% of ocean-dwelling species and 70% of land species, the new Mesozoic Era began about 250 million years ago. The first period of the era was called the Triassic Period. The first big change was seen in the types of plants that dominated the land. Most of the species of plants that survived the Permian Extinction were plants that had enclosed seeds.

When juxtaposed against the reign of the dinosaurs, human life on this earth is just a drop in the bucket. Even though the dinosaurs lived only during the Mesozoic "Middle Life" Era, which is often called the "Age of Dinosaurs," they left their indelible footprints on the landscape of this planet. All dinosaurs are classified together in a group because they evolved from a single species.[130] The Mesozoic Era (252 to 66 million years ago) is divided into three time periods: Triassic (252 to 201million years ago. the Jurassic (201 to145 million years ago) and the Cretaceous (145 to 66 million years ago) During the Mesozoic Era, life diversified rapidly, and giant reptiles, dinosaurs roamed the Earth. Non-avian dinosaurs (all dinosaurs besides birds), which are now extinct, varied greatly in shape and size. Some weighed as much as eighty tons and were more than 120 feet long. Others were the size of a chicken and weighed as little as eight pounds.[131]

The Rise of the Dinosaurs

When we look at the massive size of some dinosaurs, such as the Supersaurus, it is hard to believe that such megafauna existed. But the earth was pregnant with megaflora and these titan herbivores of the reptile world feasted off the massive vegetation of a burgeoning earth for tens of millions of years. A rich environment that produced the fossil fuels that propel our automobiles today.[132]

---

130 The Early Cretaceous Period - 146 to 100 million Years Ago
The Late Cretaceous Period - 100 to 65 Million Years Ago
https://quizlet.com/59119214/this-sets-title-has-been-removed-flash-cards/

131 Types of Dinosaurs | American Museum of Natural History (amnh.org)

132 Fossil fuels like coal and oil are made from plant material, containing carbon that was taken out of the atmosphere and buried in swamps millions of years ago. First Land Plants and Fungi Changed Earth's Climate, Paving the Way for Explosive Evolution of Land Animals, New Gene Study Suggests | Eberly

Life restoration of Supersaurus based primarily on Wyoming Dinosaur Center's more complete "Jimbo"

The Triassic period saw the rise of reptiles and the first dinosaurs.[133] The Jurassic period, from about two hundred million to 145 million years ago, ushered in birds and mammals. And the Cretaceous period, from 145 million to 66 million years ago is known for its iconic dinosaurs, such as Triceratops, and pterosaurs (avian dinosaurs) such as Pteranodon. Notwithstanding their enormous size, dinosaurs were mostly herbivores. Only 35% of the total general dinosaur population was composed of carnivores. Dinosaurs were on the earth for about 165 million years and as the earth evolved into modern continents, so did the species of dinosaurs. [134] They would continue their evolution long after their demise and become the ancestors of modern birds.[135] As of 2021, there were 1,545 scientifically described dinosaur species, according to the Paleobiology Database. Moreover, about 50 previously unknown

College of Science (psu.edu)

133 If all of Earth time from the very beginning of the dinosaurs to today were compressed into 365 days (one calendar year), the dinosaurs appeared January 1 and became extinct the third week of September. (Using this same time scale, the Earth would have formed approximately 18.5 years earlier.) Using the same scale, people (Homo sapiens) have been on Earth only since December 31 (New Year's Eve). The dinosaurs' long period of dominance certainly makes them unqualified successes in the history of life on Earth. https://www.usgs.gov/faqs/when-did-dinosaurs-become-extinct

134 https://www.usgs.gov/faqs/when-did-dinosaurs-become-extinct

135 Scientists have known for some time now that birds evolved from dinosaurs and are in fact a subgroup of dinosaurs. A rich fossil record of feathered dinosaurs discovered in China and elsewhere documents in detail the dramatic transformation of behemoth terrestrial dinosaurs into small, flight-capable birds. The bird body plan was therefore not so much a fixed blueprint but more of a Lego set that was assembled brick by brick over evolutionary time. The transition between dinosaur and bird did not happen in one fell swoop but through tens of millions of years of gradual evolution. https://www.scientificamerican.com/article/how-birds-evolved-from-dinosaurs/

species are described each year, meaning there are roughly one newfound species described each week.[136]

Early illustration of Ornithochirus umbrosus (now Pteranodon), when teeth erroneously were attributed to the species and the crest was unknown,

Types of dinosaurs[137]

136 https://www.livescience.com/3945-history-dinosaurs.html

137 By Durbed - http://durbed.deviantart.com/art/Happy-New-Year-from-Hell-Creek-345906630, CC BY-SA 3.0, https://commons.wikimedia.org/w/index.php?curid=37131266

Groups of Dinosaurs

	Group	Description
1	Ornithischia	Dinosaurs include beaked plant-eaters, such as Stegosaurus, duck-billed dinosaurs (also called hadrosaurs), as well as horned dinosaurs like Triceratops and armored dinosaurs like Ankylosaurus. Some ornithischians walked on four legs, while others walked on two.
2	Sauropodomorpha	Dinosaurs were long-necked, pot-bellied dinosaurs that had tiny heads and column-like limbs. This group includes sauropods (such as Diplodocus), their smaller antecedents (including Chromogisaurus) and extra-large sauropods known as titanosaurs (such as Dreadnoughtus and Argentinosaurus), which are among the largest land animals that have ever existed. The longest dinosaur was likely Supersaurus, a Jurassic sauropod that was at least 128 feet (39 meters) long, and possibly even reached 137 feet (42 m) in length.
3	Theropoda	A group of meat-eating dinosaurs, although some (such as Chilesaurus diegosuarezi) changed their diet to be herbivorous or omnivorous. Theropods include Tyrannosaurus rex and Velociraptor, as well as birds, which evolved from small theropods
	Archosauria includes the above	**A large subclass of Reptilia comprising the dinosaurs, pterosaurs, and crocodilians all distinguished by possessing temporal openings separated from each other by a postorbitosquamosal arch**

The Breakup of Pangea and the Extinction of Dinosaurs

From about 300-200 million years ago (late Paleozoic Era until the very late Triassic), the continent we now know as North America was contiguous with Africa, South America, and Europe. They all existed as a single continent called Pangea.[138] Pangaea existed for approximately 100 million years before it began to divide into the seven continents we know today. Pangea first began to be torn apart when a three-pronged fissure grew between Africa, South America, and North America. The breakup began about 195 million years ago in the early Jurassic period, when the Central Atlantic Ocean opened. By the beginning of the Cretaceous Period, 145 million years ago, Pangaea had completely splintered into separate continents and islands and the Atlantic had become a full-size ocean. Another shallow ocean, called the Western Interior Seaway, flooded what is now much of the midwestern United States and Canada.[139]

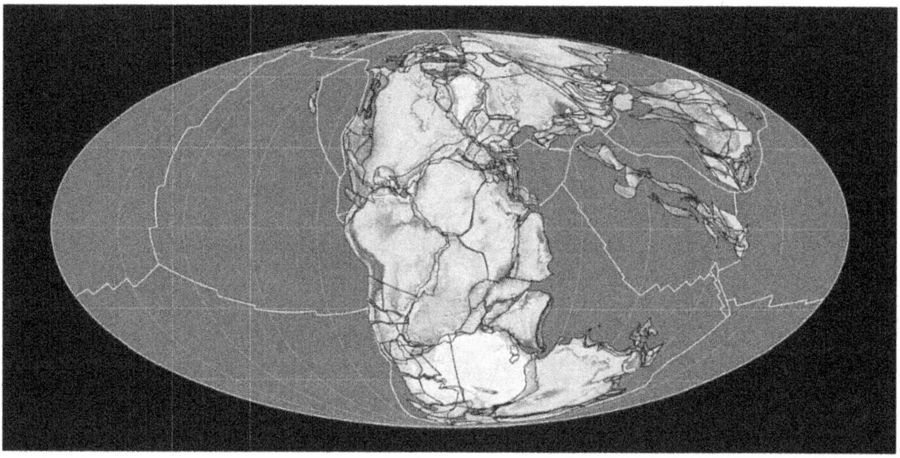

The supercontinent Pangaea in the early Mesozoic (around 200 million years ago)[140]

138 More than a century ago, the scientist Alfred Wegener proposed the notion of an ancient supercontinent, which he named Pangaea (sometimes spelled Pangea), after putting together several lines of evidence. The word "Pangaea" comes from the Greek "pan," which means "all," and "gaia" or "Earth," according to the Online Etymology Dictionary (opens in new tab). The supercontinent formed through a gradual process spanning a few hundred million years

139 Explainer: The age of dinosaurs (snexplores.org)

140 By Fama Clamosa - Own work, CC BY-SA 4.0, https://commons.wikimedia.org/w/index.php?curid=85466018

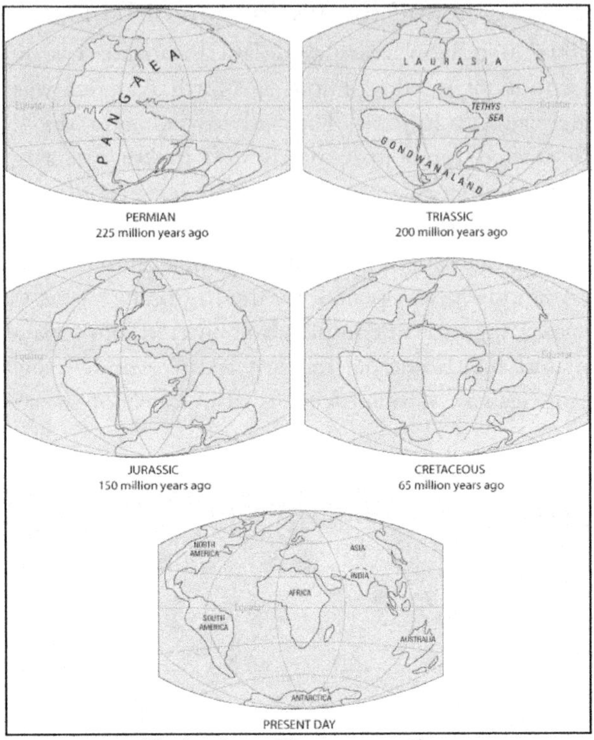

The breakup of Pangaea over time

At the beginning of the Mesozoic Era, dinosaurs lived on the one supercontinent and could walk from New York to Africa. However, as Pangea broke apart, dinosaurs became scattered on separate continents. Consequently, in each geographical area, several types of dinosaurs evolved independently, and others became extinct. For example, the Jurassic dinosaur Stegosaurus had already been extinct for approximately eighty million years before the appearance of Tyrannosaurus Rex (T. Rex) during the Cretaceous Period.[141] Renowned dinosaurs from the Cretaceous Period include: T. Rex, Triceratops, Spinosaurus and Velociraptor. The largest dinosaurs on record, including Argentinosaurus, date to the Cretaceous period.[142] Most scientists agree that an asteroid impact wiped out the dinosaurs at the end of the Cretaceous period about sixty-five million years ago. However, modern birds could be

141 https://www.usgs.gov/faqs/did-all-dinosaurs-live-together-and-same-time

142 A brief history of dinosaurs | Live Science

considered one kind of dinosaur because they share a common ancestor with non-avian dinosaurs.

There is a season for everything. A time to be born and a time to die. The demise of dinosaurs was good news for mammals, birds and insects, whose numbers burst in the aftermath. Separate continents also allowed different species of mammals to evolve on different continents. Geologically speaking, mammals are almost as old as dinosaurs. The first mammals developed toward the end of the Triassic period from a population of therapsids, the "mammal-like reptiles" that arose in the early Permian period and produced such uncannily mammal-like beasts as Thrinaxodon and Cynognathus.

Mammals differ from reptiles in several ways. For instance, mammals give birth to live young while reptiles lay eggs, and female mammals possess milk-producing mammary glands to be used to suckle their young. Moreover, all mammals have hair or fur during at least some stage of their life cycles, and all mammals are warm blooded vertebrates that can regulate their internal body temperature while reptiles are cold blooded vertebrates that cannot maintain constant body temperatures. Mammals can produce body heat while reptiles need an external heat source such as the Sun to maintain body temperature. Also, mammals have oil and sweat glands that help to regulate body temperature[143]. To date, over 5,400 species of mammals have been identified worldwide all of which have unique characteristics.

The extinction of the dinosaurs allowed large mammals to evolve and to fill many new ecological roles.[144] Ironically, the same traits that helped mammals maintain a low profile during the Mesozoic Era also allowed them to survive the extinction event that doomed the dinosaurs. As mammals evolved their behavior also changed as the first of our small furry ancestors began venturing out in the daylight. Although mammals and dinosaurs coexisted for tens of millions of years, mammals were primarily small nocturnal creatures. The first ancestral mammals to be active in the daytime perhaps lived around 65.8 million years ago, just after the mass extinction of dinosaurs. The change triggered the eventual evolution of primates and humankind that rose with the sun and slept at night. For example, the common ancestor of today's camels, hippos, and deer, probably started foraging in daylight around that time.[145]

143 Difference Between Mammal and Reptile | Compare the Difference Between Similar Terms
144 This set's title has been removed Flashcards | Quizlet

Smaller, omnivorous terrestrial animals, e.g., mammals, lizards, turtles, or birds, probably survived as scavengers feeding on the carcasses of dead dinosaurs, fungi, roots and decaying plant matter. Furthermore, smaller animals with lower metabolisms were better equipped to wait the disaster out. Our fossil fuels, along with the foliage of previous eras, are the remnants of such decay. After the dinosaurs were wiped out, some of the mammals that survived took to the trees. What is clear is that the brain size of mammals increased relative to their bodies as they struggled to contend with the dinosaurs. By this point, the brain filled the skull, leaving impressions that provide tell-tale signs of the changes leading to this neural expansion. [146]The biggest mental challenge, however, may have been keeping track of their social lives. It is important to note that fossils of the oldest known primates (our ancestors) date to around 65.9 million years old which is extremely close to the extinction of the dinosaurs.[147] Therefore, the demise of the dinosaurs led to a series of evolution that gave birth to humankind the most intelligent beings on the planet.

145 When dinosaurs went extinct, many animals literally came out of the dark | Science | AAAS

146 A brief history of the brain | New Scientist

147 When did the first primates appear? - Onpolemotorsports.com

The Mesozoic Era

1	The Mesozoic Era 252 to 66 million years ago (part of the Phanerozioc Eon)	-Referred to as the age of the dinosaurs -Many kinds of marine reptiles evolve, including the ichthyosaurs, plesiosaurs, and mosasaurs -First true turtles, corcodilians, lizards, snakes, mammals, and birds evolved at this time as well. -At Various points during the Mesozoic, North America was subdivided into two separate island subcontinents-Laramidia in the West, from which most North American Dinosaur species are known, and Appalachia in the east, from which far fewer dinosaurs are currently known. Much of the ancient sea life can now be on land throughout the American Midwest.
2	Triassic Period 252 to 201 million years ago - .	During the first ten million years of the Triassic, life gradually recovered from the mass extinction that occurred at the end of the Permian -Pterosaurs - the first vertebrates to fly. -The supercontinent Pangaea began to break apart -First representatives of the ornithiscians (like Pisanosaurus), theropods (like Eoraptor and Herrersasaurus), and Sauropodomorphs (like Panphobia) were all mostly small bipedal. -The first Ichthyosaurs evolved during this period -Plesiosaurs accompanied Ichthyosaurs in the seas during this time End of Triassic another mass extinction occurred - Cause unknown. Ichthyosaurs -Long snouts (resembles Dolphins) -Never evolved gills and requires going to the surface to breathe -Commonly called "pterodactyls" -first vertebrates to fly -have membranous wings supported by one elongated finger. -Early group belonged to a group called

		rhamphorhynchoids
3	Jurassic Period 201 to 145 million years ago -	Dinosaurs diversified. -This was the peak of sauropod Diversity, and they were the dominant terrestrial herbivores. -Coelurosaruian theropods, like Allosaurus, were the dominant terrestrial carnivores. -Stegosaurs are almost completely restricted to the Jurassic -First Ankylosaurs, ornithopods, and ceratopsians appear at this time (not abundant or diverse. -The First birds, including Archaeopteryx evolved during the Jurassic. -Rhamphorhynchoid Pterosaurs gave rise to a new pterosaur group: the Pterodactyloids. Pterodacyloids -Differed from Rhaphorhynchoids in the morphology of their tails, which were short, and the carpels in their wrists, which were elongated and made a greater contribution to the length of the wing.
4	The Early Cretaceous Period - 146 to 100 million years ago -	Dinosaurs continue to diversify, and the first flowering plants evolved. -A third of major reptilian group began patrolling Mesozoic waters. -Mosasaurs were relatives of modern monitor lizards and snakes. Mosasaurs -Like ichthyosaurs and plesiosaurs, mosasaurs had tail fins and limbs modified into paddles, but the bodies and tails of mosasaurs were more elongate. -Many mosasaurs are the right size to have preyed on small and medium sized fish.
5	The Late Cretaceous Period - 100 to 65 million years ago	-Considered the apex of non-avian dinosaur diversity. -Coelurosaurian theropods are abundant and diverse in the northern hemisphere, and include Tyrannosaurs, ornithomimids, therizinosaurs,

		oviraptorosaurs, dromaeosaurids, troodontids, and many more interesting clades. -Ankylosaurs have diverged into two groups 1. tail-clubbed ankylosaurids 2. the clubless nodosaurids. -Ceratopsians and hadrosaurs are the dominant large herbivores in the northern hemisphere. -Single lineage of sauropods remains, but the titanosaurus saruopods are the dominant herbivores in the southern hemisphere. -The first flowering plans evolved. -At the end of the Cretaceous, a large meteor collided with the earth, and this even along with its catastrophic consequences is thought to have brought about a mass extinction, which killed all non-avian dinosaurs. -much of the interior of North America was covered by a massive inland sea.
		Geologic Time Scale Flashcards | Quizlet Definition, Biota, and Facts | Britannica

OUR AFRICAN GENESIS

By all accounts, the road to the evolution of our species began in Africa. Anthropologists estimate that the human lineage diverged from other primates about 5 million years ago, with chimps being our closest living relative.[148]

Charles Darwin's theory of evolution was the first step in our understanding of how we evolved as humans. It proclaimed that species survived through a process called "natural selection," where those that successfully adapted or evolved to meet the changing requirements of their natural habitat thrived and reproduced, while those species that failed to evolve and reproduce died off. To develop his theory, he studied birds and other species in situ. Through his observations and studies, he noticed similarities among species all over the globe and he came to believe that many species had a common origin perhaps ordained by God.[149] In 1859, he documented his findings in his book entitled: *On the Origins of Species* which challenged the biblical view of creation. He wrote the following:

> The distribution of tenants of this archipelago would not be nearly so wonderful if, for instance, one island has a mocking-thrush and a second island some other quite distinct species... But it is the circumstance that several of the islands possess their own species of tortoise, mocking-thrush, finches, and numerous plants, these species having the same general habits, occupying analogous situations, and obviously filling the same place in the natural economy of this archipelago, that strikes me with wonder.[150]

Like Galileo before him, Darwin's theory of evolution shocked the religious community by suggesting that animals and humans, monkeys in particular, shared a common ancestry. However, a previous theory was proposed by the Muslim scholar Al–Jahiz a century before.[151] While in the Galapagos Islands, Darwin noticed that

148 http://geneticorigins.org/mito/intro.html

149 Darwin spent more than 500 days at sea and about 1,200 days on land during the trip. He studied plants, animals, fossils, and geological formations and wrote his observations in a series of notebooks. During long periods at sea, he organized his notes. Biography of Charles Darwin, 19th Century Naturalist (thoughtco.com)

150 What's "Natural" on the Galapagos Islands? | Origins (osu.edu)

organisms were similar from island to island but had certain features that distinguished them from each other. He attributed the unique beak shapes to the food sources available in their specific environment.[152]

Darwin's finches or Galapagos finches. Darwin, 1845. Journal of research into the natural history and geology of the countries visited during the voyage of H.M.S. Beagle around the world, under the Command of Capt. Fitz Roy, R.N. 2d edition. 1. (category) Geospiza magnirostris 2. (category) Geospiza fortis 3. Geospiza parvula, now (category) Camarhynchus parvulus 4. (category) Certhidea olivacea

Darwin's theory although scientifically sound would have social implications. The debate between science and religion continued, perhaps best exemplified by the Scopes Monkey trial that occurred around the same time that Hubble was discovering the vastness of the universe[153]. Although the religious community rejected the theory of evolution and made fun of Darwin, the scientific community embraced Darwinism. Subsequently, DNA studies provided scientific evidence for Darwin's theory of evolution. Decades of anthropological data indicated an African genesis and recent biological evidence supported the anthropological data. Today, the Out of Africa genesis is generally accepted as the prevailing view. [154]

---

151 Al-Jahiz (776-868), a Muslim polymath, theologian and scholar, was among those intellectuals whose ideas and discoveries were centuries ahead of their time. Al-Jahiz writes that "Animals engage in a struggle for existence; for resources, to avoid being eaten and to breed." And he added that "Environmental factors influence organisms to develop new characteristics to ensure survival, thus transforming into new species. Animals that survive to breed can pass on their successful characteristics to offspring." Al-Jahiz: The World's First Naturalist was Muslim (insidearabia.com)

152 Darwin's Finches: Definition, Theory & Evolution | StudySmarter

153 Like Galileo, Scopes was convicted of teaching a scientific theory that conflicted with church dogma.

154 The discussion of man's origin gained prominence from a scientific perspective with Darwin's theory of evolution in the mid-nineteenth century. There were two basic and conflicting theories: one based on a

Charles Darwin with an ape or monkey body symbolized evolution

Baby Steps Toward Humanity

The capacity of Nature to create seems boundless. Over 95 percent of the species that have existed on earth, over its 4.5-billion-year lifespan, have become extinct. Yet, Nature continues to create and destroy. Nature is not in a hurry and does not make one tree, but a series of trees, it does not make one atom, but a series of atoms or one star, but a series of stars. Science continues to reveal the evolution of sentient beings on this planet. For example, within the cells of our bodies lies the history of our earthly existence. According to anthropological biology, humans did not evolve from monkeys, but rather, shared a common ancestor. Although we share 98 percent of our DNA with chimpanzees, monkeys, and other primates, went down one path and humans down another. Our path has been long and arduous—from the early Australopithecines to the current day *Homo sapiens*. An evolution that took place over millions of years.

monogenetic paradigm, i.e., that all humans had a common ancestor, and the polygenetic paradigm of multiple and different ancestors. Although both hypotheses had their advocates and detractors, history has proven that the argument was really about race and more specifically, the vexing question: Did Caucasians have a path different from Negroes? As time progressed, this argument moved from the anthropological realm to the biological realm.

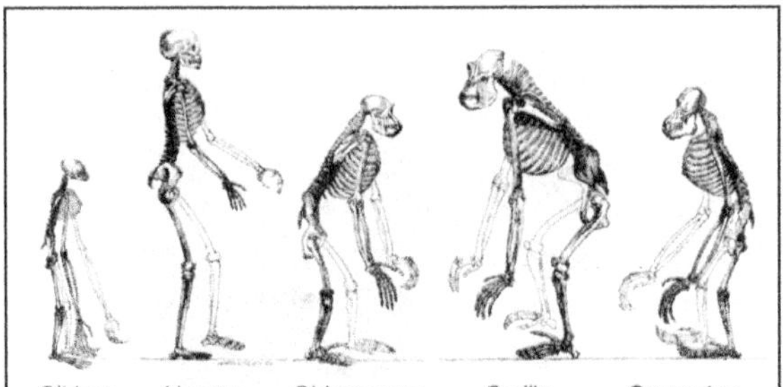

The hominoids are descendants of a common ancestor

The oldest fossils of humankind: ARDIPITHEOUS RAMIDUS: AUSTRALOPITHECUS AFARENSIS, AFRICANUS, ROBUSTUS AND BOISEI; HOMO HABILIS, HOMO ERECTUS, AND HOMO SAPIENS, were all found in Africa.[155] The discovery of a 3-million-year-old Australopithecus, named Lucy, found in Ethiopia was a major step in the understanding of the evolution of our species. This was believed to be the oldest humanoid type until the discovery and subsequent verification of an earlier humanoid type named *Ardipithecus ramidus and Ardipithecus* better known as Ardi and dating back between 5.2 and 5.8.4 million years ago. Ardi, like Lucy, was found in Ethiopia.[156] The discovery of Ardi has once again pushed back the hominoid clock, in this case by about a million years. Moreover, it has provided a genus that precedes Australopithecines. about 4 million to 6 million years ago. Geneticists have suggested

155 In presenting the history of man, based on scientific facts, it becomes obvious that history and science conflict with the Bible. For instance, the creation story diametrically opposes the archeological and biological evidence. A similar phenomenon occurred during the Age of Enlightenment when scientist such as: Galileo presented theories that conflicted with religious belief. Although they were forced to recant, science eventually prevailed. I believe that the Bible is an inspired book based on information available at the time. Therefore, it cannot be taken literally and is an earnest attempt by humankind to understand its spiritual makeup.

156 Ardi shows an unexpected mix of advanced characteristics and of primitive traits seen in much older apes that were unlike chimps or gorillas (interactive: Ardi's key features). As such, the skeleton offers a window on what the last common ancestor of humans and living apes might have been like. http://news.nationalgeographic.com/news/2009/10/091001-oldest-human-skeleton-ardi-missing-link-chimps-ardipithecus-ramidus.html Ardi was excavated between 1994 and 1997 and has been isotopically dated at 4.4 million years old. She is one of more than 100 specimens from the site that belong to Ardipithecus ramidus, a species considered by most scientists to be a very ancient hominid. Ardi possesses a small cranial cavity comparable to that of a chimpanzee (Pan troglodytes) and has long arms and fingers, opposable great toes, and relatively small canine teeth that do not project and sharpen like those in apes. Her pelvis and foot exhibit many features characteristic of later bipedal hominids. Ardi walked upright on the ground, and her foot structure suggests at least a partially arboreal existence in a woodland habitat. She weighed about 50 kg (110 pounds) and stood about 1.2 metres (3.9 feet) tall. Ardi | hominin fossil | Britannica

that the ancestors of today's humans and chimps went their separate ways about 4 million to 6 million years ago, and the ancestors of gorillas diverged about 7 million to 9 million years ago.

In more recent times, discoveries have been made in other areas of Africa which has purportedly moved the clock back to around 7 million years ago. This is awfully close to the projective date of when our species and the great apes split from a common ancestor. In addition to these watershed discoveries, 90 percent of the evidence concerning early human development comes from Africa.[157] Africa is the only continent where there is an unbroken chronological sequence of the development of human beings, from Ardipithecus to modern humans. The following explains:

> Several recent discoveries have revealed ape-like creatures that could walk upright around 5 million years ago. Perhaps the best documented is Ardipithecus, discovered in the 1990s. Around the time of Ardipithecus it seems that some apes decided to walk around on two legs. This probably happened as a result of - wait for it - climate change. Africa went through a particularly bad drought around 6 million years ago, which resulted in the drying up of the Mediterranean and seems to have set in motion some rather significant changes in the hominid - and ultimately human - lineage. Over time, these upright apes became completely bipedal, allowing them to see further and run faster, while leaving the hands free to do other things...which they started to do in earnest around 2 million years ago when we see the first evidence of tool use by our ancestors. [158]

Based on what we know today, an analogy could be drawn between the development of humankind and the development of us as individuals. Unlike other species, humans spend a lot of time developing during adolescence and this could be similar to the development of sapiens. For instance, Ardipithecus and Australopithecine were the first bipeds to stand erect and walk. However, they could not walk long distances or run very well. They were, in effect, the babies of the human species. They were not carnivores, but vegetarians. *Homo hablis* followed, they made simple tools and could eat meat, and were more adapted to walking and running. In this regard, they could be considered toddlers.

---

[157] [Grolier: Electronic Encyclopedia, ANTHROPOLOGY]

[158] https://www.greatleapforward.org/The-Great-Leap#:~:text=After%20this%20change,Jared%20Diamond%27s%20term

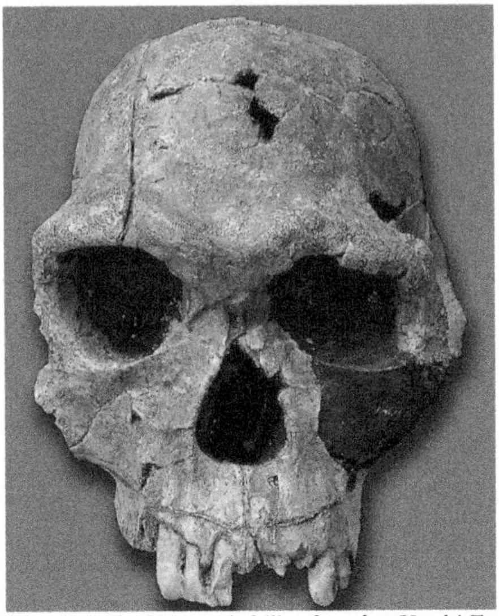

Replica of fossil skull of Homo habilis., found at Koobi Fora, Kenya

Homo erectus followed *Homo hablis*, and he was the first to leave the African continent and to use fire. In this regard, he was the wondering teenager in search of his identity. *Homo erectus* was the first humanoid to have possessed modern human-like body proportions taller than its predecessors, with longer legs and shorter arms, therefore, he had a greater capacity for long-distance running or walking, and he was the first to leave Africa. *Homo erectus* was well-equipped to thrive in a range of habitats. The most complete fossil individual of this species is known as the 'Turkana Boy' – a well-preserved skeleton (though minus almost all the hand and foot bones), dated around 1.6 million years ago. Soon after, we see evidence in the fossil record of the earliest *Homo erectus* fossils (by about 2 million years ago), and we see proof in the archeological record for the first major innovation in stone tool technology (by about 1.76 million years ago). The following explains:

> Homo erectus, the master toolmaker of the era, had a much bigger brain than his ancestors and developed many of the elements of modern human behavior, probably including the use of fire and some form of rudimentary language. He even seems to have wandered out of Africa around 1.8 million years ago - Java Man and Peking Man were both part of the Homo erectus family. He didn't stray far from the tropics, though (particularly when the world cooled down during the periodic ice ages) and was already on his way to extinction around 100,000 years ago.[159]

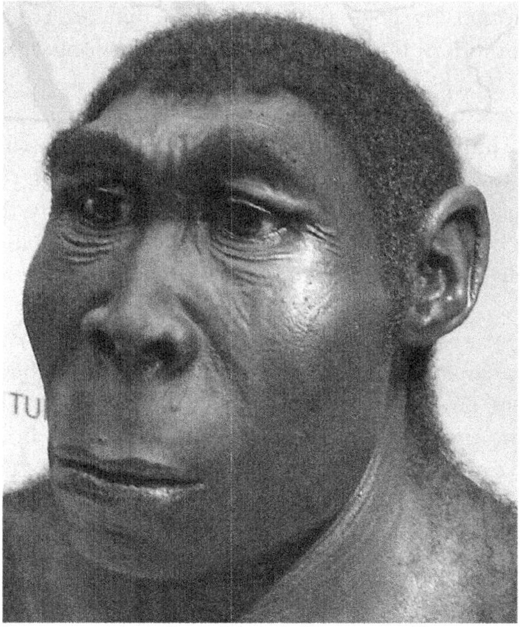

Forensic reconstruction of an adult male *Homo erectus*. Reconstruction [160]

The earliest evidence of campfires occurred during *Homo erectus*' time. *Homo erectus* coexisted in East Africa with several other early human species. Occasionally, they were even found at the same fossil sites. At the end of its lifespan, around 143,000 years ago, it coexisted with *Homo sapiens*.[161] *Homo erectus* dispersed out of East Africa about 2 million years ago as climate change triggered the expansion of East African savanna into the Southern Levant.

The proof is in the pudding, and our genes tell the story of our evolution. Fossil and genetic evidence indicate that modern humans evolved in Africa about 200 thousand years ago (kya), possibly as early as 300 kya Although several geographic regions have been proposed as potential 'birthplaces' of our species, there is now growing evidence

---

159 https://www.greatleapforward.org/The-Great-Leap#:~:text=After%20this%20change,Jared%20Diamond%27s%20term

160 by W. Schnaubelt & N. Kieser (Atelier WILD LIFE ART)Homo erectus\.JPG: photographed by User:Lillyundfreya—Homo_erectus.JPG, CC BY-SA 3.0, https://commons.wikimedia.org/w/index.php?curid=9503611

161 https://humanorigins.si.edu/evidence/human-fossils/species/homo-erectus

to support the view that our species probably evolved within structured populations, connected to each other by gene flow across the whole African continent. Thus, Africa has witnessed modern human evolution for much longer than anywhere else.

> The signatures of our recent African origin persist in our genomes today. As modern humans dispersed from Africa, probably in small populations, they experienced a dramatic bottleneck. Consequently, African populations today harbor more genetic diversity than any other population in the world, and the genetic diversity found in non-Africans represents only a subset of that found in Africa.[162]

Although many of the discoveries were found in East Africa, it is becoming increasingly clear that the entire African continent was the Garden of Eden as significant discoveries emerge from South and West Africa. These discoveries are filling in some of the gaps in the evolution of our species. For example, in South Africa a species between Homo erectus and sapiens *Homo naledi.* was found.[163]

162 https://genomebiology.biomedcentral.com/articles/10.1186/s13059-019-1740-1

163 Homo naledi is a previously unknown species of extinct hominin discovered within the Dinaledi Chamber of the Rising Star cave system, Cradle of Humankind, South Africa. This species is characterized by body mass and stature similar to small-bodied human populations but a small endocranial volume similar to australopiths. Cranial morphology of H. naledi is unique, but most similar to early Homo species including Homo erectus, Homo habili. https://www.ncbi.nlm.nih.gov/pmc/articles/PMC4559886/

Cenozoic Era[164]

Timeline	Event
55 mya	Earliest primates
8 mya	The common ancestor to African apes and hominids
5.6 mya	First hominid, Ardipithecus
4.2 mya	Frist australopithecines
2.4 mya	First hominid in the Homo genus, *Homo habilis*
1.9 mya	First Homo erectus
1.8 mya	First migration out of Africa by *Homo erectus*
500,000	First appearance of Neanderthals in Europe and Asia
125,000-200,000	First Homo sapiens (Modern man) in Africa
70,000	The cognitive revolution, language, Homo sapiens leave Africa to populate the world
30,000	Disappearance of Neanderthals; *Homo sapiens* is the only surviving hominid
12,000	The Agricultural Revolution: Domestication of plants and animals

- mya = millions of years ago

164 **Cenozoic Era**, third of the major eras of Earth's history, beginning about 66 million years ago and extending to the present. It was the interval of time during which the continents assumed their modern configuration and geographic positions and during which Earth's flora and fauna evolved toward those of the present. Cenozoic Era | Definition, Events, & Facts | Britannica

THE EMERGENCE OF MODERN HUMANS

When researchers first recovered DNA from Neanderthal bones, the available techniques for making sense of it were powerful but relatively simple. Scientists compared ancient and modern sequences, tallied up shared sites and mutations, and conducted bulk statistical analyses. That's how they discovered in 2010 that Neanderthal DNA makes up approximately 2% of the genome of people today of non-African descent, a result of interbreeding that occurred throughout Eurasia beginning 50,000-60,000 years ago[165]

OVER THIRTY YEARS AGO, I READ A GROUNDBREAKING ARTICLE IN NEWSWEEK MAGAZINE that traced the emergence of modern humans, sapiens, back to Africa. Needless to say, this article was met with the utmost skepticism by scholars and the general public. Even though the archaeological and anthropological evidence pointed in that direction, the introduction of biology into the field using mitochondrial DNA was controversial particularly when it did not yield the desired result. Much has changed in the last thirty years, and the anthropological, linguistic and biological data converged on the fact that Africa is the mother of us all. Such anthropological data combined with the recent DNA discoveries begin to paint a portrait of our early human experiences and moreover our patterns of migration. [166]

Notwithstanding, we are just beginning to understand what Africa offers us DNA wise. That is because researchers have sequenced more than 1,000 ancient human genomes from Eurasia, the Americas and even Oceania versus fewer than 80 from Africa, few of which are older than 10,000 years. Harvard Medical School geneticist David Reich states regarding Africa: We don't have a clear picture right now; Africa is the most diverse place on Earth. It's where our particular sub-lineage of humans originated.[167]

---

165 https://www.quantamagazine.org/fossil-dna-reveals-new-twists-in-modern-human-origins-20190829/

166 The breakthrough in discovering the structure of DNA came in 1952 when James Watson (1928 -), Francis Crick (1916-2004) and Rosalind Franklin (1920- 1958) discovered the double-helix.

167 https://www.nature.com/articles/d41586-020-00167-5

Sapiens

H*omo sapiens*, our species, followed in the footsteps of *Homo erectus* and according to our DNA record, we have been on earth for at least 120,000 years. This is based on Mitochondrial DNA, and Y chromosome testing (which kept a record of the beginning of our species.[168] The discovery of DNA was another major breakthrough in the understanding of our species. No longer were we solely dependent on anthropology and the study of human remains, but we could look inside ourselves for keys that would open the doors of our past. The following explains the process:

> Scientists have traced mitochondrial DNA all the way back to a woman referred to as 'mitochondrial Eve,' who is believed to have lived around 200,000 years ago, which shows the accuracy of mtDNA tests. 'Eve' is the most recent common female ancestor for everyone alive today. That means that every human being on the planet shares the DNA of this single woman, not that she was the first human woman, ever. She was just the most recent woman whose female line continued unbroken until today.
>
> Similarly, scientists were able to use Y-DNA tests to identify 'chromosomal Adam' as the most recent common male ancestor of all humanity (but again, not the first human man). Scientists have placed him as having lived around 100,000 years ago, which isn't at the same time as mitochondrial Eve, but they do have theories to explain that, including the frequency of polygamy.[169]

---

168 In the mid-1990s, the discovery of numerous Y-chromosome haplotypes enabled scientists to start building a Y-chromosome family tree, consisting of various branches all stemming back to a common trunk representing the common ancestor of all Y-chromosome haplotypes (Figure 2). This common ancestor is called Y-MRCA, for "most recent common ancestor." This is sometimes called the Y-chromosome Adam, and it tracks back to 200,000 – 300,000 years ago. Like the Y-chromosome, mitochondrial DNA exists as single copy, not paired alleles. Because our maternally inherited mtDNA is not recombined with other alleles from our fathers, it stays pure along matrilineal ancestry. Only the occasional mutation alters mtDNA haplotypes. Also, like the Y-chromosome, some mutations that occur in mtDNA are not harmful but neutral. These kinds of neutral mutations give us the mtDNA haplotypes, which are useful for tracking matriarchal ancestry.
https://www.visionlearning.com/en/library/Biology/2/Y-Chromsome-and-Mitochondrial-DNA-Haplotypes/248

169 https://www.top10.com/dna-testing/whats-the-difference-between-autosomal-mitochondrial-and-y-chromosome-dna-testing

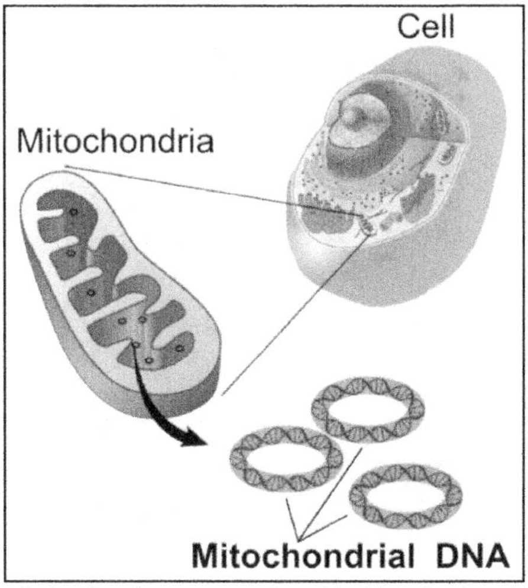

Mitochondrial DNA is the small circular chromosome found inside mitochondria. These organelles, found in all eukaryotic cells, are the powerhouse of the cell.[1] The mitochondria, and thus mitochondrial DNA, are passed exclusively from mother to offspring through the egg cell

Thanks to the Human Genome Project, not only can we discern when and where our species originated, but how we migrated out of Africa to populate other parts of the world[170]. In this regard, *Homo sapiens* are the adults. Science has proven that the diversity of races is not the result of the descendants of Noah, but rather mutations, which took place within our species. As *Homo sapiens* left Africa for Europe and Asia, they began to adapt to new environmental conditions. With the development of the Human Genome, we are just beginning to understand migration patterns and the associated mutations that occur over time.[171]

---

[170] The Human Genome Project (HGP) was one of the great feats of exploration in history. Rather than an outward exploration of the planet or the cosmos, the HGP was an inward voyage of discovery led by an international team of researchers looking to sequence and map all of the genes -- together known as the genome -- of members of our species, Homo sapiens. Beginning on October 1, 1990, and completed in April 2003, the HGP gave us the ability, for the first time, to read nature's complete genetic blueprint for building a human being. https://www.genome.gov/human-genome-project

[171] A very well-known example of the effect of mutations on a species is provided by the history of the peppered moth, a well-camouflaged species found widely distributed on the Earth, especially in cooler countries like England and Scotland. Naturalists have collected the moth for centuries, noting every now and again the appearance of a dark-colored moth among the common off-white members of the species. The off-white camouflage blends extremely well with the lichen-covered trunks of the trees in England and prevents the moth from being taken by birds while resting during the day. This advantage is not possessed by the dark-colored mutants which, because of the ease with which they are discovered by birds, rarely survive. However, this advantage was suddenly reversed in the 19th Century factory some

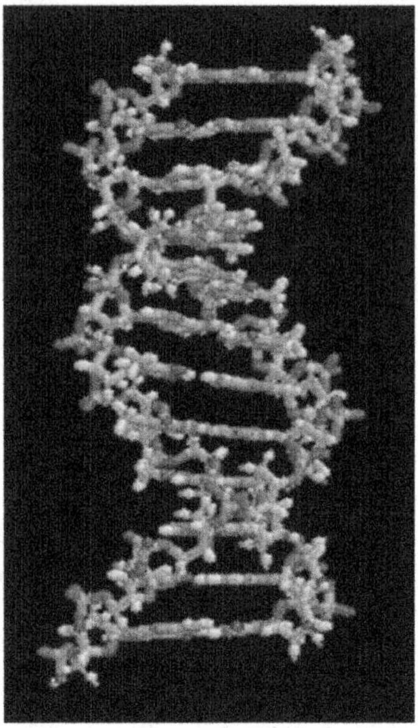

The structure of part of a DNA double helix

For example, our bodies use ultraviolet light from the sun to synthesize Vitamin D. However, there must be a balance; too much ultraviolet (UV) light could have adverse effects. Near the equator, enough ultraviolet light is present to penetrate even the darkest skin. UV rays are strongest in areas close to the equator because the sun is directly over the equator and UV rays only travel a short distance through the atmosphere to reach these areas. UV radiation is also the strongest near the equator because ozone in these areas is naturally thinner, so there is less to absorb the UV radiation. We do know that black skin has high concentrations of melanin, which protects against ultraviolet rays and therefore would be beneficial in areas of high sunlight such as the tropical equator in Africa. In contrast, white or pale skin, which absorbed ultraviolet radiation and efficiently produces Vitamin D, is beneficial in areas of low sunlight such as Ice Age Europe or northern areas such as Scotland.[172]. A

began to blacken the tree trunks and kill lichens in the English midlands, and it is now the light-colored forms that lack the protective coloring and the dark-colored moth that survives. The mutant moths produced offspring with the same characteristics and gradually these became the dominant form of the species. A better example of natural selection could hardly be found. Sickle-cell anemia is another example of a mutation. Many feel that this was a mutation required to survive in a malaria infested area, hence the rate among Africans in malarial areas of Africa. [Our Roots, pp. 60, 61]

white person in the northern areas could produce as much as five times more Vitamin D as a dark black person living around the equator. As in the case of animals and plants, it was the environment that made us different.

Therefore, depending on the environment, a mutation could be good or bad. For example, sickle cell anemia is often cited as a case where a mutation occurred based on environmental conditions. Similarly, a change that decreases the amount of melanin in the skin would increase the survivability of humans in Europe during the Wurm Glaciation period.[173] This is also the basis of Darwin's theory and evolution in general. Our early ancestors and other primates have white skin under their hair. However, as we evolved as a species, we lost hair and the amount of melanin in the skin increased. Furthermore, other mechanisms developed to help us dissipate heat such as sweating and changes in our phenotype, e.g., limps, lips, hair, etc.. However, in northern climates, with less direct ultraviolet radiation, depigmentation and other phenotypical changes occurred.

What we must understand is that nature is in no hurry, time is on her side. One of the laws of Mother Nature is that all things change in time; epitomized by the leaves on a tree in autumn, and the phases of the Moon. The laws of Nature are not restricted to things that we take for granted but are present in all things. For instance, throughout nature, we find that animals in cold environments, e.g., polar bears, tend to have a white color. If we can link the polar bear to the brown bear or white mice to brown ones, it is also feasible that Caucasians developed from Africans through a similar mutation process required to survive in post-Ice Age Europe. Not only is the white skin advantageous in cold environments with low or limited sunlight, but the phenotype of Caucasians is also advantageous in terms of being heat conservers as opposed to heat dissipaters which the first Africans that migrated to Europe were.

Based on the anthropological, biological, archeological and linguistic data, it can be categorically proven that *Homo sapiens* were present on the African continent at least 120,000 years ago. [174] Following the coastline, these native African populated the

---

[172] The Vitamin D theory was first outlined by Murray, and it has slowly gained adherents as a plausible explanation of why white skin evolved from black skin. [The African Background to Medical Science, p. 44]

[173] [Diop: Civilization or Barbarism, p. 13]

[174] The most important date, in relation to the competing evolutionary theories, is the time when all the sequences coalesce into one -- the 'mitochondrial Eve. 'From this study, a date of 171,500 years ago was obtained which fits remarkably well with that proposed in the recent African origin hypothesis. For us to accept multi-regionality, we would expect a much older date, as it would represent the common ancestor

globe. They were living in the eastern Mediterranean 90,000 years ago. They reached Asia and Australia about 40,000 years ago. The migration patterns indicate a migration from Eastern Africa south first and then across the Isthmus of Suez. The earliest European discovery of *Homo sapiens* appeared around 30,000 years ago and they crossed the Bering Straits around 15,000 years ago.[175] This gave black people (modern man) over 70,000 years to develop and walk the Earth before their Caucasian brothers.

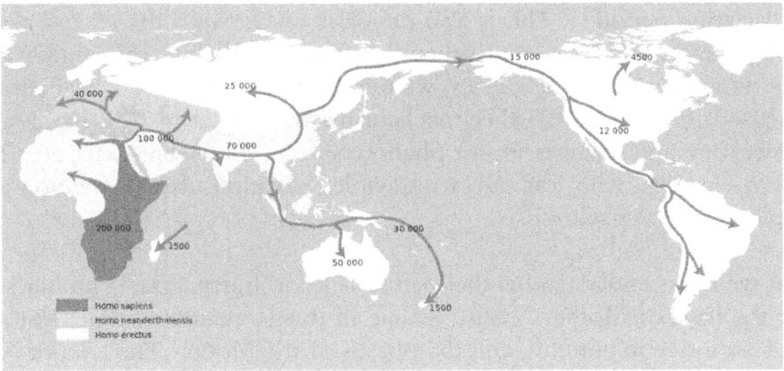

A map of early human migrations

The more and more Africa is explored using modern technology, the further the clock is pushed back, and the models adjusted. With greater investigation into West Africa, the date for the beginning of our species is being adjusted again from 200,000 to 300,000. Furthermore, the migration patterns within Africa may have been different. The following represents our latest assessment.

> Our species, *Homo sapiens*, rose in Africa some 300,000 years ago. The objects that early humans made and used, known as the Middle Stone Age material culture, are found throughout much of Africa and include a vast range of innovations. Among them are bow and arrow technology, specialized tool forms, the long-distance transport of objects such as marine shells and obsidian, personal ornamentation, the use of pigments, water storage, and art. Although it is possible that other ancestors of modern humans contributed to this material culture in Africa, some of

of Homo erectus rather than of Homo sapiens.
https://webprojects.oit.ncsu.edu/project/bio183de/Black/genetech1/genetech1_news/ingman.html

175 [John Reader: Africa, p.3]

the earliest Middle Stone Age stone tools have been found with the oldest Homo sapiens fossils found so far.[176]

Around 70,000 years ago, *Homo sapiens* in Africa started to do some very special things. The appearance of advanced tools such as harpoons in Africa, 40,000 years before Europe, supported the Out of Africa paradigm. It was the period of the great migration and they drove all previous human forms off the face of the Earth. About 45,000 years ago, they somehow crossed the open sea and landed in Australia – a continent hitherto untouched by humans. The period from about 70,000 years ago to about 30,000 years ago witnessed the invention of boats, oil lamps, bows and arrows and needles (essential for sewing warm clothing).[177]

The first objects that can reliably be called art date from this era as does the first clear evidence for religion, commerce and social stratification. Sapiens had the ability to transmit larger quantities of information about the world surrounding them. Along with the ability to transmit information about things that do not really exist as well as advanced social techniques.[178] All of these things, so vital to human existence emerged in Africa and were spread to the rest of the world. Still, the gene pool in Africa remains strong. According to a recent report:

> Africans are more diverse genetically than the inhabitants of the rest of the world combined, according to a sweeping study that carried researchers into remote valleys and mountaintops to sample the bloodlines of more than 100 distinct populations.[179]

The diversity of genes in Africa is consistent with the out of Africa data. Africa was the tree, and the dispersal of DNA throughout the world only represents a branch on that tree. Before biological anthropology and radioactive decay of the surrounding environment, scientists had to rely on Carbon 14 dating techniques.[180]

---

176 https://www.sapiens.org/archaeology/west-africa-evolution/

177 Scientists speculate that a random genetic mutation enabled Sapiens' brains to function differently, causing a massive cognitive leap forward. Harari calls this the Cognitive Revolution.

178 [Yuval Noah Harari, Sapiens]

179 https://www.washingtonpost.com/wp-dyn/content/article/2009/04/30/AR2009043002485.html

180 C14 is an artifact of nature that is based on the decade of C14, an isotope of Carbon 12.

When living things die, they stop assimilating C^{14} and the rate of the decline is known as its half-life. Based on this precise half-life, scientists can date the age of things that once lived from wood to bones.[181] The half-life of C^{14} is approximately 5,600 years. This significantly reduces the accuracy of some of our far distant past. However, it is important to note, that wherever modern humans appeared, the megafauna became extinct due to the advanced hunter-gatherer techniques used.

The Neanderthal Controversy

Neanderthals were an extinct species of humans that was widely distributed in ice-age Europe between 120,000–35,000 years ago, with a receding forehead and prominent brow ridges. What happened to the Neanderthals remains a topic of discussion and inquiry. The reasons for the demise of the Neanderthals some thirty thousand years ago, only a few millennia after the first appearance of modern humans in Europe, remain controversial and are a focus of Neanderthal research. Currently, there are two prominent theories: One is based on interbreeding, the other on replacement through massive genocide. However, both theories can be true. In the first theory, the Neanderthals did not disappear but were absorbed by the more advanced African sapiens. This theory is supported by the fact that some inhabitants of Eurasia appear to have small amounts of Neanderthal DNA.[182] In his book *Sapiens*, Yuval Noah Harari explains:

---

181 Libby, Anderson and Arnold (1949) were the first to measure the rate of this decay. They found that after 5568 years, half the C14 in the original sample will have decayed and after another 5568 years, half of that remaining material will have decayed, and so on (see figure 1 below). The half-life (t 1/2) is the name given to this value which Libby measured at 5568 30 years. This became known as the Libby half-life.

182 The Neanderthal disappearance is viewed by some as a true extinction. Others, however, contend that Neanderthals did not become extinct, but instead were assimilated into the modern human gene pool. The fossil record is ambiguous on this point: a few European Upper Paleolithic modern human specimens have been proposed as potential Neanderthal-modern human hybrids, but this interpretation has been questioned (e.g., see Smith 2005, Harvati et al. 2007). Analysis of Neanderthal and Upper Paleolithic modern human mitochondrial DNA shows no indication of interbreeding (e.g., Ghirotto et al. 2011). However, recent research on Neanderthal nuclear DNA has found evidence for limited admixture: a small portion (up to ~4%) of the genomes of non-Africans so far examined may derive from Neanderthals, suggesting that interbreeding probably occurred in the Near East during the earliest dispersal of modern humans out of Africa, but prior to their arrival in Europe (Green et al. 2010). Demographic modeling of admixture combined with territorial expansion, however, indicates that this level of introgression would be produced under very low (<2%) interbreeding rates and strong barriers to reproduction between Neanderthals and modern humans, arguing against assimilation (Currat & Excoffier 2011). Pending the completion of the Neanderthal genome and ancient DNA analyses of early modern Europeans dating to the Upper Paleolithic and following the recent discovery of a third possibly coexisting species from Denisova cave (Krause et al. 2010), it is premature to conclude that the currently observed level of admixture constitutes assimilation. Regardless of this small contribution to the modern human gene pool,

It seems that about 50,000 years ago, Sapiens, Neanderthals and Denisovans were at that borderline point.[183] They were almost, but not quite, entirely separate species. Sapiens were already very different from Neanderthals and Denisovans not only in their genetic code and physical traits, but also in their cognitive and social abilities, yet it appears it was still just possible, on rare occasions, for a Sapiens and a Neanderthal to produce a fertile offspring. So the populations did not merge, but a few lucky Neanderthal genes did hitch a ride on the Sapiens Express. It is unsettling – and perhaps thrilling – to think that we Sapiens could at one time have sex with an animal from a different species and produce children together.

According to the Interbreeding Theory, when Sapiens spread into Neanderthal lands, Sapiens bred with Neanderthals until the two populations merged._If this is the case, then today's Eurasians are not pure Sapiens.

The replacement theory is more widely accepted as the reason for the massive extinction[184]. In his book Sapiens, Yuval Noah Harari also explains the clashing of Neanderthals with sapiens:

Neanderthal populations across Europe vanished abruptly in the fossil record, and several scenarios have been proposed to account for this observation. Most invoke a degree of competition, either direct or indirect, with modern humans, or alternatively, deteriorating environmental conditions, as major factors. https://www.nature.com/scitable/knowledge/library/what-happened-to-the-neanderthals-68245020/

183 Denisovans were an extinct species of human of robust build, distributed from Siberia to Southeast Asia in the Upper Paleolithic. Denisovans, Neanderthals and modern humans descend from the same population of ancestors, who most likely lived in Africa between 550,000 and 765,000 years ago. Some of these early humans spread to Eurasia, where they split into Neanderthals in Europe and Denisovans in Asia. Denisovans were "like an Eastern cousin" of Neanderthals. For some period after that initial dispersal, the human groups — now divided geographically — were evolving into distinct species. Had they stayed separated longer, they would have naturally accumulated too many genetic differences to mate successfully. https://www.discovermagazine.com/the-sciences/meet-the-denisovans

184 Putting this all together, the Replacement model asserts that Homo sapiens originated or evolved only in Africa, then spread out over Europe and Asia. When speaking of this theory, the word 'only' is probably one of the most important words since the hallmark of the theory is that Homo sapiens only evolved in Africa. Then as these Homo sapiens migrated out of Africa, the model asserts that all lesser-evolved human species somehow became extinct, leaving Homo sapiens as the only remaining human species. For this reason, the Replacement model is also referred to as the Out of Africa model. https://study.com/academy/lesson/replacement-models-of-human-origin-attributes-claims-arguments.html

> In a one-on-one brawl, a Neanderthal would probably have beaten a Sapien. But in a conflict of hundreds, Neanderthals wouldn't stand a chance. Neanderthals could share information about the whereabouts of lions, but they probably could not tell – and revise – stories about tribal spirits. Without an ability to compose fiction, Neanderthals were unable to cooperate effectively in large numbers, nor could they adapt their social behavior to rapidly changing challenges. While we can't get inside a Neanderthal mind to understand how they thought, we have indirect evidence of the limits to their cognition compared with their Sapiens rivals. Archaeologists excavating 30,000-year-old Sapiens sites in the European heartland occasionally find their seashells from the Mediterranean and Atlantic coasts. In all likelihood, these shells got to the continental interior through long-distance trade between different Sapiens bands. Neanderthal sites lack any evidence of such trade. Each group manufactured its own tools from local materials.

The following describes the cognitive ability of a Neanderthal:

> After this change in abilities that led to the erectus African exodus, a long period of stasis seems to have set in. The hominids in the ensuing 1.7m years grew larger brains; several species appeared and then became extinct, and some even wandered out of Africa - giving rise to the Neanderthals in Europe about a quarter of a million years ago. During this time there were no big innovations. The tools used a million years ago in Africa do not differ much from those used by the Neanderthals 900 millennia later.[185]

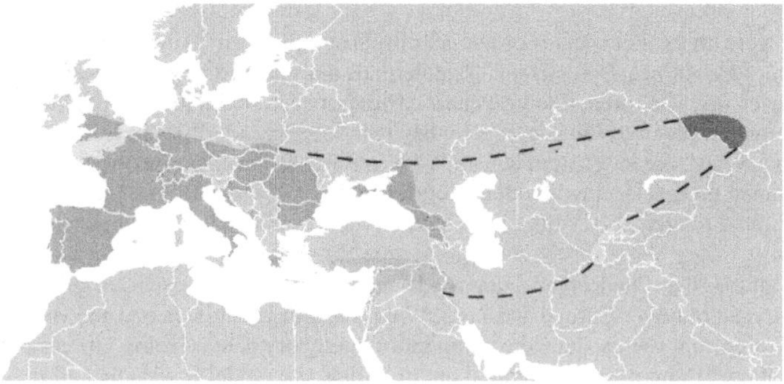

[185] https://www.greatleapforward.org/The-Great-Leap#:~:text=After%20this%20change,Jared%20Diamond%27s%20term

Stretching across all of Portugal, Spain, Switzerland, Italy, England, southern Germany and Austria, all of the Czech Republic, Hungary, Romania, Croatia, Montenegro, the Peloponnesian Peninsula, the Crimean Peninsula, the Black Sea–Caspian Steppe west of the Caucasus, southern Turkey, northern Syria, the Levant, northern Iraq spilling over into Iran, the east end of Uzbekistan, and in Russia just northeast of Kazakhstan.[186]

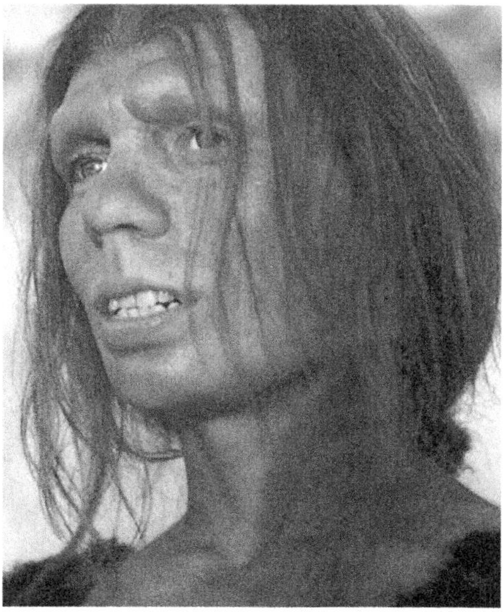

Reconstruction of a Neanderthal woman[187]

It turns out, that having black skin and wooly hair represents modern humankind in its earliest form. The proverbial Adam and Eve. The theory that the Neanderthals living in ice age Europe were white is based on the Vitamin D hypothesis. However, this was not the only source of white genes in Europe. The following is from a report from the American Association for the Advancement of Science (AAAS)[188]

[186] By Nilenbert, Nicolas Perrault III - File:Range_of_Homo_neanderthalensis.png, CC BY-SA 3.0, https://commons.wikimedia.org/w/index.php?curid=60634820

[187] By Bacon Cph - Cro-Magnons Conquered Europe but Left Neanderthals Alone" From PLOS ONE, an open-access article distributed under the terms of the Creative Commons Attribution License, which permits unrestricted use, distribution, and reproduction in any medium, provided the original author and source are credited."© 2004 Public Library of Science. https://commons.wikimedia.org/w/index.php?curid=78873692

[188] Founded in 1848, the American Association for the Advancement of Science (AAAS) is an international organization dedicated to advancing science around the world by serving as an educator, leader, spokesperson, and professional association. AAAS fulfills its mission to advance science and serve society through initiatives in science policy; international programs; science education; and more. https://www.charitynavigator.org/ein/530196568

> Based on 83 human samples from Holocene Europe as analyzed under the 1000 Genomes Project, it is now found that for the majority of the time that humans have lived in Europe, the people had dark skin, and the genes signifying light skin only appear within the past 8,000 years. This recent and relatively quick process of natural selection suggests to researchers that the traits which spread rapidly were advantageous within that environment.
>
> AAAS reports that the "modern humans who came out of Africa to originally settle Europe about 40,000 years are presumed to have had dark skin, which is advantageous in sunny latitudes. And the new data confirm that about 8500 years ago, early hunter-gatherers in Spain, Luxembourg, and Hungary also had darker skin: They lacked versions of two genes—SLC24A5 and SLC45A2—that lead to depigmentation and, therefore, pale skin in Europeans today.
>
> Then, the first farmers from the Near East arrived in Europe; they carried both genes for light skin. As they interbred with the indigenous hunter-gatherers, one of their light-skin genes swept through Europe, so that central and southern Europeans also began to have lighter skin. The other gene variant, SLC45A2, was at low levels until about 5800 years ago when it swept up to high frequency." [189]

It has now become apparent that wherever humans encountered each other, they interbred and shared their DNA. This not only includes the sharing of DNA between Neanderthals and Sapiens but between Neanderthals and Denisovans, Sapiens and Denisovans and possibly others humanoid forms yet to be discovered.

Creativity and Collective Learning

Since early hominids emerged 5 million years ago, humans have evolved large brains to support higher cognitive functions. The brains of the earliest humans were similar in size to those of chimpanzees. However, over time, human and chimpanzee

---

[189] DNA taken from the wisdom tooth of a 7,000-year-old human found in Spain in 2006 overturned the popular image of light-skinned European hunter-gatherers. The study revealed that the individual had dark hair and the dark-skinned genes of an African. However, the man had blue eyes, an unexpected find by researchers. The hunter-gatherer is the oldest known individual in Europe found to have blue eyes. https://www.ancient-origins.net/news-evolution-human-origins/white-skin-developed-europe-only-recently-8000-years-020287

brains evolved in several different and important ways. In particular, the human cerebral cortex is greatly expanded, allowing accommodation of the evolutionary increases in the number of cortical areas, the functional modules that subserve perception, attention, motor control, cognition, memory, and learning. Duplicated genes specific to the Homo lineage have played key roles in human speciation, particularly in the development of the highly complex human brain and the circuits of the cerebral cortex.[190] The growth in the size of our brains ended around 200,000 years ago.[191] Our brains burn 20 percent of our energy at a rate of about 15 watts. In terms of weight and energy usage, the human brain is an efficient organ. The reason behind human intelligence, in part, is neurons and folds in the brain. Sapiens have more neurons per unit volume than other animals, and the only way they can all fit within the brain's layered structure is to make folds in the outer layer, or cortex.[192]

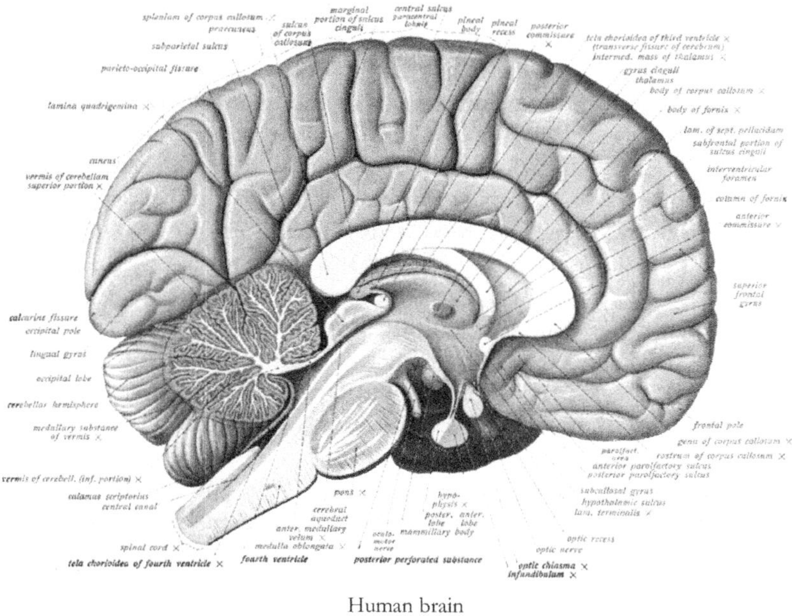

Human brain

The human brain is divided into two hemispheres. The left hemisphere is the "logical brain" and is involved in language and analysis and the right hemisphere

---

190 Evolution of the human brain | Science

191 A brief history of the brain | New Scientist

192 https://www.livescience.com/29365-human-brain.html

is the "creative brain," involved in daydreaming and imagination. The left hemisphere controls the right side of the body while the right hemisphere controls the left side.[193]

Although other creatures on this Earth have displayed some level of ingenuity, creativity is one of the features that demarcates us from them and could be one of the main reasons why *Homo sapiens* survived and dominated over Neanderthals and chimpanzees.[194] The ability to imagine lies at the core of our creativity. For instance, the young of our species engage extensively in pretend play in infancy and childhood. From the age of about eighteen months, all normal children, in all human cultures, start to do something which (when viewed from an external perspective, at least) appears very odd indeed – they begin to pretend.

They engage in imaginary conversations with make-believe characters (talking to a doll; inventing an imaginary companion) and they pretend to be engaging in a wide variety of adult or fictional activities (talking into a banana as if it were a telephone; pretending to cook and eat mud pies; pretending to be a bird or an airplane). Horses and other animals know how to run right from the womb. Such things seem to be hardwired in their DNA. But humans must go through certain stages, crawling, standing, baby steps, walking, followed by running. A mini exercise in collective learning. This uniqueness is a presage to much greater collective learning processes. The emergence of art is one of the earliest indicators of our imagination and creativity, and children learn to draw and color at a very early age.

In search of what makes our species different from the others are two distinct features: The use of language and the efficient control of energy. Although *Homo erectus* was the first species to use and control fire, *Homo sapiens* were the first to harness the full range of potential energy sources, e.g., fire, wind, fossil fuels, etc.. Language and energy usage enabled our species to embark on a collective learning experience that allowed each generation to transfer the knowledge gained to successive generations. Language was an important step, but it was preceded by the ability to cognitively understand and control fire and other tools.

193 Language and the Human Brain (news-medical.net)

194 The idea that creativity may have given Homo sapiens a survival edge over Neanderthals has been around a long time, said senior author Dr. Claude Robert Cloninger, a professor emeritus in the psychiatry and genetics departments at Washington University in St. Louis. But that's a tricky case to prove, as we still don't know how creative Neanderthals actually were, he said. In the new study, the researchers analyzed DNA previously taken from Neanderthal (Homo neanderthalensis) fossils, modern humans (Homo sapiens), and chimpanzees (Pan troglodytes). They found that the genes related to the oldest network — emotional reactivity — were identical among Homo sapiens, Neanderthals and chimpanzees. But the chimpanzees completely lacked the genes that led to self-awareness and self-control in humans. https://www.livescience.com/creativity-genes-gave-homo-sapiens-edge.html

The Control of Energy

Animals need food, water, and shelter; humans need all those things, but they also need fire or better yet, the control of energy. Fire is, and has always been, a central part of the human experience. Fire helped humans to shape their environment, and this was an important step in human development. If we were cold, it kept us warm, if it was dark, it gave us light. We are the only species that cook our meat and reap the nutritional and digestive benefits that cooking has to offer. The use of fire was essential to the migration and survival of the human species across the globe. Without fire, humans would have never left the womb of Mother Africa. Although it took almost a million years from the time *Homo erectus* first mastered fire to the emergence of modern humans, it was a time well spent. [195]

It was with fire and language that the seeds of civilization were planted. Humans became hunter-gathers and communities banded together to better control the energy required to sustain the community. Women and men begin to take on specific domestic roles and we began to develop social structures that would form the bedrock of civilization. Men for the most part were the hunters and the women the gathers.[196] The ancillary effects of fire cannot be overstated and even the advent of the computer does not compare with this nascent discovery. Fire has often been associated with knowledge and it was indeed the control of fire that became man's first formidable tool and led to subsequent forms of knowledge.

It is our quest for and use of energy that draws a clear distinction between us and other species. Even though other species are dependent on the entropic forces of nature, and therefore, have developed means of extracting energy to survive, we are the greatest consumers of energy in the form of light emanating from the Sun. Although fire and light are often used as metaphors for knowledge, they could also be used as symbols of our unbounded quest for energy. In fact, the amount of energy we use is directly proportional to the amount of collective learning that we acquire, leading to a greater complexity of life.

---

195 Some scientists contend that before Homo sapiens arrived in Europe, our Paleolithic cousins didn't just spend a few months or years in a cold land without fire—they spent entire lifetimes, many generations even, without the warm glow of a hearth to take the chill off their toes, cook their meat, and lift their spirits https://www.sapiens.org/archaeology/neanderthal-fire/.

196 This is not intended to place women in a more subservient role. Gathering was more amiable to caring for the infants as it did not represent the same level of danger. Gathering was just as important as hunting, and some have argued that it was more important.

Language and Art

Another critical ingredient was the development of symbolic language. This enabled humans to share their ideas very efficiently. Human communities grew and interacted, creating the perfect conditions for something new, the ability to learn collectively instead of just as individuals. Ideas and knowledge accumulated generation after generation, and human technologies became more and more prolific. Just as genetics has confirmed the anthropological data, linguistics has also confirmed that the earliest languages are of African origin. The world's several thousand languages have been grouped into twenty linguistic groups. Of these groups, four have the most distant relationships and all four are African with points to an early origin. They are Khoisan, Niger-Congo, Nilo-Saharan and Afro-Asiatic.[197] All four are representative of the languages spoken today in the corresponding regions.[198] Human spoken language is unprecedented among animals both in information content and rate of transmission. It is a clear indication of a very sophisticated brain function and to some degree, it represents the presence of higher *Consciousness* and cognition.

---

[197] Africa contains more human genetic variation than any other continent, but the majority of the population-scale analyses of the African peoples have focused on just two of the four major linguistic groups, the Niger-Congo and Afro-Asiatic, leaving the Nilo-Saharan and Khoisan populations under-represented. https://www.sciencedirect.com/science/article/pii/S0002929720302378
Atkinson discovered that the further humans traveled from Africa, the fewer number of phonemes survived. To put this into perspective: Many African click languages or "click consonants," found in all three Khoisan language families, have more than 100 phonemes while the languages of Oceania, the spoken language of the Pacific Islands, Papua New Guinea, and New Zealand – the latter being the furthest migration route out of Africa, have only 13. The Modern English language has approximately 45 phonemes.
https://www.dictionary.com/e/origin/#:~:text=The%20origin%20of%20spoken%20language%20has%20stumped%20linguistics,experiment%20conducted%20by%20a%20Pharaoh%20named%20Psammetichus%20I.

[198] [John Reader: Africa pp, 109,110] A researcher analyzing the sounds in languages spoken around the world has detected an ancient signal that points to southern Africa (meaning Sub-Sahara) as the place where modern human language originated. The finding fits well with the evidence from fossil skulls and DNA that modern humans originated in Africa. The detection of such an ancient signal in language is surprising. http://faculty.ucr.edu/~legneref/igbo/westafricanorigin.htm#igbooriginhumankind

One does not necessarily need language to make a tool, but one needs language to explain to others how to make a tool. Thus, it was a natural evolution as toolmaking became more sophisticated. Language, therefore, was the key that opened the door to a greater and shared imagination. Reading, writing, mathematics, etc. are all tools that are the result of a spoken and shared language. Throughout the ages, language, art and science have served as conduits of the human imagination and collective experiences. It is through such voices that the ancients speak.

Language and art reflected a new type of thinking; one in which the human mind began associating things with symbols orally and artistically. These antecedents of writing were fundamental to the development of higher thinking that enabled humans to emerge from the realm of beasts and humans to become the cynosure of God's eye. Such cognitive thinking created an unprecedented species in which symbols became germane to social gatherings that ultimately led to civilizations. It seems almost improbable that of the hundreds of thousands of species that have graced the confines of this Earth, only one developed the uniqueness of the spoken word, with its grammar and syntax which makes this form of communication unique and exceptional.

It is true that other animals make sounds and communicate, however, that is far from a language. Only animals in Aesop's fables or Disney movies, express themselves in a uniquely human manner. It is as if we humans are hardwired for language and the association of symbols as a form of expression.[199] Language and art are not only the harbingers of civilization but of science, religion and philosophy that give us the power to question our place in the universe. Wherever we go on this planet, and we encounter fellow members of our species, we intuitively develop means to communicate with each other and to exchange languages using symbols and names that we assign for things and events. Even though each language is unique, the

---

199 How can babies learn the melodies and sounds of their mother tongue even before they are born? The answer is as simple as this: about 3 months before birth, while still in their mother's womb, babies start to hear. At that time, their ears are developed enough and start working. Usually, it will mostly be the mother's voice that reaches the baby's ears inside the womb, but other loud sounds or voices as well. Consequently, every day of the last few months before birth, the baby can hear people speaking – this is the first step in language learning! This first step, in other words, is to learn the melody of the language. Later, during the next few months and years after birth, other features of language are added, like the meaning of words or the formation of full sentences. Although infants already possess an important groundwork for language acquisition, more advanced language learning becomes possible as the brain continues to develop during childhood. The Brain and Language: How Our Brains Communicate · Frontiers for Young Minds (frontiersin.org)

common threads are woven into a fabric of understanding of our environment and experiences.

When language appeared has been a matter of debate and conjecture among scholars of various disciplines. However, we do know that language is anatomically unique among our species although many species make and respond to sounds. For instance, packs of wolves communicate and cooperate during hunting. Chimpanzees communicate to warn others of danger. However, words, sentences and syntax comprise a realm of expression that is uniquely human. Most intriguing is the position of the larynx in adult humans. In most animals, and infants, the larynx sits high in the throat and serves as a valve. This allows animals and infants to breathe and drink at the same time. However, as infants become mature, their larynx is repositioned halfway down the throat, the so-called Adam's apple prevalent in adult males. Because of this shift, our species can no longer drink and breathe at the same time without choking.

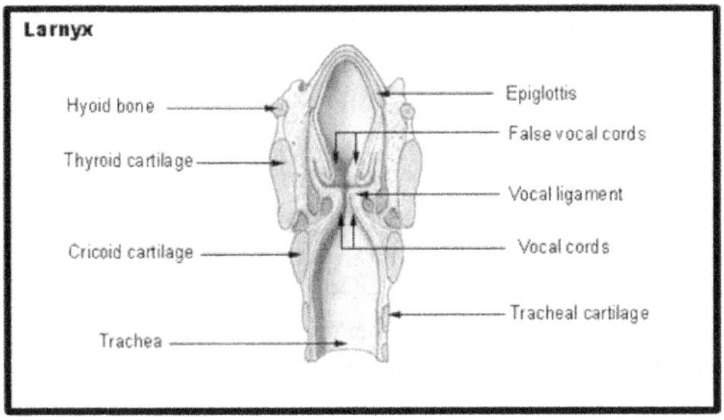

The position of the larynx, along with bipedalism, is an aspect that makes humans unique among the creatures of this Earth. Bipedalism freed the hands, which facilitates writing, a key feature in the collective learning process. Writing allowed the transmission of general and specific information to be transferred from generation to generation. From these two anatomical features, the diverging roads traveled alone by our species throughout time can be traced. Therefore, the communication skills of which language is a central part represent a quantum leap along the road not only to civilization but to an understanding of our place in the universe. Language and its derivatives, e.g., reading and writing are the links in our human chain connecting us not only to each other but to past and future generations to come. Language is, in many ways, the catalyst of our consciousness and innovation. In hearing or reading words, we are merging our consciousness with others forming an invisible chain that links our species together from generation to generation and from one idea to another.

Art and language are the expressions of higher intellect, and both originated in Africa. Much has been made of the rock paintings found in Europe, particularly, rock art located in the Franco-Cantabrian region and has subliminally been used to suggest that such findings were a sign of the innate white European intelligence. However, it appears that these skills migrated to Europe and the rest of the world, along with everything else, with sapiens out of Africa. The following explains:

> Scientists determined the age of a cave painting by dating the lumpy mineral deposits that formed on top of the paintings after they were made. The deposits contained small amounts of uranium. Uranium is an unstable radioactive element. It decays into thorium at a known rate. If you can measure the ratio of uranium to thorium in the deposits, then it's possible to calculate their age. They found they were at least 40,000 years old. That's at least as old as the oldest known art in Europe. It would mean that humans probably didn't first invent art in Europe. Our species, *Homo sapiens,* first appeared in Africa, and then migrated outward into Europe and Asia. It might mean that art was invented much earlier than we thought, in Africa, and that humans carried the tradition with them as they spread.[200]

Along with language came the ability to sing and to praise a deity. From those early flames of yesteryear, embers of understanding ignited and reignited our imagination. We should be forever mindful that whatever series of *Consciousness* and development that separated humans from the realm of beasts took place first among Africans. Perhaps Immanuel Kant said it best:

> The history of humanity, viewed as a whole may be regarded as the realization of a hidden plan of nature to bring about a political constitution, internally, and for this purpose, also externally perfect, as the only state in which all the capacities implanted by her in mankind can be fully developed.[201]

200 How Art Began | A Moment of Science - Indiana Public Media

201 https://tuningjohn.com/book/kant/

Sapiens are the most advanced species *known* in the universe. As we gaze into a cosmos, with stars more abundant than grains of sand on our Earth, we are left to ponder: if such steps were taken by other sentient beings on planets orbiting around distant stars? Or are we alone: orphans on a blue planet revolving around a mediocre star in a mediocre galaxy in a universe among universes? By looking into our past, perhaps we are, in a strange way, gazing into our future. Our ability to imagine and to collectively learn has not only enabled us to trek around the globe, but to place our footprints on another world.

Astronaut Buzz Aldrin, lunar module pilot, walks on the surface of the Moon near the leg of the Lunar Module (LM) "Eagle" during the Apollo 11 extravehicular activity (EVA). Astronaut Neil A. Armstrong, commander, took this photograph with a 70mm lunar surface camera. While astronauts Armstrong and Aldrin descended in the Lunar Module (LM) "Eagle" to explore the Sea of Tranquility region of the Moon, astronaut Michael Collins, command module pilot, remained with the Command and Service Modules (CSM) "Columbia" in lunar orbit.

LET THERE BE LIGHT

Then God said, "Let there be light"; and there was light!

Genesis 1:3

The ability of the human mind to comprehend the vastness and to explore the idiosyncrasies of the cosmos is unprecedented in the evolution of species on this planet. In 1845, Michael Faraday discovered that a magnetic field influenced polarized light, a phenomenon known as the *magneto-optical effect* or the Faraday effect. He noticed that the plane of vibration of a beam of linearly polarized light incident on a piece of glass rotated when a magnetic field was applied in the direction of propagation of the beam. This was one of the first indications that electromagnetism and light were related. Hence, the laws of electricity and magnetism are due much more to Faraday's experimental discoveries than to any other scientist. [202]

Faraday, Maxwell and Electromagnetism

Using the empirical discoveries of Faraday, in 1867, James Clerk Maxwell predicted that electromagnetic waves traveled at the speed of light and opened the door to a new room of Physics.[203]. He was able to calculate the speed of these waves in a vacuum based on the properties of electricity and magnetism. He found that this speed was the experimentally measured speed of light.

$$c = \frac{1}{\sqrt{\varepsilon_0 \mu_0}}$$

Where ε_0, the permittivity of free space, has an experimentally determined value of 8.85×10^{-12} square coulomb per newton square meter, and μ_0, the magnetic permeability of free space, has a value of 1.26×10^{-6} newton square seconds per

202 Faraday and the Electromagnetic Theory of Light | OpenMind (bbvaopenmind.com)

203 In all Michael Faraday is cited six times and mentioned three times in Maxwell's 1865 paper. However, this is not surprising considering that a large amount of Maxwell's work is based on Faraday's work and Maxwell mathematically modelled most of Faraday's discoveries on electromagnetism into the theory that we know today. Faraday and the Electromagnetic Theory of Light | OpenMind (bbvaopenmind.com)

square coulomb.[204] The speed of an electromagnetic wave, c, is equal to 3 × 108 ms, the same as light. Although the complete electromagnetic spectrum was unknown, Maxwell predicted that there should be light with even longer wavelengths than infrared light.

Portrait of Faraday in 1842 by Thomas Phillips and James Clerk Maxwell by Jemima Blackburn [205]

Maxwell recognized that electromagnetism's processes could be determined using mathematics. Maxwell predicted the connections of magnetism and electricity leading directly to the prediction of electromagnetic waves. He also predicted that the speed of light was constant. Maxwell worked in mathematical physics until his death in 1879

204 light - Light as electromagnetic radiation | Britannica

205 Two of the boldest and most creative scientists of all time were Michael Faraday (1791-1867) and James Clerk Maxwell (1831-1879). This is the story of how these men, separated in age by forty years, discovered the existence of the electromagnetic field and devised a radically new theory that overturned the strictly mechanical view of the world that had prevailed since Newton's time. Faraday was an autodidact, who overcame class prejudice and a lack of mathematical training to become renowned for his acute powers of experimental observation, technological skills, and prodigious scientific imagination. Maxwell was highly regarded as one of the most brilliant mathematical physicists of the age and made numerous advances in his own right. But when he translated Faraday's ideas into mathematical language, thus creating field theory, this unified framework of electricity, magnetism and light became the basis for much of later, 20th-century physics. Their collaborative efforts gave rise to many of the technological innovations we take for granted today - from electric power generation to television, and more. Told with panache, warmth, and clarity, this captivating story of their greatest work, in which each played an equal part, and their inspiring lives will bring new appreciation to these giants of science. https://books.apple.com/us/book/faraday-maxwell-and-the-electromagnetic-field/id1479483011?l=vi

and formulated what is now known as Maxwell's Equations. They describe, through mathematics, the functions of electricity and magnetism and launched a new branch of Physics. He also predicted the existence of electromagnetic waves. His equations demonstrate that moving electric charges create magnetic fields, while moving magnets create electrical fields (which is how power plants generate electricity). Maxwell's equations are empirically sound and are based on the meticulous work of Gauss, Faraday, Ampere, Oersted, Coulomb and others that came before him. The four basic laws of electricity and magnetism had been discovered experimentally through the work of the aforementioned physicists. However, Maxwell discovered logical inconsistencies in these earlier results and identified the incompleteness of Ampère's law as their cause. By consolidating their work, Maxwell mathematically quantified the electromagnetic spectrum. The following are his equations.

Maxwell's Equations[206]

Integral form in the absence of magnetic or polarizable media:

I. Gauss' law for electricity $\oint \vec{E} \cdot d\vec{A} = \frac{q}{\varepsilon_0}$

II. Gauss' law for magnetism $\oint \vec{B} \cdot d\vec{A} = 0$

III. Faraday's law of induction $\oint \vec{E} \cdot d\vec{s} = -\frac{d\Phi_B}{dt}$

IV. Ampere's law $\oint \vec{B} \cdot d\vec{s} = \mu_0 i + \frac{1}{c^2} \frac{\partial}{\partial t} \int \vec{E} \cdot d\vec{A}$

Maxwell's first equation is built on Gauss' Law for electricity, which states that "when a closed surface integral of electric flux density is always equal to charge

206 http://hyperphysics.phy-astr.gsu.edu/hbase/electric/maxeq.html

enclosed over that surface." Maxwell's second equation is based on Gauss' law for magnetism which states that "closed surface integral of magnetic flux density is always equal to total scalar magnetic flux enclosed within that surface of any shape or size lying in any medium." Maxwell's third equation is derived from Faraday's laws of Induction. It states that "Whenever there are n-turns of conducting coil in a closed path placed in a time-varying magnetic field, an alternating electromotive force gets induced in each coil. Maxwell's fourth equation is based on Ampere's circuit law which states: that "The closed line integral of magnetic field vector is always equal to the total amount of scalar electric field enclosed within the path of any shape", which means the current flowing along the wire is equal to the magnetic field vector.[207]

Maxwell also predicted that the electrical and magnetic components were perpendicular to each other and shared a symbiotic relationship.[208]

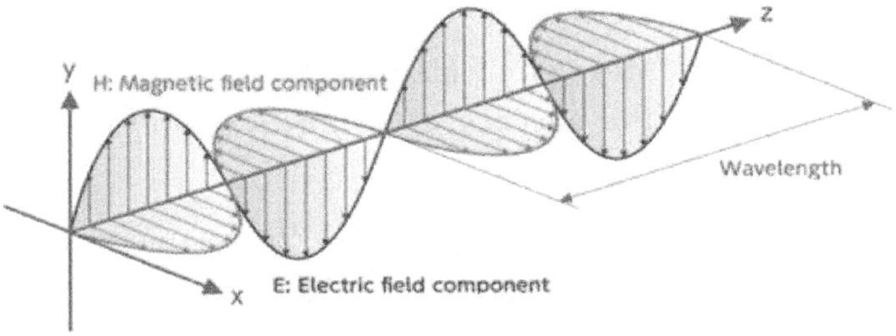

Basic Electromagnetic Theory - Emagtech Wiki by Unknown Author is licensed under CC BY-SA-NC

German physicist Heinrich Hertz proved Maxwell's electromagnetic wave theory was correct, and in the process, generated and detected electromagnetic waves. [209] Between 1879 and 1889, he did a series of experiments that used electrical and magnetic fields to produce waves that could be measured. He confirmed that the velocity of the waves was equal to the speed of light. He measured the characteristics of the fields by measuring their magnitude, polarization, and reflections.[210] Hertz's

207 Maxwell's Equations: Maxwell's 4 Equations And Their Derivations (byjus.com)

208 Maxwell's paper On Physical Lines of Force—written over the course of two years (1861-1862) and ultimately published in several parts—introduced his pivotal theory of electromagnetism. Among the tenets of his theory were (1) that electromagnetic waves travel at the speed of light, and (2) that light exists in the same medium as electric and magnetic phenomena. https://www.thoughtco.com/james-clerk-maxwell-inventor-1991689

209 https://www.thoughtco.com/electromagnetism-timeline-1992475

work exposed the electromagnetic spectrum and radio waves in particular. He also focused on a concept called the photoelectric effect which Einstein later refined earning him a Nobel prize in 1921 and it became the basis of physics called Quantum Mechanics.[211]

Heinrich Rudolf Hertz (1930)

In 1887, while Hertz was experimenting with electromagnetic waves, the implication that the speed of light was constant led to the Michelson-Morley experiment and the subsequent response by Lorenz and Fitzgerald to their conclusion.[212] It was assumed that light, like sound, traveled, on some medium called the aether, but this was not the case. As a result of the Michelson-Morley experiment, the speed of light was empirically measured to be constant at 186,350 miles per second with a likely error of around 30 miles per second. The constant speed of light meant that light was not like sound and that the so-called aether did not exist.

210 Biography of Heinrich Hertz (thoughtco.com)

211 Einstein resolved this problem using Planck's revolutionary idea that light was a particle. Since light is bundled up into photons, Einstein theorized that when a photon falls on the surface of a metal, the entire photon's energy is transferred to the electron Einstein's Explanation Of Photoelectric Effect - Threshold Frequency | BYJU'S (byjus.com)

212 They found absolutely no evidence of the relative motion bias they were looking for. No matter which path the beam took, light seemed to be moving at precisely the same speed

Mathematicians Lorenz and Fitzgerald tried to explain the conclusions of the Michelson-Morley experiment which resulted in the concept of time dilation and length contraction. The equations developed by Lorenz and Fitzgerald form the basis for space and time dilation and length contraction and also established the speed of light as the limit. Einstein was the beneficiary of the ideas and brilliance of the aforementioned men. He pointed to the Michelson-Morley experiment as the basis for the development of Special Relativity. He said:

> I was familiar with the strange results of Michelson's experiment while I was still a student pondering these problems, and instinctively realized that, if we accepted his result as a fact, it would be wrong to think of the motion of the Earth with respect to the ether. This insight actually provided the first route that led me to what we now call the principle of special relativity. I have since come to believe that, although the Earth revolves around the Sun, its motion cannot be ascertained through experiments using light.[213]

Albert A. Michelson and Edward Williams Morley in 1907, Michelson became the first American scientist to earn the Nobel Prize: he did so in physics

[213] Michelson-Morley Experiment - The Flat Earth Wiki (tfes.org)

Special and General Relativity

Like those that came before him, Einstein did not arrive at his conclusions by clinging to conventional wisdom, instead, by stretching his imagination into the realm of the unknown using science as a tool to chisel out a sculpture of the universe that displays its natural beauty. Hence, he initiated a new kind of expression for the artisans of his craft to follow. In Einstein's universe, space and time were no longer constants but malleable variables that were related to the speed of light. We, therefore, lived in the realm of spacetime of which Newtonian physics was a part. A relativistic universe in which gravity was linked to acceleration.[214] Mass and energy were two sides of the same coin.

Albert Einstein in 1947

In 1905, Albert Einstein presented one of the most important papers ever published in the field of physics. Like Maxwell's Equations, the *Theory of Special Relativity* stood on the shoulders of empirical and theoretical observations that came before it. Special

214 A century ago, Einstein's powerful field equations of gravitation showed the way forward. His uncanny intuition about the universe prevailed despite temporary reversals, and his decades-old insights are now at the cutting edge of modern physics and cosmology, helping us shed light on the greatest mysteries of all: the nature of matter, gravity, time, space, and the mysterious dark energy pushing it all https://www.discovermagazine.com/the-sciences/einsteins-lost-theory-describes-a-universe-without-a-big-bang

relativity is an explanation of how speed affects mass, time and space. The theory includes a way for the speed of light to define the relationship between energy and matter — small amounts of mass (m) can be interchangeable with enormous amounts of energy (E), $E = mc^2$.[215] On the most basic level, the equation says that energy and mass (matter) are interchangeable; they are different forms of the same thing. Under the right conditions, energy can become mass, and vice versa. Before Special Relativity, Newton's laws were unmalleable, and time and space were constants. In everyday life, things move at a slow pace whether it's running, trains, rockets or bullets. It is only when the velocity approaches the speed of light when Newtonian physics breakdown.

Special relativity is based on two postulates. The first postulate was that the laws of physics are the same everywhere, regardless of the frame of reference, as long as the frame was moving with a constant velocity. Zero acceleration is therefore equal to a state of rest. This means that physics operates the same whether you are on an orbiting Moon, floating alone in outer space, standing on planet Earth, or flying in an airplane. For example, the Earth is moving about 1,000 mph yet we do not feel its motion because its acceleration is zero.

The second postulate was that the speed of light was constant regardless of the motion of the light source or the observer. The thing that Einstein knew (that both Newton and Galileo had no way of knowing) was that: the speed of light in a vacuum must be exactly the same for everyone. This had been proven empirically by the. Michelson-Morley experiment. "Light is always propagated in empty space with a definite velocity c which is independent of the state of motion of the emitting body." Special relativity applies to "special" cases. It is mostly used when discussing huge energies, ultra-fast speeds and astronomical distances, all without the complications of gravity. Einstein officially added gravity to his theories in 1915, with the publication of his paper on general relativity.[216]

Lorenz and Fitzgerald could be considered the true primogenitors of Special Relativity and many scientists believed that if Einstein did not develop the principles that another physicist would have, based on the mathematics of Lorenz and Fitzgerald.[217].

215 Michelson-Morley Experiment - The Flat Earth Wiki (tfes.org)

216 Einstein's Theory of Special Relativity | Space

Germane to their equations are length contraction and time dilation. One of the many implications is that time moves relative to the observer. An object in motion experiences time dilation, meaning that when an object is moving very fast it experiences time more slowly than when it is at rest.[218] The following formulas describe length contraction and time dilation relative to the speed of light. The faster you move through space; the shorter distance becomes according to the following formula:

$$L_o = \frac{L}{\sqrt{1 - \frac{v^2}{c^2}}}$$

Similarly, the faster you move through space, the slower you move through time according to the following formula:

$$\Delta t' = \frac{\Delta t}{\sqrt{1 - \frac{v^2}{c^2}}}$$

The following best explains time dilation:

> When astronaut Scott Kelly spent nearly a year aboard the International Space Station starting in 2015, he was moving much faster than his twin brother, astronaut Mark Kelly, who spent the year on the planet's surface. Due to time dilation, Mark Kelly aged just a little faster than Scott — "five milliseconds," according to the earth-bound twin. Since Scott wasn't moving near lightspeed, the actual difference in aging due to time dilation was negligible.[219]

Furthermore, as an object approaches the speed of light, the object's mass becomes infinite and so does the energy required to move it. This means it is impossible for

---

217 The concept of the contraction was proposed by the Irish physicist George FitzGerald in 1889, and it was thereafter independently developed by Hendrik Lorentz of the Netherlands https://www.britannica.com/science/Lorentz-FitzGerald-contraction

218 Einstein's Theory of Special Relativity | Space

219 https://www.space.com/36273-theory-special-relativity.html

any matter to go faster than light. This cosmic speed limit inspires new realms of physics and science fiction, as people consider travel across vast distances. As you can see, in the following equation, when v =c, m is infinite.[220]

$$m = \frac{m_0}{\sqrt{1 - \frac{v^2}{c^2}}}$$

Albert Einstein and Hendrik Lorentz, photographed by Ehrenfest in front of his home in Leiden in 1921

In Special Relativity, Einstein described the relationship between space and time when an object moved at a constant velocity, however, acceleration and gravity remained another problem to be solved. Although Einstein is most famous for his work on Special Relativity, he was merely following in the footsteps of Lorenz and Fitzgerald. Notwithstanding, General Relativity is by far a more original work and took seven years of work by Einstein, the final two to three being years of intense and exhausting labor. Germane to Einstein's Theory of General Relativity is the concept of spacetime. To understand the concept, it is important to juxtapose Newton's laws of gravity with Einstein's.

220 Einstein's Theory of Special Relativity | Space

In Newton's classical account of gravitation, the Earth and the planets want to move inertially, that is, uniformly in a straight line. The gravitational force from the sun deflects it and causes it to move in an elliptical orbit around the Sun. In Einstein's theory, the presence of the Sun curves the very fabric of spacetime. The Earth then merely moves inertially in this new disturbed spacetime. It follows an inertial trajectory, but that trajectory has been distorted so that it ends up as an ellipse in the space around the Sun; or, more precisely, a helical trajectory winding around the Sun's curvature of spacetime.[221] Spacetime remains undistorted only if a unit of spacetime has less energy density. Newton's concept is weaker but not incorrect, however, Einstein's concept is stronger and more precise. In Einstein's universe, like space and time, gravity and acceleration are two sides of the same coin. Einstein noted, every kind of acceleration should be describable by the equations for gravity. If not, "the theory refutes its own starting point; then it has no foundation whatsoever." Einstein showed that gravity did not just make what goes up always come down, it made the universe go around.

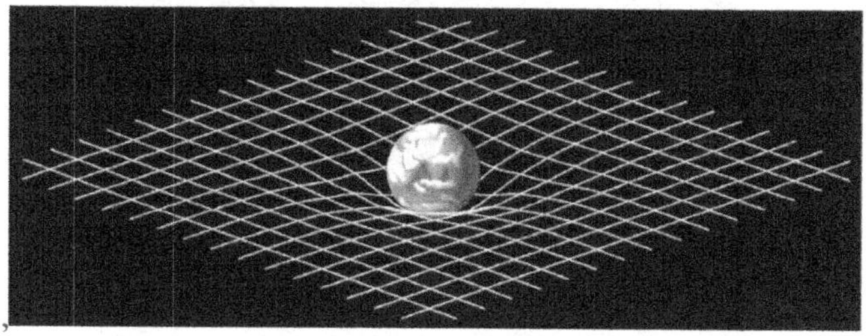

Spacetime curvature schematic[222]

We increasingly find, thanks to Maxwell and others, that Newtonian physics is a subset of Einsteinian physics. Einstein's theory proposes that gravity is not an actual force but is instead a geometric distortion of spacetime not predicted by ordinary Newtonian physics. The more mass you have, the more it distorts spacetime. This distortion changes the path of objects moving through space, and even the paths of light rays, as they pass close-by a massive body. Even so, this effect is very feeble for an object as massive as our own sun, so it takes enormous care to even discover that it is happening![223] Over the last century, general relativity's predictions have been

221 https://sites.pitt.edu/~jdnorton/teaching/HPS_0410/chapters/general_relativity/

222 By Mysid - Own work. Self -made in Blender & Inkscape., CC BY-SA 3.0, https://commons.wikimedia.org/w/index.php?curid=45121761

repeatedly verified by modern precision measurements. General relativity explains how the universe can obey physical laws that apply to any form of motion. [224]

Einstein's idea of general relativity had to incubate and coalesce over years. During a solar eclipse, such a shift could be photographed and measured. Such measurements, made during an eclipse in 1919, confirmed Einstein's calculation.[225] By bending light, masses act like a lens; such "gravitational lensing" alters the apparent position of a distant object, creating multiple images of it, or (if the images overlap) appearing to brighten it. Such effects can be used to probe matter's distribution in space or detect the presence of unseen masses.[226] One hundred years after Albert Einstein published his *Theory of General Relativity*, the Hubble Space Telescope has provided a demonstration of the theory at work: a picture of a distant galaxy so massive that its gravitational field is bending the light from an even more distant supernova.[227]

This blue horseshoe is a distant galaxy that has been magnified and warped into a nearly complete ring by the strong gravitational pull of the massive foreground luminous red galaxy.

223 Testing General Relativity | Total Solar Eclipse 2017 (nasa.gov)

224 Einstein's genius changed science's perception of gravity (sciencenews.org)

225 On May 29, 1919, Einstein's four-year-old Theory of General Relativity was put to its first test during a total solar eclipse. Testing General Relativity | Total Solar Eclipse 2017 (nasa.gov)

226 Einstein's genius changed science's perception of gravity (sciencenews.org)

227 Hubble's 'Einstein Cross' Marks the Space-Warping Spot (nbcnews.com)

The Cosmic Horseshoe is one of the best examples of an Einstein Ring. This blue horseshoe is a distant galaxy that has been magnified and warped into a nearly complete ring by the strong gravitational pull of the massive foreground luminous red galaxy. Einstein realized that his new theory of gravity was really a theory of the cosmos. In 1917, he wrote a famous paper applying general relativity to the entire universe. The physicist Max Born wrote: Einstein's ideas," over half a century ago, "have given the physical sciences the impetus which has liberated them from outdated philosophical doctrine and made them one of the decisive factors in the modern world of man.[228]

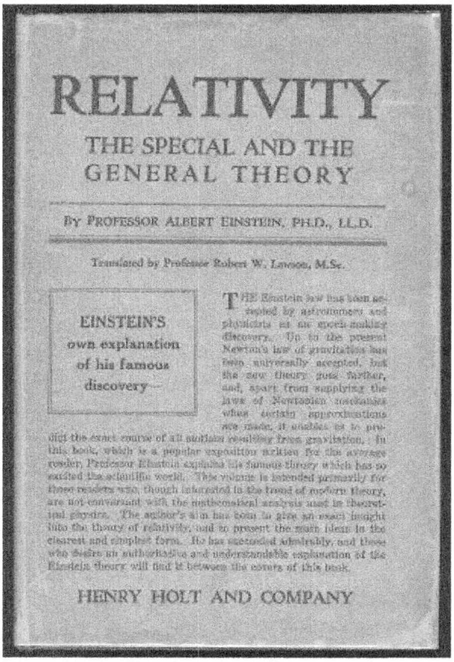

The original 1920 English publication of the paper.

228 Einstein's genius changed science's perception of gravity (sciencenews.org)

EINSTEIN'S UNIVERSE

God does not play dice with the universe

Albert Einstein

On April 26,1920, at the Smithsonian Museum of Natural History a historic debate took place between two astronomers, Heber Doust Curtis and Harlow Shapley regarding the size and scope of the universe. However, what makes the *Great Debate* a true classic in the history of science is the way it ended. At the time, its outcome seemed inconclusive, and despite the drama heralded, it raised more questions than answers. [229] In the following years, new empirical evidence showed that Shapley and Curtis were both half right — and half wrong. Shapley believed that the Milky Way contained all the known and unknown celestial objects – that it represented the whole universe. He calculated that if Andromeda were not in the Milky Way, then it should be as big as 100,000,000 light-years an unfathomable size in 1920.

On the other hand, Curtis claimed that the Milky Way and Andromeda were independent galaxies, just two of many such bodies. In retrospect, Shapley was correct in his estimation of the size of galaxies, and Curtis was correct in his assertion of their independence. It all came down to a matter of scale. Before 1920, many scientists believed that the Milky Way galaxy was the universe. With the Curtis/Shapley debate came the first generally accepted assumption that our galaxy may not be the only galaxy in the cosmos.[230]

[229] The Shapley-Curtis Debate Questioned the Size of the Universe (northropgrumman.com)

[230] Shapley believed that the Milky Way contained all the known and unknown celestial objects – that it represented the whole Universe. He argued that galaxies like Andromeda were simply part of our own galaxy. He backed up his claim with a dimensional argument. Looking into former studies, he calculated that if Andromeda were not in the Milky Way, then it should be as big as 100,000,000 light-years. The Curtis-Shapley Debate And The Discovery Of The Universe. (secretsofuniverse.in) The irony of the Shapley-Curtis debate is that it advanced our knowledge of the cosmos, while simultaneously revealing our insignificance on the grandest scales.

Heber Doust Curtis and Harlow Shapley

The Expanding Universe

In 1927, a Belgian mathematician, Catholic priest and astronomer named Georges Lemaître had a big idea. He said that a very long time ago, the universe started as just a single point. He said the universe stretched and expanded to get as big as it is now, and that it could keep on stretching. [231] The Russian physicist Alexander Friedmann created the mathematical theory of a real expanding universe, based on the idealized earlier work of de Sitter. Thus, began the idea of an infinite expanding universe.

Lemaître used Friedmann's equations and theory to create the final realistic model of an expanding universe. He was the first to foresee a specific redshift-distance relationship, directly from theory.[232] However, the empirical proof was deficient. Two years later, Edwin Hubble turned the 100-inch Mount Wilson telescope to the sky and opened a new chapter in the book of the cosmos. Meanwhile, at the Lowell Observatory, Vesto Slipher independently observed the spectra of galaxies that Hubble later observed himself. Slipher noted a redshift in the spectrum of light which he interpreted as a Doppler effect, due to the nebulae moving away from us. [233]

231 https://spaceplace.nasa.gov/big-bang/en/

232 Mount Wilson Observatory | Discovering Mount Wilson Chapter 9: Hubble's Law (mtwilson.edu)

233 Mount Wilson Observatory | Discovering Mount Wilson Chapter 9: Hubble's Law (mtwilson.edu)

Georges Lemaître and Alexander Friedmann

All the pieces of the galactic puzzle began to fit. We are one of the myriad stars that grace the sky, and our galaxy is one of the billions of galaxies included in our universe. Not only are there billions of galaxies, but the universe is expanding into spaces unknown.[234] Like all discoveries, this one ended not with an answer but another question. As in most cases, when we peer inside the new known, we find a plethora of unknowns. The nature of the expanding universe bemused many scientists, including Einstein that viewed it as static.[235] He went as far as to introduce a cosmological constant to explain the static universe. Einstein heard about Hubble's results, and in the early 1930s, he traveled to California and met with Hubble. At the Mount Wilson Observatory, he saw the massive data set on distant galaxies that had led to "Hubble's law" describing the expansion of the universe. The redshift of starlight could only be explained by an expanding universe.[236] The following explains:

234 The theory of an expanding universe pre-dates Hubble's observations. Dutch mathematician and astronomer Willem de Sitter proved that Albert Einstein's theory of general relativity did not allow a static universe (1917). Russian physicist Alexander Friedmann created the mathematical theory of a real expanding universe, based on the idealized earlier work of de Sitter (1921, 1924). Mount Wilson Observatory | Discovering Mount Wilson Chapter 9: Hubble's Law (mtwilson.edu)

235 In 1917 Einstein applied his theory of general relativity in the universe, and suggested a model of a homogenous, static, spatially curved universe. However, this interpretation has one major problem: If gravitation was the only active force, his universe would collapse – an issue Einstein addressed by introducing the cosmological constant

236 In 1929, Edwin Hubble observed that all but the closest galaxies surrounding our own had a red shift in their hydrogen spectra that was proportional to their distance from us. Applying the Doppler Effect, Hubble recognized that this meant that all galaxies were receding from our own, with those farther away

After Hubble had established an observational redshift-distance relationship in 1929, the expanding universe model created by Lemaître became more than mathematical speculation and was accepted as a valid model of the real universe. Hubble's redshift-distance relationship was not remarkably different from the one theoretically predicted by Lemaître. The rate of expansion quickly became known as the 'Hubble Constant,' and the linear equation describing the redshift distance became known as 'Hubble's Law.' Although it is now sometimes called the Hubble-Lemaître Law, Lemaître himself always attributed it solely to Hubble.[237]

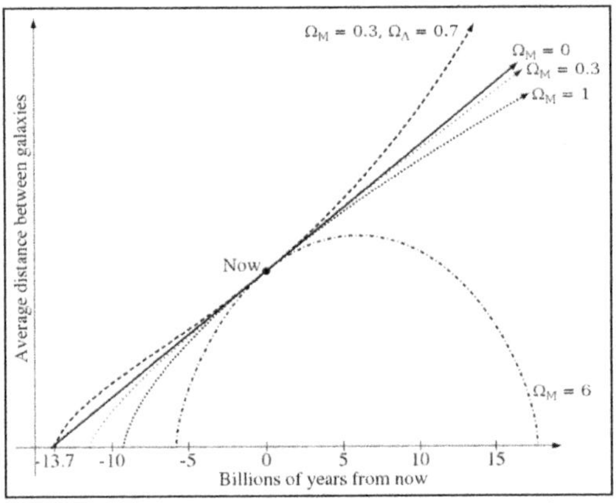

The age and ultimate fate of the universe can be determined by measuring the Hubble constant today and extrapolating with the observed value of the deceleration parameter, uniquely characterized by values of density parameters (ΩM for matter and ΩΛ for dark energy). A "closed universe" with $\Omega M > 1$ and $\Omega\Lambda = 0$ comes to an end in a Big Crunch and is considerably younger than its Hubble age. An "open universe" with $\Omega M \leq 1$ and $\Omega\Lambda = 0$ expands forever and has an age that is closer to its Hubble age. For the accelerating universe with nonzero ΩΛ that we inhabit, the age of the universe is coincidentally very close to the Hubble age.[238]

receding even faster. Knowing that our place in the universe was no more unique than any other, the implication was clear: The space within the universe itself was expanding. Just like pen marks on an expanding balloon, everything in the universe was accelerating away from everything else. 23.3 The Unification of Forces - Physics | OpenStax

237 Mount Wilson Observatory | Discovering Mount Wilson Chapter 9: Hubble's Law (mtwilson.edu)

238 Hubble's Law only works for distant galaxies. For nearby galaxies (in the Local Group), stars inside the Milky Way, and for objects in our Solar System, the relationship between distance and velocity does not hold. The reason for the discrepancy for nearby galaxies is the "peculiar velocity" of the galaxy, that

Light emanating from distant stars and galaxies has an encoded message within. The color of light is related to its wavelength. In a spectrum of what we view as white light, blue has the shortest wavelength and red the longest. Consequently, if starlight from a galaxy appears with a blue shift, such as Andromeda (M31), it is moving toward us and if it appears redshift like most galaxies, it is moving away from us.[239] Blue shifted galaxies (e.g., Andromeda) are only seen in our local neighborhood, not far away. All distant galaxies show a redshift.

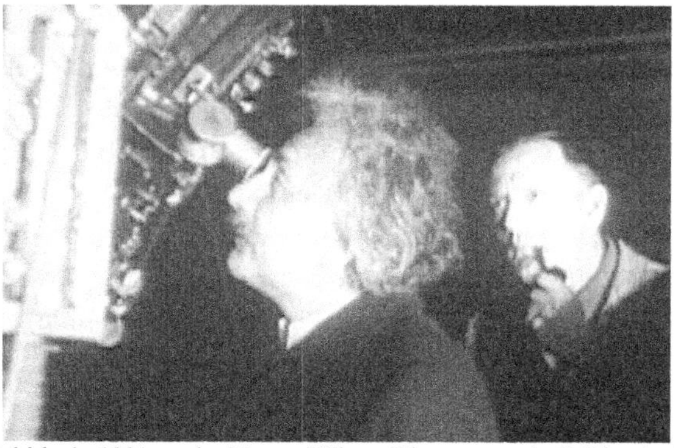

Einstein with Edwin Hubble, in 1931, at the Mount Wilson Observatory in California, looking through the lens of the 100-inch telescope through which Hubble discovered the expansion of the universe in 1929. (Credit: Courtesy the Archives, CalTech)[240]

---

is, its real velocity through space that is unrelated to the expansion. For distant galaxies, their peculiar velocities are small enough that they still lie on or near the line for Hubble's Law. For nearby galaxies, though, their peculiar velocity is larger than their velocity from the expansion, so their peculiar velocity dominates their total velocity, causing them to lie far from the line relating velocity to distance. For example, the galaxy M31 does not even show a redshift; it is blue shifted, showing that its peculiar velocity is pointed towards us, rather than away from us. Hubble's Law | Astronomy 801: Planets, Stars, Galaxies, and the Universe (psu.edu)

239 The blue shift is a result of moment within our local cluster of galaxies in in the future our galaxy and Andromeda will combine to form one larger galaxy.

240 https://www.discovermagazine.com/the-sciences/einsteins-lost-theory-describes-a-universe-without-a-big-bang

Einstein came up with mathematical formulas to describe the universe. The formulas only worked if the universe were growing or shrinking. At the time, though, he and others believed the universe was doing neither. So, he added an extra term to the formulas. He called it a "cosmological constant." With that the formulas described a universe that did not change. Einstein soon dropped the term, however, when he found out later that the universe is expanding. He called the whole idea his "greatest blunder. "Einstein's idea may not have been a blunder after all. Scientists have found that the universe isn't just expanding. It is expanding faster and faster, they say. So, what is causing this? One way to explain this is "dark

When Alexander Friedmann, developed an explanation of an expanding or contracting universe from Einstein's original equations, Einstein first thought Friedmann to be in error. Nonetheless, he eventually yielded, although still viewing the "expanding universe" as of mathematical interest only. But a few years later, when Edwin Hubble's analysis of light from distant galaxies confirmed the universe's expansion, Einstein capitulated and concurred with Hubble's analysis.[241]

Hubble used Cepheid variable stars, discovered by Henrietta Leavitt to measure the distances to a sample of galaxies thereby establishing that or galaxy was just one of billions of galaxies in the universe. A Cepheid variable star's pulse rate reveals the star's true, fundamental brightness. The amount by which the star's brightness is dimmed by distance allows the star's distance from the Earth to be calculated. In 1929, Hubble demonstrated that the distance to a galaxy (measured from Cepheid variables) and the velocity of the galaxy (measured by the shift in the spectral lines), are inextricably connected.[242]. This was the missing link, and the watershed moment of astronomy in the 20th century. The following is known as Hubble's Law:

$$v=H_0d$$

energy." This hidden force may be Einstein's cosmological constant. NASA - Einstein and Beyond

241 At larger distances (as measured e.g., with Cepheids) the redshift is larger. The reddening of the light from distant galaxies indicates that the wavelength has been stretched somehow. By determining the redshift of an object, scientists can calculate the speed at which the object is moving away from us (as a result of the universe expanding) and the distance to the object from Earth Einstein's genius changed science's perception of gravity (sciencenews.org)

242 The intrinsic brightness of Cepheid variable stars is strongly tied to their period. This is the famous period-luminosity relationship discovered by Henrietta Leavitt more than a century ago. So, astronomers can predict the absolute magnitude (i.e., the mean intrinsic luminosity) of any given Cepheid by measuring the time it takes to rhythmically change its brightness. By observing the apparent luminosity, dimmed by the inverse square law of light traveling across the vast reaches of space, and comparing this with the predicted luminosity, astronomers can calculate the distance to that star. How do astronomers use Cepheid variables to measure distances? | Astronomy.com

H_0 is known as the Hubble constant and is the slope of the line that connects the distance of a galaxy to its velocity. If H_0 is known, and if you can calculate the velocity, v, from the spectrum, then you can use this equation to calculate the distance, d, to that galaxy. [243] Hubble continued measuring distances and velocities in deep space. He discovered that the further apart galaxies are from each other, the faster they move away from one another. His findings, published in 1929, led to the widely accepted belief that the universe is expanding.[244]

It is important to acknowledge that the Hubble constant could not have been derived without the painstaking work of Henrietta Leavitt. In a constellation dominated by men, she was a bright star that shined.

Henrietta Swan Leavitt (July 4, 1868 – December 12, 1921)

Because of her gender, like many women of her day, she was relegated to the mundane work of a computer, and the light of her genius, like the Cepheid variable star, was dimmed by the hubris of men. It is most ironic but apropos that a two-part (or diptych) quilt depicts stars as leaving on the morning of Leavitt's birth in

243 https://www.e-education.psu.edu/astro801/content/l10_p3.html

244 When did Einstein Meet Hubble? - Sohoplayhouselv.com

Lancaster, Massachusetts, on July 4, 1868, and returning on the evening of her death in Cambridge, Massachusetts on December 12, 1921. Between these events is a life about which little is known or remembered.[245]

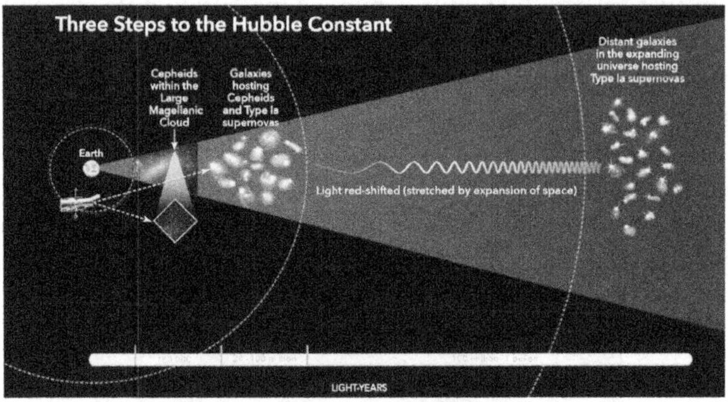

This illustration shows the three basic steps astronomers use to calculate how fast the universe expands over time, a value called the Hubble constant.[246]

As with Hertz, Hubble, and others discerned that deep inside the electromagnetic spectrum was more than meets the eye. Tales of distant stars, stories of our beginnings, and the vastness of our universe. Missions such as the James Webb Space Telescope take advantage of this phenomenon by viewing light in the near- and mid-infrared wavelengths – the wavelengths to which light has shifted after traveling from the most distant, and therefore oldest, stars and galaxies.[247] This allows astronomers to peer into the early history of the universe.

245 Remembering Astronomer Henrietta Swan Leavitt | National Air and Space Museum (si.edu)

246 All the steps involve building a strong "cosmic distance ladder," by starting with measuring accurate distances to nearby galaxies and then moving to galaxies farther and farther away. This "ladder" is a series of measurements of different kinds of astronomical objects with an intrinsic brightness that researchers can use to calculate distances. Among the most reliable for shorter distances are Cepheid variables, stars that pulsate at predictable rates that indicate their intrinsic brightness. Astronomers recently used the Hubble Space Telescope to observe 70 Cepheid variables in the nearby Large Magellanic Cloud to make the most precise distance measurement to that galaxy. Astronomers compare the measurements of nearby Cepheids to those in galaxies farther away that also include another cosmic yardstick, exploding stars called Type Ia supernovas. These supernovas are much brighter than Cepheid variables. Mystery of Universe's Expansion Rate Widens With New Hubble Data | NASA: the Hubble constant.
Credits: NASA, ESA and A. Feild (STSc

247 Ibid

The Electromagnetic Spectrum

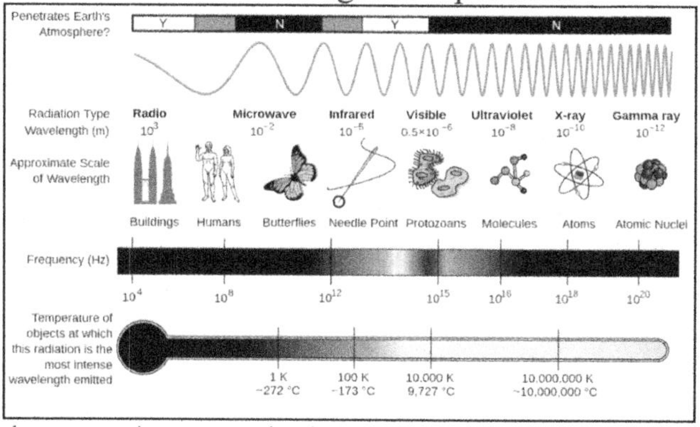

A diagram of the electromagnetic spectrum, showing various properties across the range of frequencies and wavelengths[248]

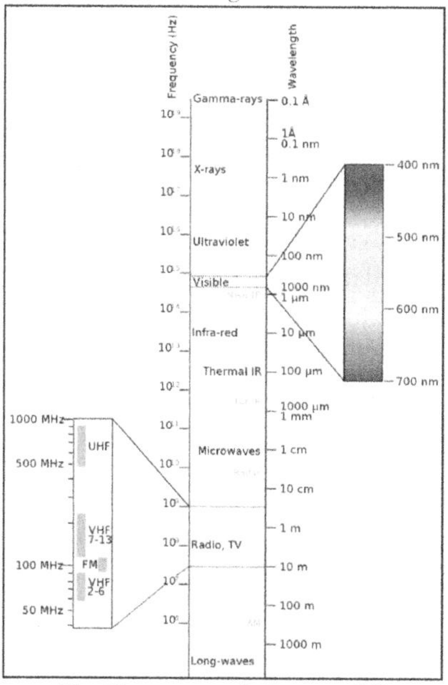

The Electromagnetic Spectrum[249]

---

248 CC BY-SA 3.0, https://commons.wikimedia.org/w/index.php?curid=2974242

249 By Original: Penubag Vector: Victor Blacus - Own work based on: Electromagnetic-Spectrum.png, CC BY-SA 3.0, https://commons.wikimedia.org/w/index.php?curid=22428451

Astronomers believe that Hubble's law is a direct consequence of the ongoing expansion of the universe and that the evidence suggests that the universe began in an explosion, which we call the Big Bang. The Big Bang theory has been challenged by scientists most noted was the late Fred Hoyle that coined the phrase in sarcasm. The following explains the assumption behind the Big Bang.

> The universe was born with the Big Bang as an unimaginably hot, dense point. When the universe was just 10^{-34} of a second or so old — that is, a hundredth of a billionth of a trillionth of a trillionth of a second in age — it experienced an incredible burst of expansion known as inflation, in which space itself expanded faster than the speed of light. During this period, the universe doubled in size at least 90 times, going from subatomic-sized to golf-ball-sized almost instantaneously.[250]

Therefore, the Big Bang rests on three pillars: Hubble expansion, cosmic background heat radiation, and the relative abundance of chemical elements (most noted are Hydrogen and Helium).[251] The Big Bang suggests that the universe started from a single point and that as it expands, the radiation within it cooled. Scientists believe that the Cosmic Microwave Background (CMB) is leftover radiation from the Big Bang or the time when the universe began. As the theory goes, when the universe was born it underwent rapid inflation, expansion and cooling. The universe is still expanding today, and the expansion rate appears different depending on where you look. The CMB represents the heat leftover from the Big Bang.[252] Observations made about the energy distribution of the CMB map indicates that predictions of the spectrum by the Big Bang Theory were correct.[253]

250 Our expanding universe: Age, history & other facts | Space

251 [Davies: The Last Three Minutes, p.27]

252 What is the cosmic microwave background? | Space

253 The discovery of Cosmic Microwave Background Radiation was basically an accident, Radio astronomers, Robert Wilson and Arno Penzias, were building a radio receiver when they picked up a peculiar buzzing that was coming from all parts of the sky. At a loss, Wilson and Penzias did everything they could think of to expel any possible interference before considering what it could mean. https://astronomywebproject.weebly.com/cosmic-background-radiation.html

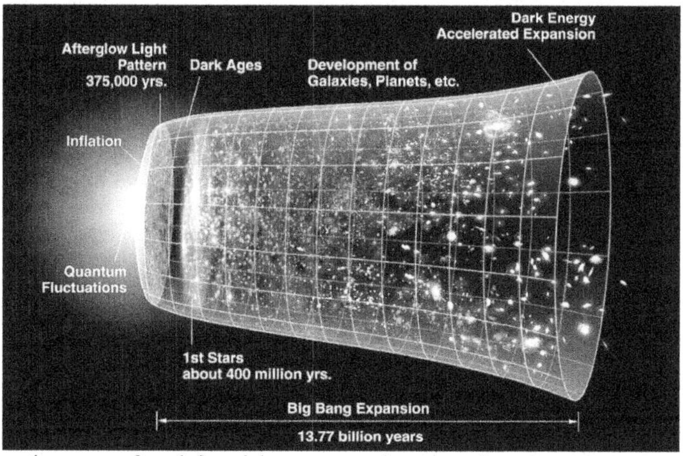

In this diagram, time passes from left to right, so at any given time, the universe is represented by a disk-shaped "slice" of the diagram

When we extrapolate backward in time, the Big Bang would have occurred around 13.8 billion years ago, when all matter would have been at a single point. We do not know what caused the explosion nor can we infer what happened before it. However, the James Webb Telescope will be able to observe the light from around the time of the Big Bang.

Einstein and Black Holes

Einstein's *Theory of General Relativity* introduced the concept of black holes and baby universes. It describes the relationship between space, time, gravity and matter. The scions of Einstein continued in his footsteps applying General Relativity to scenarios that lay at the boundaries of Astrophysics and Quantum Mechanics. For example, Karl Schwarzschild, formulated that an astrophysical singularity would warp spacetime so severely that even light would not be fast enough to get out of the spacetime black hole created.[254]

---

254 One of the key points of interest for Karl Schwarzschild's mathematics career were the publications of none other than Albert Einstein. Schwarzschild took great interest to the works published by Einstein, especially those relating to the Theory of Relativity. Without getting into the specifics of this theory (which is very complex), Schwarzschild used aspects of this theory to recognize and prove the existence of cosmic black holes. This was a scientific breakthrough and shaped the way we understand the universe today. https://www.outerspaceuniverse.org/man-who-discovered-cosmic-black-holes-karl-schwarzschild.html

Furthermore, a point of no return, a spherical region surrounding the singularity, would become known as the "event horizon" and the point of singularity would be called a black hole. A singularity is formed when a gigantic star, at the end of its life cycle, collapses on itself so that all its mass is contained within a singular point of zero radius and infinite density. Although the mass of the dying star is now contained within a singular point, the mass remains the same. So, the mass of a black hole is the same as that of the dying star from which it was born. Known physics breaks down beyond the event horizon and, no information can escape, we can have no experience as to what lies inside. As described by American physicist John A. Wheeler: general relativity governs the nature of spacetime, particularly how it reacts in the presence of matter: "matter tells spacetime how to curve, and spacetime tells matter how to move. [255] When we enter an event horizon, we have therefore crossed the Rubicon into a new universe beyond our comprehension.

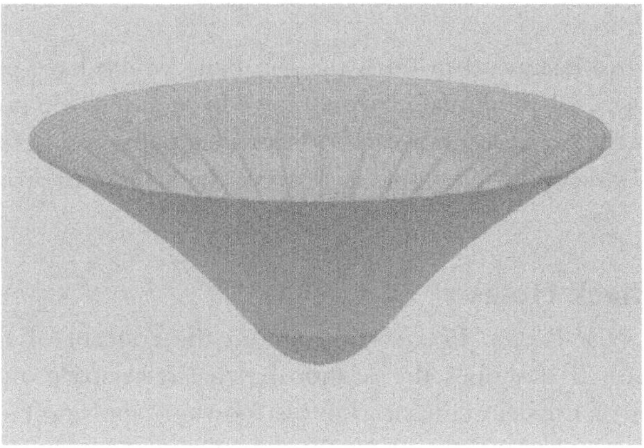

Boundary region of Schwarzschild interior and exterior solution

The Schwarzschild radius is given by the following formula:

$$Rg = 2GM/c^2$$

Where (Rg) is the radius of an object, M is its mass, G is the universal gravitational constant and c is the speed of light:[256] However, it was not until 1935, when the

255 Even Albert Einstein Had His Doubts About Black Holes - HISTORY

256 Schwarzschild radius | Definition, Equation, & Facts | Britannica

Indian astrophysicist Subrahmanyan Chandrasekhar realized that, if a massive star ran out of fuel, the absolute gravitational pressure of that mass would be reduced to a point, causing spacetime to collapse in on itself. He, therefore, provided a physical scenario that supported Schwarzschild's hypothesis.[257]

Karl Schwarzschild and Subrahmanyan Chandrasekhar

Even with Chandrasekhar's contributions toward the modern understanding of the nature of black holes, astrophysical singularities were assumed to be, at best, extremely rare. However, Stephen Hawking and Roger Penrose proved that, far from being rare, singularities were a part of the cosmic ecosystem, and are a part of the natural evolution of massive stars after they run out of fuel and die.[258] John Wheeler wrote in his 1999 autobiography the discovery of black holes demonstrates that spacetime can be crumpled like a piece of paper into an infinitesimal dot, that time can be extinguished like a blown-out flame, and that the laws of physics that we regard as 'sacred,' as immutable, are anything but.[259]

---

257 Subrahmanyan Chandrasekhar (1910-1995) was one of the giants of modern astronomy and astrophysics in the 20th Century. His work connected the study of physics to the structure and evolution of stars and helped astronomers understand how stars live and die. Without his forward-thinking research, astronomers might have labored far longer to comprehend the basic nature of stellar processes that govern how all stars radiate heat to space, age, and how the most massive ones ultimately die. Biography of Subrahmanyan Chandrasekhar (thoughtco.com)

258 Black holes and Einstein's Theory of Relativity - Sciworthy

259 Even Albert Einstein Had His Doubts About Black Holes - HISTORY

Until recent times, these concepts were theoretical. However, the announcement of the first-ever picture of a black hole by the Event Horizon Telescope (EHT) in April 2019 not only confirmed Einstein's original theory, but also provided indisputable evidence that black holes are real, and moreover, exist at the center of most galaxies.[260] For instance, at the center of our Milky Way galaxy lies a black hole with a mass about 4 million times that of our Sun. It's called Sagittarius A*, after its location in the constellation Sagittarius.[261] Einstein's Theory of General Relativity was again confirmed.

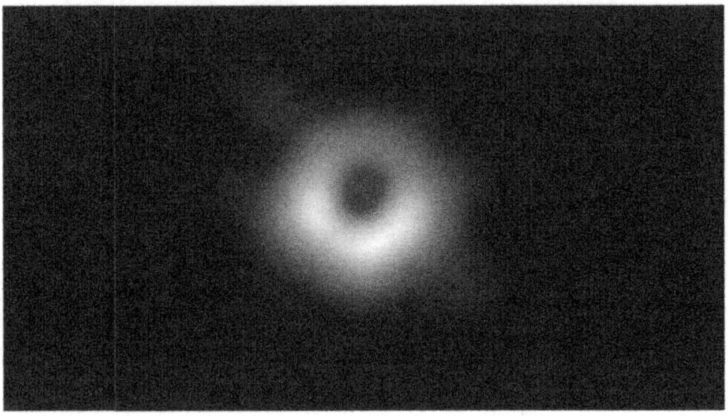

The first image of the shadow of the black hole in the center of M87 taken with the Event Horizon Telescope in 2019.[262]

260 The Event Horizon Telescope (EHT) Collaboration is a group of observatories united to image the emission around supermassive black holes. In 2019, the EHT produced iconic images of the emission around the black hole in M87. The images showed a ring, as predicted by general relativity. The size of the ring indicates that the black hole has a mass that is 6.5 billion times that of the Sun. The images provide evidence that the structure of the emission in M87 changed over the course of six days of observing. Images of M87 obtained by the Event Horizon Telescope Collaboration. Another prime target of the EHT is Sagittarius A* (Sgr A*), the black hole in the center of the Milky Way. Estimates of the mass of the black hole in Sgr A* are much smaller than for M87, weighing in at only 4 million times that of the Sun. Due to this much lower mass, the timescale of variability in Sgr A* is mere minutes, making imaging much more challenging. Event Horizon Telescope - MIT Haystack Observatory

261 Einstein's theory was put to the test. Guess what happened (Opinion) | CNN

262 EHT Collaboration By Event Horizon Telescope, uploader cropped and converted TIF to JPG - https://www.eso.org/public/images/eso1907a/ (image link) The highest-quality image (7416x4320 pixels, TIF, 16-bit, 180 Mb), ESO Article, ESO TIF, CC BY 4.0, https://commons.wikimedia.org/w/index.php?curid=77925953

THE FORCES OF THE COSMOS

No great discovery was ever made without a bold guess.

Isaac Newton

Scientists have initially identified four fundamental forces of Nature. They are the Strong Nuclear force, the Weak Nuclear force, the Electromagnetic force, (recently combined with the Weak Nuclear force) and lastly, the Gravitational force (which seems to be the hardest force to unify)[263] In the last decades, scientists have tried to unify all of the known forces into one as they once were after the big bang. To date, they have combined the Electromagnetic force with the Weak Nuclear force. It is most ironic that the strongest force is associated with the tiniest things, e.g.., atomic nuclei operating at the smallest distance, and the weakest force is associated with larger objects such as stars and galaxies operating at large distances. Although it may seem counterintuitive, gravity, by far, is the weakest force.

The Fundamental Forces of Nature

The strong nuclear force holds protons and neutrons together, the weak nuclear force is responsible for things such as radioactivity and is critical for the nuclear fusion reactions and Carbon -14 dating. The electromagnetic force holds atoms and molecules together. It consists of two parts: the electric force and the magnetic force. The electric component acts between charged particles whether they're moving or stationary, creating a field by which the charges can influence each other. But once set into motion, those charged particles begin to display the second component, the magnetic force. The electrons create a magnetic field around them as they move. So, when electrons zoom through a wire to charge your computer or phone or turn on your TV, for example, the wire becomes magnetic.[264] Gravity is the chief architect of celestial bodies. It holds solar systems and galaxies together and keeps terrestrial bodies from floating into space.

263 Sometime the fundamental forces are referred to as four not three with the weak nuclear force viewed separately.
264 The Four Fundamental Forces of Nature | Space

To give you some idea of the disparity between forces. When juxtaposed against each other the nuclear force or the strong force, reigns supreme. It is ten-trillion-trillion trillion (10^{-37}) times the force of gravity. It binds the fundamental particles of matter together to form larger particles. Both neutrons and protons are subject to the strong force, but electrons are not. It holds together the quarks that make up protons and neutrons, and part of the strong force also keeps the protons and neutrons of an atom's nucleus together.[265] Yet, gravity is the force that creates black holes and is so powerful that it can bend light by warping the fabric of spacetime. The following is a chart that compares the fundamental forces,

Comparison of the Fundamental Forces

Force	Approximate relative strength	Range	Carrier particle
Gravity	10^{-38}	∞	Graviton(conjectured)
Electromagnetic	10^{-2}	∞	Photon (observed)
Weak Nuclear force	10^{-13}	$<10^{-18}$m	W+, W–, Z0 (observed)
Strong Nuclear force	1	$<10^{-15}$m	Gluons (conjectured)

The Four Basic Forces | Physics (lumenlearning.com)

However, at the molecular and atomic levels, gravity has almost no effect relative to the other fundamental forces. It is most ironic that the life and death of the universe is dependent upon nature's weakest force. Nevertheless, if we view the micro and macro worlds as two different worlds with different spheres of influence, we begin to see why nature made its choices. The nuclear force must be the strongest because it is

265 The strong force is odd, though, because unlike any of the other fundamental forces, it gets weaker as subatomic particles move closer together. It reaches maximum strength when the particles are farthest away from each other. The Four Fundamental Forces of Nature | Space The strong force is carried by eight proposed particles called gluons, which are intimately connected to a quantum number called color—their governing theory is thus called quantum chromodynamics (QCD). Taken together, QCD and the electroweak theory are widely accepted as the Standard Model of particle physics. 33.6 GUTs: The Unification of Forces – College Physics (ucf.edu)

only within the atom that the known forces compete against each other and for the universe to work at the micro and the macro levels the nuclear force must prevail in this arena. These forces form the laws that govern our universe and the baryonic matter within it. Maxwell's unification of electricity and magnetism was a work of genius as profound as Einstein's Theories of Relativity. In 1979, nearly a century later, Physicists Sheldon Glashow and Steven Weinberg from Harvard University with Abdus Salam from Imperial College London won the Nobel Prize in Physics for unifying the electromagnetic force with the Weak Nuclear force. These two forces have been merged into a common force called "Electroweak Force". One of Einstein's goals was to unify all the known forces. Although such exercises have not been fruitful in unifying all the known forces, they have given us greater insight into the nature of fundamental forces. The search for a correct theory linking the forces is called the Grand Unified Theory (GUT). The following explains:

> A new theory for the strong force was developed known as quantum chromodynamics (QCD). After that researchers discovered the theory that says the energy of the electromagnetic and weak forces can combine at some point. Moving further at higher energies these two forces electromagnetic and weak forces combine with strong forces, this is known as grand unified theory (GUT). So, we can say that to combine electromagnetic and weak forces with strong forces we have to go at very high energies in the range of Giga electron volts (GeV). In this theory quarks and leptons are also unified, as quarks are responsible for strong forces and leptons are responsible for weak forces. Unified theories are based on group theories and are started after the Maxwell equations. Gravitational force is still not included in the grand unification theory instead when we unify gravitational force with all the other three fundamental forces it becomes the theory of everything (TOE).[266]

Moreover, the interaction between the known forces is germane to how our universe works. Although the strength of the forces differs considerably, the ratio of forces must be as they are for the universe to function as observed. For instance, a small change in their relative strengths could permit the Electromagnetic force to overwhelm the Strong Nuclear force, inhibiting the formation of atoms. If gravity was

266 Learn About Grand Unification Theory (Gut) | Chegg.com

any different, stars would use their nuclear fuel much faster, and matter may not coalesce. If the strong force was not strong enough, the adjacent protons would repel each other. Not to mention how the forces independently act with gravity and the electromagnetic forces operating at infinite distances. Furthermore, there is a distinctive messenger particle associated with each of the four fundamental forces called bosons. Among the most familiar is the photon which is the messenger particle of the Electromagnetic force. All the above with additional subatomic particles are the unseen denizens of the quantum world and without their interaction, and moreover their fine tuning, the cosmos would not be possible. Thus, the values of the fundamental forces are fixed and constitute the nature of our universe.

The Standard Model of Particle Physics

Although much of the universe consists of dark matter and dark energy that do not fit into the Standard Model, all ordinary (baryonic) matter, including every atom on the Periodic Table of elements, all the stars and life as we know it consists of only three types of matter particles: up and down quarks, which make up the protons and neutrons in the nucleus, and electrons that surround the nucleus. It describes three of the four fundamental forces that govern the universe:[267] Leptons and Quarks are the building blocks which build up matter, i.e., they are seen as the "elementary particles". Unlike Quarks, Leptons do not take part in strong interactions. There are six leptons in the present structure, the electron, muon, and tau particles and their associated neutrinos. In the present standard model, there are six "flavors" of Quarks. They can successfully account for all known mesons and baryons. These quarks combine to produce composite particles called hadrons, the most stable of which are neutrons and protons which are the components of atomic nuclei.

We can define the flavors of Quarks as follows: up, down, strange, charm, bottom, and top. Protons consist of two up quarks and one down quark, whereas a neutron is made up of two down quarks and one up quark. Up quarks are the lightest among all the quarks. The mass of the up quark ranges from 1.7 – 3.1 MeV / c2. Its electronic charge is +2/3 e. They have maximum stability due to the lowest mass. The down

---

[267] The existence of quarks was first theorized in 1964 in the work of two physicists, Murray Gell-Mann (opens in new tab) and George Zweig, who were both at the California Institute of Technology (CalTech) but who came to the conclusion that quarks exist independently of one another. Contrary to how science is often portrayed in the media, Gell-Mann and Zweig's conclusions were not an "a-ha!" moment but were instead built on the back of many years of hard work and careful discoveries by the particle physics community. Quarks: What are they? | Space

quark comes next to up quarks regarding its light mass. The mass of the down quark ranges from 4.1 – 5.7 MeV / c2. Its electric charge is -1/3 e. [268]

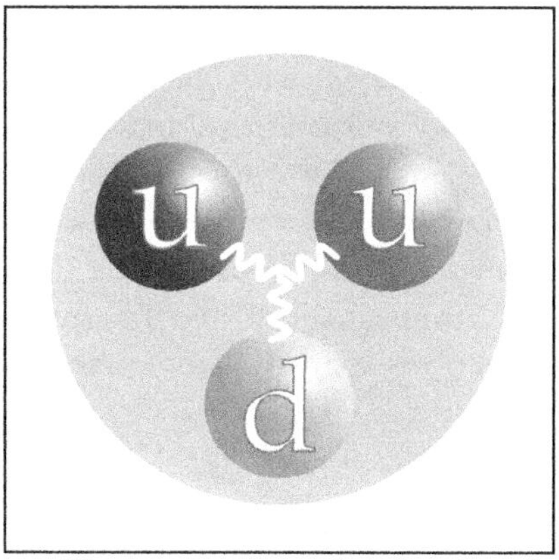

The quark structure of the proton. There are two up quarks in it and one down quark. The strong force is mediated by gluons (wavey). The strong force has three types of charges, the so-called red, green and the blue. Note that the choice of green for the down quark is arbitrary; the "color charge" is thought of as circulating among the three quarks.[269]

The strange quark is the third lightest of all. Its electric charge is -1/3 e. The charm quark. The meson which is called a J/Psi particle is an example of the charm quark. The mass of the top quark is 172.9 – 1.5 GeV/c2. Its electric charge is +2/3. The mass of the bottom quark is approximately 4.1 GeV/c2. Its electric charge is -1/3 e. A property of quarks labeled *color* is an essential part of the quark model. The force between quarks is called the color force. Since quarks make up the baryons, and the strong interaction takes place between baryons that are colorless, you could say that the color force is the source of the strong interaction, or that the strong interaction is like a residual color force that extends beyond the proton or neutron to bind them together in a nucleus.[270]

Fermions such as quarks, electrons, neutrinos, protons, and neutrons are the foundation of matter. However, bosons are responsible for acting as the 'carriers' of

268 What is Quark? - Definition, Meaning, Types, Model, Example (byjus.com)
269 By Arpad Horvath - Own work, CC BY-SA 2.5, https://commons.wikimedia.org/w/index.php?curid=637353
270 Quarks (gsu.edu)

three of the four fundamental forces: electromagnetism, the strong nuclear force, and the weak nuclear force. That means that fermions interact with each other via the exchange of bosons.[271] According to the Standard Model, there are several fundamental bosons, which are not made up of smaller particles. Photons carry all electromagnetic energy, gluons interact with the strong nuclear force, which binds together quarks to form protons and neutrons and holds the protons and neutrons together within an atom's nucleus. A boson is a "force carrier" particle that comes into play when particles interact with each other, with a boson exchanged during this interaction. For example, when two electrons interact, they exchange a photon, the force-carrying particle of electromagnetic fields. The W Boson and Z Boson work with the weak nuclear force. The Higgs boson, sometimes referred to as the *God particle,* is the fundamental force-carrying particle of the Higgs field, which is responsible for giving other particles their mass.[272]

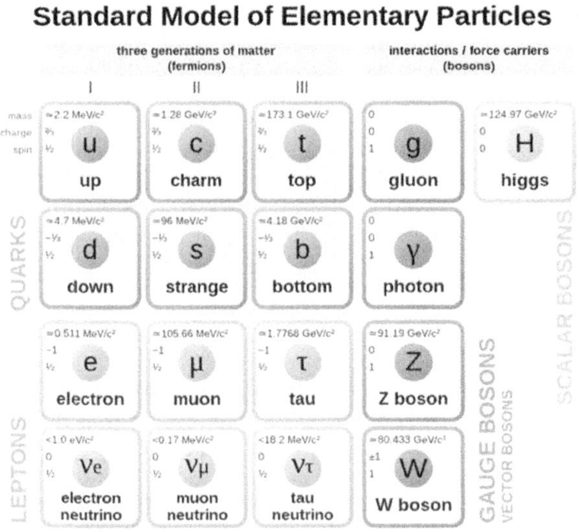

---

271 What are bosons? | Space

272 Higgs boson: The 'god particle' explained | Space Business Insider says that when Lederman authored a book on the Higgs boson in the 1990s the title was to be "The Goddamn Particle" but the publishers changed this to "The God Particle" and a troublesome connection with religion was drawn, one which bothers physicists to this day. Higgs boson: The 'god particle' explained | Space

Because quantum field theory explains the microscopic world and the quantum fields that fill the universe with wave mechanics, a boson can also be described as a wave in a field.[273] One of the most recent additions to the Standard Model of Particle Physics is the Higgs boson. First proposed in 1964 and discovered in 2012, the Higgs boson is the fundamental particle associated with the Higgs field, a field that gives mass to other fundamental particles such as electrons and quarks. A particle's mass determines how much it resists changing its speed or position when it encounters a force. However, not all fundamental particles have mass. The photon, which is the particle of light and carries the electromagnetic force, has no mass at all.[274] The Higgs boson gets its mass just like other particles from its own interactions with the Higgs field. The Higgs boson can be a unique portal to finding signs of dark matter due to its own distinctive characteristics and properties. [275]

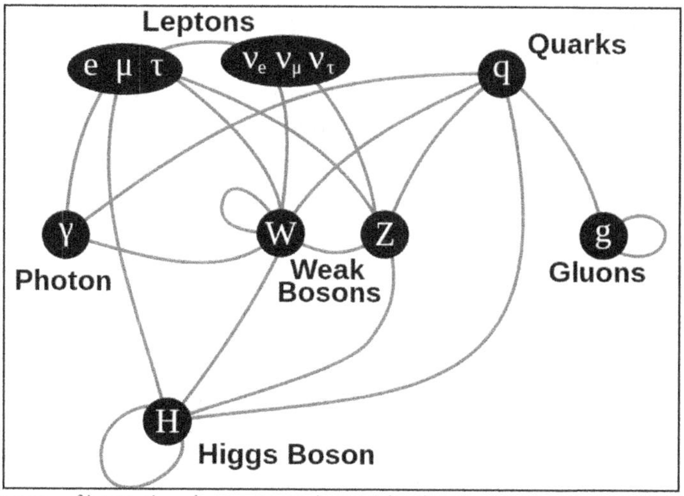

Summary of interactions between certain particles described by the Standard Model

The Higgs boson has a mass of 125 billion electron volts,meaning it is 130 times more massive than a proton. It is also chargeless with zero spin, the quantum

273 Higgs boson: The 'god particle' explained | Space

274 A photon is a particle and a wave that arises from an excited electromagnetic field and the Higgs boson is the particle or "quantized manifestation" that arises from the Higgs field when excited. That field generates mass via its interaction with other particles and the mechanism carried by the Higgs boson called the Brout-Englert-Higgs mechanism.

275 DOE Explains...the Higgs Boson | Department of Energy

mechanical equivalent to angular momentum. It is the only elementary particle with no spin. The discovery of the Higgs boson completed the Standard Model of Particle Physics. In a wider sense, the discovery of the Higgs boson sets a new paradigm for particle physics to build on. Many questions remain in the field, most of which have some relation to the Higgs sphere including the quagmire of dark matter and energy. Several models have been recently proposed where the only link of dark matter with regular matter would be through the Higgs boson.[276] Therefore, we are just beginning to scratch the surface of the Higgs boson and its moniker *the God particle* may not be a misnomer.

As far as gravity is concerned, it is not part of the Standard Model of Particle Physics. Since subatomic masses are so small, the gravitational attraction is minuscule when compared to the other forces. Furthermore, unlike the other fundamental forces, we do not understand the boson that is associated with gravity. However, at the macro level, gravity defines our spacetime because anything that has mass has an acceleration component that emerges in our cosmos as gravity. Perhaps one day the fundamental forces will be combined into a theory of everything bringing us one step closer to an understanding of our beginning and the secrets of the cosmos.

276 The Higgs boson: the hunt, the discovery, the study and some future perspectives | ATLAS Experiment at CERN

ENTROPY AND ENERGY

Entropy is often times described as the measure of chaos, randomness, or disorder within a closed system. While that is true, to an extent, entropy may play a much bigger role in our everyday lives. In fact, our mere existence may depend on it[277]

The *First and Second Laws of Thermodynamics* define energy and its flow throughout the universe. Therefore, we cannot have a conversation about the constructs of the cosmos without discussing the epic battle between entropy and energy which are integral to both laws. The relationship can be summed up in a single statement: Although the total energy content of the universe is conserved, and therefore, cannot be created or destroyed in accordance with *The First Law of Thermodynamics,* the total entropy increases in accordance with *The Second Law of Thermodynamics.*[278] These two laws are the most fundamental of all laws governing our universe. Energy and entropy literally drive the birth, lives, and deaths of stars, which are vessels that store and convert enormous amounts of energy into light, heat, and other energy forms.[279]

The concept of entropy is complex and applies to our reality on different levels. *The Second Law of Thermodynamics* also states that systems, if left undisturbed, tend to progress towards equilibrium with their surroundings, e.g., hot coffee cools and ice cream melts. These and all other *natural* systems, which are not influenced by outside agents, progress towards a state of equilibrium with the surroundings. Therefore, the entropy of such systems (including their surroundings) must also be increasing as these systems move toward equilibrium.

---

277 Does Entropy Determine the Direction of Time? | by Devin Gates | Medium

278 The concept of entropy was first introduced in 1850 by Clausius as a precise mathematical way of testing whether the second law of thermodynamics is violated by a particular process. Second law of thermodynamics | Definition, Formula, & Entropy | Britannica

279 The entropic process is tending to change from a state of order to a state of disorder. The entire universe and everything in it are entropic and requires the insertion of energy to maintain a state of equilibrium. Time is the measurement of entropy and entropy is the measurement of the amount of usable energy.

The Dynamic Duo

Energy and entropy are like the ones and zeros of a computer, complementary in nature. Together, they create the motion which lies at the heart of our cosmos that has sustained us throughout time. Energy and entropy are measures of different characteristics of a system, and both must be considered in order to know how much work can be extracted from a system. The following explains:

> The implications of the *Second Law of Thermodynamics* are considerable. The universe is constantly losing usable energy and never gaining. We logically conclude the universe is not eternal. The universe had a finite beginning – the moment at which it was at "zero entropy" (its most ordered possible state). Like a wind-up clock, the universe is winding down, as if at one point it was fully wound up and has been winding down ever since. The question is who wound up the clock.[280]

In this battle, mass and energy are allies and so are time, energy and entropy. In the battles between two cosmic nemeses, energy and entropy, there are no true victories only Pyrrhic ones because when all the usable energy is exhausted entropy can no longer exist and the disorderly chaos converges on its antithesis which is order. Thus, the phrase: out of chaos comes order proves true. To acknowledge chaos is not only to acknowledge the entropic nature of our universe but also its randomness which is also germane to the procreation and destructive process. Therefore, we owe our lives not only to the birth of our sun but also to the death of many suns and the slow sacrifice of the one that we cherish and revolve around. Without the energy produced by stars, life as we know it could not exist and all living forms must find ways to extract and use energy or fall to the perils of entropy.

The entropic process is the change from a state of order to a state of disorder. When we think of disorder, we often think of something awful or unstable. However, the disorder imbued in entropy is required for the universe to operate. In this regard, we can think of entropy as the dynamics of the universe. Without it, we would have no sense of time. No daylight or nighttime, no flowers in spring or autumn leaves and no wind to blow across desert sands. The entire cosmos, and everything in it, is entropic and require the insertion of energy to maintain a state of equilibrium. Time is the

280 Second Law of Thermodynamics (allaboutscience.org)

measurement of entropy and entropy is the measurement of the amount of usable energy.

However, there is a difference between the energy being dissipated or consumed and energy being destroyed; the latter being a violation of the *Conservation of Energy*. Since energy must be conserved, the amount of useable energy, i.e., the amount of energy available to do work decreases as entropy increases, although the amount of total energy remains the same. For instance, the molecules in a glass of water or an ice cube are all in continual motion, but at different rates and so each molecule has a tiny amount of kinetic energy. The temperature of an object reflects the kinetic energy of the atoms or molecules that make it up. To sum up, fast molecules equal high kinetic energy or high temperature, but they are more disordered. The following explains:

> The motion of molecules in liquids is in turn less vigorous than in gases. When thermal energy is added to solid ice, it melts from ice at zero degrees Celsius into liquid water that is also at zero degrees Celsius. When the liquid water is further heated it can evaporate, becoming a gas, water vapor. The more thermal energy a molecule has, the faster it moves and the more disordered the motion of a collection of such molecules becomes. In other words, as the thermal energy of the system of molecules increases (solid to liquid to gas), there is a progression from less disordered to more disordered behavior, that is, from a lower entropy state to a higher entropy state. Recall that the random motion of gas molecules is an example of dynamic disorder. Dynamic arrangements depend on time, but static arrangements do not.[281]

Furthermore, water is warmer than ice, which means that the molecules in water generally have more kinetic energy (and are thus moving faster) than the ones in ice. When ice cubes are placed in a glass of water, the slow-moving molecules in the ice are bombarded by the faster molecules in the water. As a result, the molecules in the ice end up faster while the molecules in the water end up slower. Hence, the ice gets warmer, and the water gets colder! And this process will continue until a state of equilibrium is reached. This transfer of energy from the warmer substance to the colder substance is called "heat."[282] The process of entropy always moves in one direction from the hotter to the colder.[283]

281 Entropy.pdf

The fact that matter cannot be created or destroyed is perhaps best explained by the following anecdote. When Sir Walter Raleigh returned to England after his sojourn to America, he brought a package of tobacco to Queen Elizabeth I. He made a bet with the Queen that he could measure the amount of smoke that was in the tobacco. And she was foolish enough to take it. He then weighed the tobacco before he smoked it, then after he smoked it, he weighed the ashes, the difference must be attributed to the smoke. However, the wise Queen, never outdone, said that she has heard of money going up in smoke, but never smoke turning into money and that is what she wanted to do. Not only was the *Law of Conservation of Energy* at play here but also the direction of entropy because the ashes and the smoke could not be reversed to form the initial orderly state of the tobacco.

The battle between entropy and energy has raged since the beginning of time. In fact, the creation of nuclei during the Big Bang sequence is a contest between entropy and energy.[284] Large entropy requires many states, but the *Law of Conservation of Energy* limits the number of states available. Immediately after the Big Bang sequence, within the first 10^{-43} seconds which we call Planck time, the temperature was high, energy was high and entropy was the dominating factor because the number of possible states was maximized.[285] At very low temperatures, energy becomes more dominant, in fact at Absolute Zero when energy reaches its lowest point, entropy no longer exists, and matter is totally ordered as it was before the Big Bang sequence. Between the highest state of energy and the lowest state of energy, the *Second Law of Thermodynamics* prevails but within all states of existence, it is the *Conservation of Energy* that dictates all the laws of our cosmos.

---

282 Entropy and energy (morningside.edu)

283 Wind is caused by uneven heating of the earth's surface by the sun. Because the earth's surface is made up of different types of land and water, the earth absorbs the sun's heat at different rates. One example of this uneven heating is the daily wind cycle. Gases move from high-pressure areas to low-pressure areas. And the bigger the difference between the pressures, the faster the air will move from the high to the low pressure. That rush of air is the wind we experience. Why Does Wind Blow? | NOAA SciJinks – All About Weather

284 [Longing for Harmonic, p. 80]

285 In the era around one Planck time, 10^{-43} seconds, it is projected by present modeling of the fundamental forces that the gravity force begins to differentiate from the other three forces. This is the first of the spontaneous symmetry breaks which lead to the four observed types of interactions in the present universe. http://hyperphysics.phy-astr.gsu.edu/hbase/astro/planck.html#c2

Without the *Conservation of Energy,* nuclei could not have formed after the Big Bang sequence. Without nuclei, there would be no atoms and no nuclear fusion, no stars, no photons and no speed of light to govern the universe and subsequent forces such as gravity. The fact that the cosmos has moved from an ordered state to a disordered state and that all the energy during the Big Bang was not burned, gave rise to the stars and its residual energy which spun life in the universe through its conversion of mass into energy by the nuclear fusion process. In this regard, mass is nothing more than a storage "battery" that energy uses to combat entropy which in turn is governed by the *Conservation of Energy* and the vector of energy flow, best exemplified by the unidirectional and irreversible flow of heat.

When you view the cosmos through the lens of the symbiotic relationship between energy and entropy, the nature of our existence and the universe we live in comes into focus. The universe is a slave to the entropic process on all levels. It could be said that entropy is the 42 in *The Hitchhiker's Guide to the Galaxy.*[286] The quest to maintain local equilibrium is the driving force of Nature. Our Sun burns its fuel, and thus sacrifices itself to maintain its equilibrium. The radiation pressure that comes from nuclear reactions in its core balances the enormous gravity from the star's mass. The following explains:

> The energy released through gamma rays and the kinetic energy of the particles contributes to generation of thermal pressure in the solar interior, effectively balancing it with the gravitational pressure to maintain an overall steady state. It takes 10,000 to 170,000 years for a photon to travel from the Sun's core to its surface. On its way, the gamma ray photons emitted in the fusion reactions are converted into visible light, infrared and ultraviolet photons, as they reach the photosphere.[287]

Survival of the fittest succinctly put is entropy which is the nature of the universe. The *Second Law of Thermodynamics* does not rule out local reductions in the entropy of a system. To locally reduce entropy, a system is taken from an equilibrium situation to a nonequilibrium one, and energy must be transferred to the system. Each localized entropy decrease is accompanied by a greater increase in the entropy of the

---

286 A note to non-Geeks. In Douglas Adams' The Hitchhiker's Guide to the Galaxy, 42 is the number from which all meaning ("the meaning of life, the universe, and everything") could be derived.

287 Nuclear Fusion in the Sun Explained Perfectly by Science - Universavvy

surroundings, thereby causing an increase in the total entropy of the universe. For instance, humans band together to prevent or slow down the entropic process at the expense of their surroundings. Once a nexus is developed, i.e., being of the same race, religion or even sex, one group searches for and wants to maintain its local equilibrium at the expense of another group. Wars are fought and alliances are made to exploit people outside a given group and shibboleths are developed to keep some people in and others out of the group. Order within one group is achieved through the chaos or disorder of the other. This could be viewed as social entropy.

Slavery is a perfect example of this as one group is ostracized so that the other group can control (maximize the resources) thus minimizing the effects of entropy on their group while maximizing it among the other group. The rich coalesce around the rich forcing the poor to coalesce around the poor. On this basis, privileges are given to one group yet denied to others. These are local and temporal states of equilibrium. However, in the long run, empires rise and fall, and Nature will establish equilibrium among both groups, perhaps best exemplified when hot water meets cold water which ultimately converges on a state of warm water. In the final stages both groups, over time, fall prey to the entropic process. This is also seen in other natural phenomena. For example, if the population of rabbits increases dramatically, it will attract predators that will ultimately decrease the population. Once the population decreases significantly, predators will move on, and the population of rabbits will start to increase until it reaches a state that attracts a new group of predators. Thus, creating an endless feedback loop controlled by the vicissitudes of Nature. Such feedback is a necessary condition of any controlled "closed looped" system (S).

In each case, energy must be introduced to maintain local states of equilibrium and to counterbalance the disorder created by entropy. It is through the local equilibrium process that order is manifested and maintained in the universe. And this is a function of the disorder of the universe and the increase in entropy. Out of disorder comes order, and out of order comes disorder, bound together by the motion of the universe, in a manner akin to an electric field and a magnetic field that are orthogonal to each other and moving at the speed of light.

Although many are familiar with $E=mc^2$, few are familiar with $S = k \log W$: where S is the amount of Entropy, k is Boltzmann's constant and W is the number of possible microstates within a closed system (S). Perhaps, each one of the microstates has the potential to generate local equilibrium states thus creating temporal orderly systems that over time succumb to the entropic process. In this regard, the Sun may be considered a microstate, which establishes its local equilibrium and in turn, the Earth is a subset of the Sun's microstate with its established local equilibrium and life on Earth is a subset of the Earth's microstate.

The summation of these microstates constitutes a macro-state or orderly system that is completely ordered and or hierarchically stacked upon each other. If local equilibrium is not maintained at the higher level of the system, it cannot be maintained at any state lower in the system. In this regard, each point of local equilibrium may not be aware of its function in the macro -state and is only driven by its need to maintain its state of local equilibrium. This state is achieved by extracting energy from the macro-state. For example, when a beaver builds a dam, it is not aware of the extent and impact of the dam on the ecology of a system. Yet, a beaver is equipped with the tools to build such a dam. The beaver's main focus is to maintain its local equilibrium. In fact, every point within W is so equipped including the human mind.

The Transfer of Energy

However, the transference of energy rages on and is essential to life. At the terrestrial level, herbivores eat plants to transfer their energy, and carnivores eat animals to transfer their energy but eventually we die because of the nature of entropy. In this regard, we are microcosms of the stars that made us, burning the residual energy made in our sun to maintain our local equilibrium. For us humans, it is measured in calories per day. There are only two forms of usable energy available for our use; terrestrial energy embedded or stored in the Earth which bears the seeds of stars that met their demise billions of years ago and celestial energy available from the Sun's battle with its local equilibrium. And as each day goes by, the amount of energy available to us has decreased in both arenas.

As far as life on this Earth is concerned, every living being is a dissipater and consumer of energy. Not only is humankind a consumer of energy, but so are the machines they create. At the living organism level, the paramount goal is to acquire energy from its surrounding thus increasing the overall entropy of its surrounding because at each stage energy is lost at a higher percentage in a similar manner that each nuclear phase of fusion in the Sun produces less energy. The Earth itself is set up to support the chemical process necessary for the extraction and transference of energy. Life as we know it is a symbiotic chemical process; hence, the proverbial food chain. Each organism must compete for a limited energy supply.

Trophic levels are the feeding positions of all organisms in a specific ecosystem. Every living thing must compete for a limited supply of energy generated by the Sun while obeying the laws of thermodynamics. You can think of them as food chains or as a trophic level pyramid. At the first tropic level, plants get their energy from the Sun via photosynthesis. At the second tropic level, herbivores eat the plants, and at the third tropic level, carnivores eat herbivores to extract the precious energy bestowed to us by our nearest star. At each stage, about 90% of the energy is lost and thus unavailable for the next stage.[288] However, this quest for energy is a must to maintain local order and to keep the local entropy negative enough to offset the positive entropy of the environment. In this regard, every living thing is like a sun, a storage battery of energy. Therefore, every living thing is equipped with tools to maintain its local state of order for a period but not forever. Its anatomy, ability to reproduce, and perhaps its *Consciousness* was structured for this sublime purpose. And when they need to evolve or mutate, they contain within their DNA the ability to do so.

---

288 The fourth trophic level includes carnivores and omnivores which eat the animals that belong to the third level. Omnivores are animals that eat both plants and animals. Omnivores consume both primary producers and secondary consumers. Animals on this level are called tertiary consumers. These animals tend to receive less energy from their food than animals in the third level. This is because the energy created by primary producers has been transferred and converted at least twice by animals in the previous groups. Each time you go up a trophic level, available energy is reduced by at least one magnitude. https://sciencing.com/trophic-levels-ecosystem-8205653.html
Almost all life on Earth relies on solar energy for food, either directly or indirectly.
Producers rely directly on solar energy. They absorb sunlight and convert it into nutrients through a process called photosynthesis. Producers, also called autotrophs, include plants, algae, bacteria, and fungi. Autotrophs are the foundation of the food web. Consumers rely on producers for nutrients. Herbivores, carnivores, omnivores, and detritivores rely on solar energy indirectly. Herbivores eat plants and other producers. Carnivores and omnivores eat both producers and herbivores. Detritivores decompose plant and animal matter by consuming it. Fossil Fuels. Photosynthesis is also responsible for all of the fossil fuels on Earth. Scientists estimate that about 3 billion years ago, the first autotrophs evolved in aquatic settings. Sunlight allowed plant life to thrive and evolve. After the autotrophs died, they decomposed and shifted deeper into the Earth, sometimes thousands of meters. This process continued for millions of years. https://education.nationalgeographic.org/resource/solar-energy1

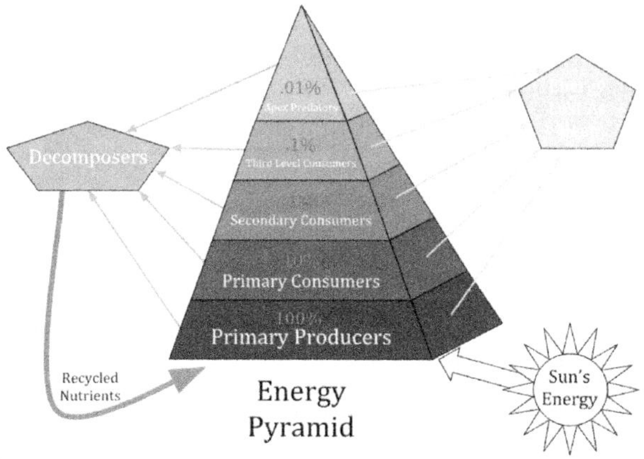

An energy pyramid illustrates how much energy is needed as it flows upward to support the next trophic level. Only about 10% of the energy transferred between each trophic level is converted to biomass[289]

For instance, predators have the tools to kill prey, prey the tools to escape predators, etc... As part of the chaos of the universe, some become extinct. Evolution, mutations and the presence of the known forces, universal constants, etc., exist to assure that every piece of usable energy is dissipated. Because entropy wins all battles of local equilibrium, living things must have the ability to reproduce, to perpetuate their DNA so that future generations could continue to consume energy. Therefore, entropy is the shadow of death that lurks beyond the stars and into the realm of everything it gave life to. However, Nature has imbued in us the ability to combat entropy by passing our genes which is the essence of us to our progeny.

Trophic levels can also be extended to people and nations. Those with the greatest resources are best equipped to deal with the entropic aspects of nature, i.e., better food, health, access to education, and so forth. In this regard, they can succeed where others fail and therefore able to impose their will on others less fortunate. This often leads to a quest for power and has been the leitmotif of history. For example, in ancient times the Egyptians were at the top of the food chain for nearly 3,000 years before Christ. Although they experienced ebbs and flows throughout their history, they were the keepers of arcane knowledge and wisdom during that period and their accomplishments are renowned best exemplified by their pyramids and temples that

289 By Swiggity.Swag.YOLO.Bro - Own work, CC BY-SA 4.0, https://commons.wikimedia.org/w/index.php?curid=40660207

we still adore. They were able to feed their people and devote resources toward the advancement of their nation and civilization and exert their influence over the entire Mediterranean region whose remnants we see today. The Greeks, Romans, and others were able to follow in their footsteps often at the expense and exploitation of those on the lower trophic level. Hence the development of totem poles such as first and third world countries, princes and paupers, and everything in between. However, at the heart of it all, and the core of our reality is entropy.

The only things that are constant throughout the cosmos are change and the presence of energy in various forms. That, in a nutshell, is the essence of getting old and constitutes the life of a star and a person. It is entropy that gives us gray hair and wrinkles on our faces so that others may know that we had many trips around our closest star. The cosmos is driven by the transference of energy. It is not only the essence of life but is what constitutes our universe from Planck time to our current day and from Planck length to the edges of the cosmos.[290]

The Arrow of Time

The tendency of a system, entropy to increase defines the arrow of time. However, the speed of the entropic process is not constant and although time seems to be constant, in reality, it is not as Einstein's formula illustrates. Entropy determines the arrow of time, but it does not determine the rate of time. Hence, from one perspective, entropy is a way of differentiating the past from the future. It is the only quantity in the physical sciences that requires a particular direction for time. Everyone has an intuitive sense of the effects of entropy. Ice melting, smoke coming out of a chimney, things wearing out or dying. All these things are the result of entropy. The following explains:

> The explanation that is usually given as to why we don't see broken cups gathering themselves together off the floor and jumping back onto the table is that it is forbidden by the second law of thermodynamics. This says that in any closed system disorder, or entropy, always increases with time. In other words, it is a form of Murphy's law: things always tend to go wrong! An intact cup on the table is a state of high order, but a

[290] A hydrogen atom is about 10 trillion trillion Planck lengths across or about 10^{-20} times larger than an electron.[290]http://www.askamathematician.com/2013/05/q-what-is-the-planck-length-what-is-its-relevance/[290]

> broken cup on the floor is a disordered state. One can go readily from the cup on the table in the past to the broken cup on the floor in the future, but not the other way around. The increase of disorder or entropy with time is one example of what is called an arrow of time, something that distinguishes the past from the future, giving a direction to time. [291]

Therefore, in a non-entropic realm such as paradise, where entropy does not exist in this ordered system, time does not exist, and energy does not exist or is not dissipated. Although we can imagine such a state, it is not a part of the physical universe or our reality. There once was a story told about a scientist that climbed the mountain of knowledge only to find a priest atop praying. In the local battle between energy and entropy, entropy wins every time, because energy is on the offense while entropy holds its ground. Entropy depends on the nature of the star, i.e., energy stored as mass, to eventually be its downfall because it is the mass of the star that produces the gravity that eventually kills it. Of all the fundamental forces, gravity is the servant of entropy.

Perhaps on a more esoteric level, death could be defined as the end of the entropic process and birth as the beginning of it. It is at birth that we become consumers of energy and at death when we cease to consume energy. Our beliefs of the world beyond life are one of a non-entropic world be it heaven or hell. One in which our souls are immortal and condemned to either a life of eternal bliss or one of eternal damnation based on the performance of the body (matter) in the entropic world. It is from such constructs that religions and myths came to realization. The human quest to see what is unknowable and unseen. This conceived existence lies in an unknown world, outside of spacetime and one of fear and wonderment. It is in such a world that the scales of the entropic world are held, and judgments rendered.

Time does not exist in such a world because entropy does not exist. A world where energy is not consumed but is omnipresent, a massless world submerged in a field of conscious energy that we conceive but cannot comprehend. It is the driving force of the entropic world we call spacetime and lies beyond the boundary that we call infinity. From time immemorial, humans have searched for immortality and looked to the stars as if they held the key to our existence. However, to be immortal is not to be

291 [Hawking: A Brief History of Time, p.149]

a slave of entropy and in order to achieve such a goal one must be outside of this thing we call a universe and beyond the stars that made us human.

Spacetime is a product of our universe and as Einstein's equation has shown, space and time could be altered and are the result of reference frames. Within our universe, entropy cannot be reversed, but it could be impeded or negated locally through the process of temporal equilibrium systems or by varying space and time. If time is relativistic, so is local entropy. On the other hand, universal entropy is a function of the rate of change and the *Conservation of Energy* and is not altered in the same way as local entropy. For example, if we view aging as a measurement of a person's local entropic state, and if you send that person off at a certain percentage of the speed of light and keep her twin within the current system of equilibrium, i.e., unaltered Earth, when she returns, she will be younger than her sister, in accordance with how close to the speed of light she traveled and how long she was away from Earth.

Therefore, you have changed the local entropies of the one that returned by changing its environment relative to space and time. In this particular case, time has an effect on entropy's clock slowing it down according to its reference frame where time is also slower. Although Einstein's theories address space, time and gravity, he failed to address Heraclitus's assertion that you can't step into the same place on the river twice. Hence, they fall short of addressing the direction of time and entropy something that Gödel tried to address. Within our dimension, entropy and time have the same arrow but that need not be the case in all dimensions.

Within our universe, entropy and time also go hand in hand not only in their direction but also in the battle between entropy and energy. You cannot reverse time or entropy because they are related to the fact that energy is moving from an available state to an unavailable state thus the direction of events or vectors of our cosmos is fixed. Therefore, the three are inextricably linked directionally. In our dimension, local equilibrium can only be maintained by an influx of energy required to combat entropy. Just as a battery disperses its energy over time, so does our sun which is required to maintain its local equilibrium.

It is gravity that makes a star by coalescing the available hydrogen in a region and then compressing the hydrogen until its mass can fight back. At this point, the star ignites and produces enough heat and pressure to hold off gravity and to establish local equilibrium. To do this, it must burn its energy thus, decreasing the amount of available energy as time goes on thereby increasing the entropy of its surroundings. The bigger the star the faster the energy is burned or fused and the smaller the star the slower the energy is burned. Although our Sun will burn for about 10 billion years, a star that is 10 times the mass of the Sun will only burn for about 35 million years. Stars that are too small, i.e., a brown dwarf which is less than 8 percent of the Sun's mass, will not ignite and the nuclear process will not commence.[292] And all while this is happening, the battle between energy and entropy remains a draw and a state of equilibrium is reached.

However, just as a battery cannot run forever, neither can a star because its mass and albeit its energy is finite. Entropy is thus the instrument of death, but it is constrained not by energy but by the *Conservation of Energy* and this is an important distinction and lies at the essence of this nuclear battery that we call a star. All the building blocks, of what we define as life and the planet that we dwell on, are a result of this life and death battle raging throughout the cosmos and throughout time. A battle that we estimate that has been raging for about 14 billion years and the life of our particular star that we revolve around has been fighting this battle with entropy for about 5 billion years or half its estimated life. Stars that are between 8 and 80 percent of a solar mass have been burning since the beginning of the universe and have experienced little change since their birth. Some of these stars may burn for trillions of years.

Thus, life and death are fundamental, and the *Conservation of Energy* is always coeval with the motion or transference of energy the latter, often referred to as the *Second Law of Thermodynamics* replete with its definition of entropy could be viewed as a corollary to the *Law of Conservation of Energy.* The Second Law is different from the First because it deals with the direction in which the process takes place in nature. Not only does it define the direction, but it makes it clear that the process is irreversible. Therefore, the *First and Second laws of Thermodynamics* are inextricably linked. The notion of direction, flow or motion is alien to the First Law and can only be explained by the Second which assumes the existence of the First.

---

292 [Livio: The Accelerating Universe, p.202]

The transference of heat from a hotter body to a colder one, the increase in the disorder of the universe through the entropic process, and the arrow of time are all consequences of entropy and the *Second Law of Thermodynamics.* Our cosmos took on its current form because it must conserve energy and the acceleration within it must always be less than the speed of light. If one of these rules is broken, then the other rule is broken and hence you are no longer a part of the spacetime continuum. In fact, every rule in our cosmos is broken. Therefore, it is the presence of energy that must be transformed into mass that brought the fundamental forces into existence in our nascent cosmos. Almost all irreversible processes (e.g., life) on Earth are fueled (ultimately) by sunlight. The evolution of stars obeys the *Second Law of Thermodynamics*, i.e., entropy always increases in a burning star. This means entropy is increasing everywhere in universe.

One may ask a poignant question; how can the universe be only about us when stars like ours were here before us and will be here after us? What would be the point of a universe so designed if, in the Eye of Providence, we are the only cynosure? Hence, an obvious paradox exists. Such a Creator is far too great and oh too wise to have only considered the existence of life on this planet in His grand plan. Everywhere in the cosmos, Nature repeats itself. For life to exist on this planet alone is not only improbable but inconsistent with the replicating nature of Nature. There are stars in our universe that are billions of years older than our Sun. For all we know, there could be sentient beings in our universe millions of years ahead of us scientifically and the gap between us and them is much greater than us and chimpanzees. We must be forever mindful that our species has only been on this planet for about three hundred thousand years which is a very short time in the history of the cosmos.

A STAR IS BORN

The first stage in the birth of a star is called a protostar. This is where the majority of the stellar material has collected together in ball in the center, but there is a huge disk of gas and dust obscuring it from our view. ...The protostar stage takes about 100,000 years to complete[293]

The birth, life and death of stars represents the greatest example of the dichotomy between energy and entropy. There are trillions of stars in the universe. However, they all were born the same way and must continuously burn their energy in order to survive. In the past century, scientists have learned that stars, like people, are born and from time to time, they have witnessed the death of these heavenly bodies.

The Birth of Stars

The birth of the first stars took place billions of years ago and scientists for the first time have constructed a scenario of our cosmic genesis. Today, in the Orion nebula, we have witnessed the birth of stars. The intense wind from the newly formed star at the heart of the Orion Nebula is creating the bubble and preventing new stars from forming in its neighborhood. At the same time, the wind is pushing molecular gas to the edges, creating a dense shell around the bubble where future generations of stars can form. [294]

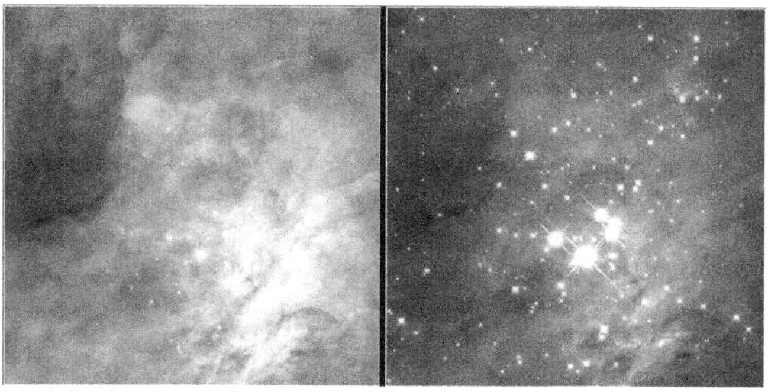

Optical images reveal clouds of gas and dust in the Orion Nebula; an infrared image (right) reveals the new stars shining within.

293 Birth of Stars - Universe Today

294 Star Formation in the Orion Nebula | NASA

Although all stars are born the same way, they live and die differently. Some fall on their swords, others hang themselves and their choice of death is predetermined by its mass. The more massive the star, the faster it burns its mass and the faster it dies. [295] But the executioner is always gravity the force that gave them life. In his seminal book: *A Brief History of Time*, Stephen Hawking explains the birth of stars:

> The hydrogen and helium gas in the galaxies would break up into smaller clouds that would collapse under their own gravity. As these contracted, and the atoms within them collided with one another, the temperature of the gas would increase, until eventually, it became hot enough to start nuclear fusion reactions. These would convert the hydrogen into more helium, and the heat given off would raise the pressure, and so stop the clouds from contracting any further. They would remain stable in this state for a long time as stars like our sun, burning hydrogen into helium and radiating the resulting energy as heat and light. More massive stars would need to be hotter to balance their stronger gravitational attraction, making the nuclear fusion reactions proceed so much more rapidly that they would use up their hydrogen in as little as a hundred million years. [296]

When the hydrogen supply in the core begins to run out, the core becomes unstable and contracts. The outer shell of the star, which is still mostly hydrogen, starts to expand. As it expands, it cools and glows red. The star has now reached the red giant phase. It is red because it is cooler than it was in the main sequence star stage, and it is a giant because the outer shell has expanded outward. All stars evolve the same way up to the red giant phase.

A main sequence star chart, or Hertzsprung-Russell (HR) diagram, is a chart that indicates the relationship between a star's luminosity and its stellar classification, or effective temperature. The following is an example of an HR diagram. The horizontal axis represents stellar surface temperature, which corresponds to the spectral type.

---

295 Our Sun is approximately 5 billion years old. On the average, about ten new stars are born in our galaxy per year. [Lightman, Alan, Time for the Stars, New York, Warner Books,1992, p.35]

296 [Hawking: A Brief History of Time, p.119]

The stellar spectral types are classified using letters O, B, A, F, G, K, and M, where O is the hottest and M is the coolest.

A Hertzsprung-Russell Diagram

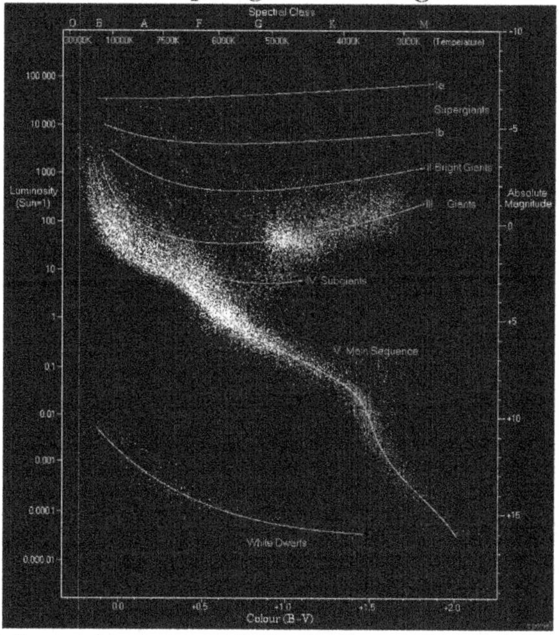

A Hertzsprung–Russell diagram plots the luminosity (or absolute magnitude) of a star against its color index (represented as B−V). The main sequence is visible as a prominent diagonal band that runs from the upper left to the lower right. This plot shows 22,000 stars from the Hipparcos Catalogue together with 1,000 low-luminosity stars (red and white dwarfs) from the Gliese Catalogue of Nearby Stars.[297]

The spectral types can also be organized by color, where O is the bluest and M is the reddest. For this reason, the temperature decreases from left to right on the HR diagram. The vertical axis represents stellar luminosity, or absolute magnitude, and is generally given in units of the Sun's luminosity. The Sun is a main sequence star with a G spectral type, or a yellow dwarf star. The color of a star tells us much about its history.

---

297 By Richard Powell - The Hertzsprung Russell Diagram, CC BY-SA 2.5, https://commons.wikimedia.org/w/index.php?curid=1736396

As shown in the main sequence chart, the colors of stars are like the colors of the rainbows they produce, from blue to red depending on their temperature. As the main sequence star glows, hydrogen in its core is converted into helium by nuclear fusion. When the hydrogen supply in the core begins to run out, and the star is no longer generating heat by nuclear fusion, the core becomes unstable and contracts. The outer shell of the star, which is still mostly hydrogen, starts to expand. As it expands, it cools and glows red. The star has now reached the red giant phase. In the core of the red giant, helium fuses into Carbon. Our red giant Sun will still be eating up helium and cranking out Carbon. But when it's finished its helium, it isn't quite hot enough to be able to burn the Carbon it created.

Regarding the brightness of stars, it is not linear but logarithmic with a star 25 times the mass of our Sun having a brightness of 4,000 times that of the Sun. Each time the size of a star doubles, the luminosity increases tenfold. Therefore, a star such as Sirius B, whose initial mass was five times that of the Sun must have been extremely luminous when it was burning its hydrogen.[298]

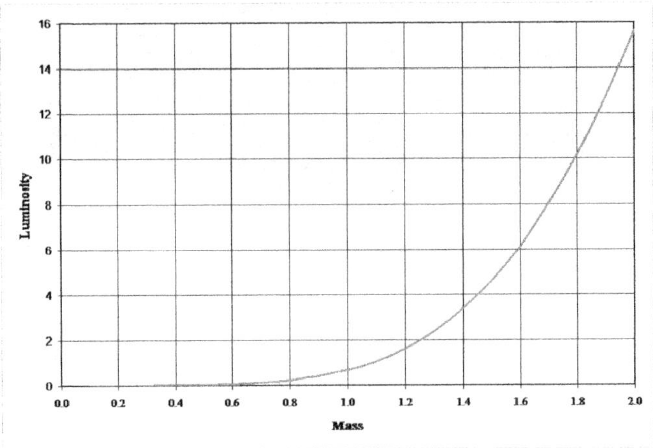

This plot gives an example of the mass-luminosity relationship for zero-age main-sequence stars. The mass and luminosity are relative to the present-day Sun

[298] http://arxiv.org/abs/astro-ph/0507523
[298] Bartusiak: Archives of the Universe, p.259]

The Death of a Star

Once a star has reached the red giant stage, it is on its deathbed and all that remains is how it will die. Since our Sun won't be hot enough to ignite the Carbon at its core, it will succumb to gravity again. When the core of the star contracts, it will cause a release of energy that makes the envelope of the star expand. Now the star has become an even bigger giant than before! Our Sun's radius will become larger than Earth's orbit! All stars evolve the same way up to the red giant phase. However, the star's initial mass has determined its future life cycle path.[299] For instance, a main sequence star, like our Sun; could not die in a supernova because its Chandrasekhar limit is less than 1.4 solar masses. Stars much larger than our own, and above 1.4 solar masses burn their fuel faster, and as they die, they create heavier elements such as gold and silver before exploding in a supernova, spreading these elements throughout the universe. After they have exhausted all their fuel, for low- and medium-mass stars, like Sirius B and our sun, do not explode, but are compressed by gravity into a white dwarf. The amount of mass a star has determines which of the following life cycle paths it will take after the red giant phase[300]

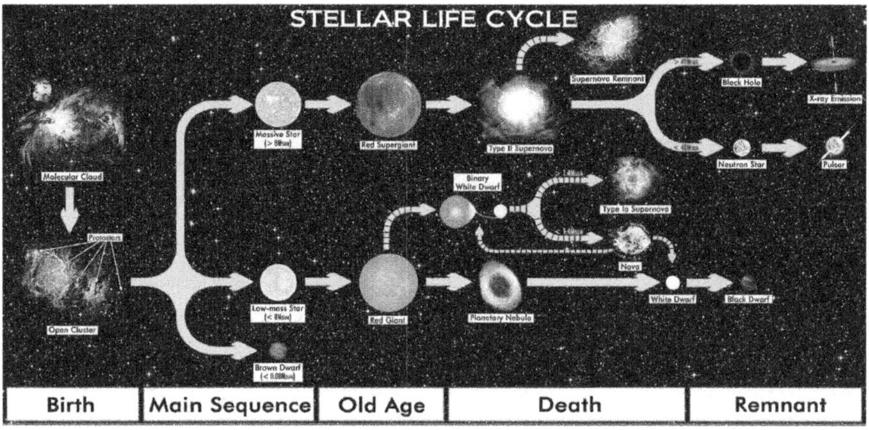

Our Sun will follow the path of the Average Star[301]

---

299 Background: Life Cycles of Stars (nasa.gov)
Imagine the Universe! (nasa.gov)

300 https://imagine.gsfc.nasa.gov/educators/lifecycles/SC_main_p1.html

301 By R.N. Bailey - Own work, CC BY 4.0, https://commons.wikimedia.org/w/index.php?curid=59672008

White dwarfs are dense, dim, stellar corpses in the last visible stage of evolution. They are stars about the size of our Earth, however, the gravity on the surface of a white dwarf is 350,000 times that of the Earth. Furthermore, the mass is so heavy that a teaspoon of it would weigh several tons. If a white dwarf happens to orbit another star in a binary system, e.g., Sirius B orbiting Sirius A, it will gradually steal material from the other star, increasing its mass in the process.[302] Since Sirius B was five times as massive as our Sun, its lifespan was measured in millions and not billions of years as in the case of our Sun.

Image of Sirius A and Sirius B taken by the Hubble Space Telescope. Sirius B, which is a white dwarf, can be seen as a faint point of light to the lower left of the much brighter Sirius A.

When massive stars burst during a supernova, they can explode with the brightness of 10 billion suns. Supernova remnants consist of material from the exploded star and any interstellar material it swept up in its path. The Crab Nebula is an example of a supernova remnant. The explosion that created it in the year 1054 was so bright that for weeks it could be seen even in the daytime sky, and it was

302 Since every white dwarf has the same upper mass limit, the energy generated from every Type 1A Supernova will be the same, allowing astronomers to use these supernovae as standard candles, which can be used to determine the distances to other galaxies.

recorded by astronomers across the world.[303] The material from the star is still rushing outward at around 3 million mph (4.8 million kph).[304]

The Crab Nebula as seen by Herschel and Hubble telescopes.

When a giant star dies in a supernova, it has three forms it can end in: entirely dead, transform into a neutron star or manifest as a black hole.[305] The outcome will depend on several factors that mainly rely on the dying star's total mass. neutron stars, and black holes are very similar in the ways they are formed and even their core characteristics such as a strong magnetic field and overall density. The main difference

303 During a supernova, the amount of energy produced is so high that Iron can be fused into even heavier elements, such as gold and silver. If stars did not cease to exist one day, the elements they fuse would remain locked within their cores. In order for the very ingredients of planets and life to find their way into space, stars must explode. Our very existence is only made possible due to the laws of physics that govern stellar life cycles. If it were not for the stars, the universe would be nothing but a cloud of hydrogen. The atoms that compose nearly everything we see were once located within the cores of massive stars. Are We Made Of Stardust? - WorldAtlas

304 Discoveries - Hubble's Nebulae | NASA

305 A supergiant star that has burnt its elements develops an Iron core at which point the radiation pressure holding the gravitational force in place collapse. Gravity consumes the core and collapses leaving a black hole. Gravity and Pressure | Stellar Evolution | Space FM

between a neutron star and a black hole is the size of the star that created them.[306] The following explains:

> The initial explosion of a supernova has so much energy that it can split atoms apart at the core, sending protons and neutrons flying into the universe. In the moments following the explosion, these particles crash into each other with enough energy to fuse back together. Light elements continue colliding with protons and neutrons in this way, constantly growing larger and larger. This process, which is similar to nuclear fusion, is called nucleosynthesis. The nucleosynthesis that occurs during the explosion of a supernova produces elements with a higher atomic number than iron, which cannot be created by nuclear fusion. When the first stars died out this way, brand new elements, including gold, were formed. Eventually, those elements ended up here on Earth.[307]

After a supernova explodes, the whole core of the star becomes nothing, but a dense ball of neutrons, and its remains become a neutron star. Neutron stars are white dwarfs on steroids. Of all heavily condensed objects in the universe, the neutron star is the densest. They are called neutron stars because they are primarily composed of neutrons. Because of the intense pressure resulting from the immense gravity at their cores, protons and electrons are fused together turning them into neutrons without any charge. Because of their density, they have intense gravitational and magnetic fields. The gravity of a neutron star is about a thousand billion times stronger than that of the Earth and a teaspoon of a neutron star would weigh about a billion tons. Moreover, they have a magnetic field that is a trillion times greater than Earth's magnetic field.

For the largest stars of all, the remaining core is so massive and has such a strong gravitational pull that not even light can escape. This is called a stellar black hole.[308] Although the birth of stars is vital to maintaining and creating life, the death of stars is required to create it. The life and death of stars form the chemical elements that make up Earth, making stars critical to life as we know it.

306 Neutron Star vs Black Hole (Similarities And Differences) | Scope The Galaxy

307 The Life and Death of Stars | Let's Talk Science (letstalkscience.ca)

308 The Life and Death of Stars | Let's Talk Science (letstalkscience.ca)

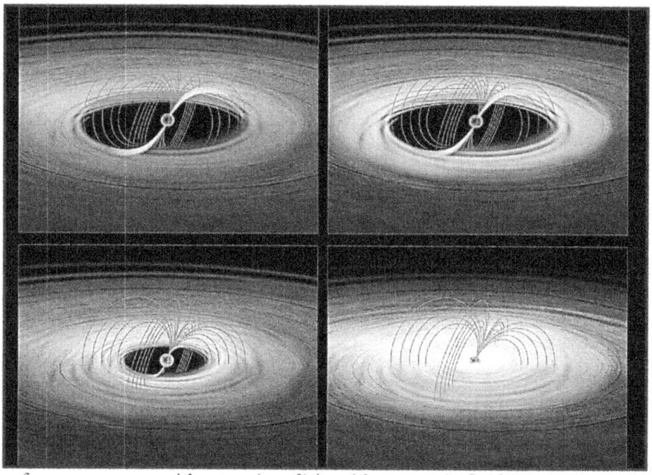

Computer renders of a neutron star with accretion disk, with magnetic field lines projected, showing bursts of powerful X-rays and Radio Waves. The simulations are taken from 2017 data from NASA's NuSTAR and Swift, and ESA's XMM-Newton observatories.

Some neutron stars have jets of materials streaming out of them at nearly the speed of light. These beams flash like the bulb of a lighthouse and are called pulsars. Normal pulsars spin between 0.1 and 60 times per second, while millisecond pulsars can result as much as 700 times per second.[309]

The following are the typical phases of a star's life:

1. Giant Gas Cloud
2. Protostar
3. T-Tauri Phase
4. Main sequence star
5. Expansion into Red Giant
6. The fusion of Heavier Elements
7. Supernovae and Planetary Nebulae
8. Neutron Star and Black Hole

309 Neutron Stars: Definition & Facts | Space

WE ARE ALL MADE OF STARDUST

The astronomer Carl Sagan once famously said that "we are made of starstuff." Interestingly, this is actually an accurate statement about the universe. Not only are humans made of starstuff, but everything we see around us is, along with everything in the solar system. The elements that make up our bodies and all the planets were cooked in the cores of massive stars.[310]

It is most ironic for our universe to be as massive as it is that most of the conversion of mass to energy is performed within the micro world of a star through nuclear fusion. This process is conducted at its core and radiated to the surface. Thus, the arduous journey of a life-giving photon begins deep within the core of the Sun. With each stage of fusion, the amount of available energy decreases. The lower the atomic number of an atom, the greater the differential in lost mass and the maximum amount of energy is produced. After the Big Bang, the universe consisted mostly of Hydrogen with some Deuterium, Helium and Lithium. [311]

Nuclear Fusion of Elements

In 1957, Fred Hoyle proved that all the elements from which our world is made, from Carbon to Uranium had been cooked inside stars eons ago from a basic fuel of hydrogen. The energy from the Sun, both heat and light, originates from a nuclear fusion process that is occurring inside its core. These heavy elements were then blasted into space in great stellar explosions called supernovae, where they later congealed into planets, mountains – and humans. In other words, we are all made of stardust. The nuclear fusion conversion process for a star with a large enough mass is as follows: Hydrogen, Helium, Carbon, Oxygen, Neon, Silicon and Iron.[312] A star like our sun will burn through its Hydrogen and Helium and end up as a Carbon/Oxygen hybrid. When the nuclear fusion of Iron is achieved, the fusion process is no longer autonomous and requires an external energy source such as gravity to fuse heavier elements.

310 https://www.worldatlas.com/space/are-we-made-of-stardust.html

311 Hydrogen is the most abundant element in the universe, however, until 1925 no one knew this. [Archives of the Universe, p.251]

312 [Lederman, Schramm: From Quarks to the Cosmos, p.83]

All the known elements were formed from Hydrogen. These elements formed molecules and life as we know it began. Hydrogen is the only element that does not contain a neutron because it has only one proton. When an additional proton is added to the nucleus to form higher elements, these protons repel each other, and it is the neutron (with zero electrical charge) that helps to keep the nucleus together via the Strong and Weak nuclear forces. Therefore, beyond Hydrogen, there must be at least one neutron for every proton. Without neutrons protons would repel each other and for every proton there must be an electron to keep the atom electrically neutral. However, the atomic number and weight is determined by the number of protons and neutrons.

Furthermore, when four Hydrogen atoms are fused together to form one Helium atom. The mass of the Helium atom is less than the sum of the masses of the four Hydrogen atoms. This delta is equal to .07 percent, and it is precisely this number (no greater or no less) that brought our universe into being. [313] Every second, the Sun fuses 620 billion Kg of Hydrogen nuclei (protons) into Helium, to produce 384.6 trillion trillion Joules of energy per second. This is equivalent to the energy released in the explosion of 91.92 billion megatons of TNT per second. About 75% of the Sun is made up of Hydrogen, while the rest is mostly all Helium.[314] Today, the Sun continues to fuse Hydrogen atoms to make Helium in its core. It fuses about 600 million tons of Hydrogen every second, yielding 596 million tons of Helium. The remaining four million tons of Hydrogen are converted to energy, which makes the Sun shine.[315]

The fusion process continues in the following manner: When the temperature in the core reaches about 100 million degrees, the Helium will begin to fuse into Carbon by a reaction known as the triple-alpha process.[316]

313 [Rees: Just Six Numbers, p.47] The efficiency with which hydrogen fuses to helium, which has to do with the strength of the attraction between particles in an atomic nucleus. It's about 0.007. Raise it to 0.008 and the hydrogen created in the Big Bang would have turned to helium almost immediately. Reduce it to 0.006, on the other hand, and helium would never form, the stars wouldn't have ignited to give energy for life." [New Scientist: The Universe Next Door]

314 Nuclear Fusion in the Sun Explained Perfectly by Science - Universavvy

315 https://brainly.com/question/24706656

316 What is the fusion of hydrogen into helium? - TimesMojo

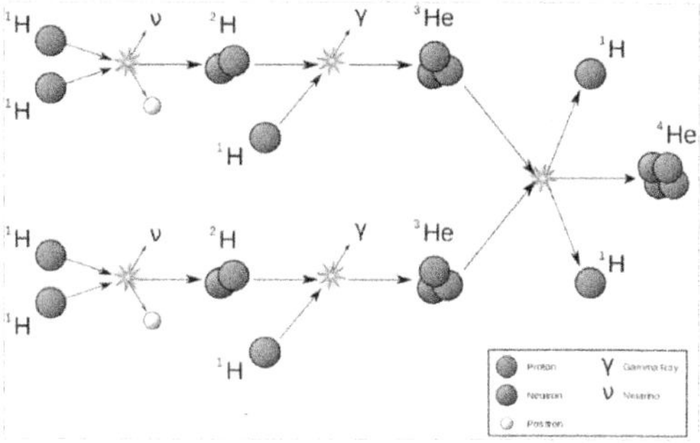

Nuclear Fusion of Hydrogen into Helium[317]

In the triple alpha process, two alpha particles combine to form the nucleus 8Be which then captures another alpha particle to form the Hoyle state in 12C. This state can then decay to the ground state of ^{12}C by emitting gamma rays and thereby making ^{12}C, or decay back into three α's. The production of ^{12}C in nature is proportional to the probability of decay of the Hoyle state to the ground state of ^{12}C.[318]

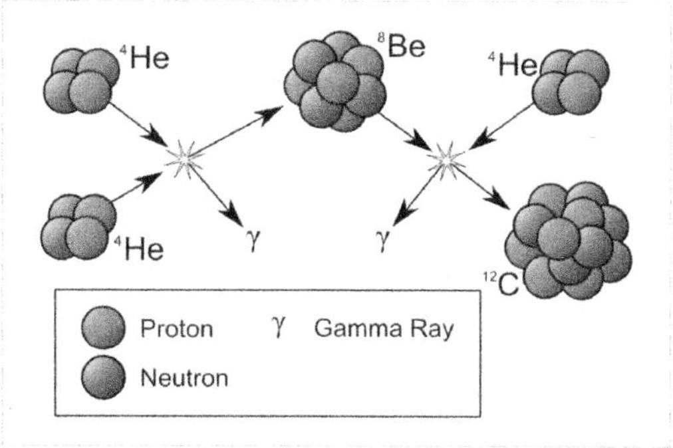

Additional elements made via the fusion process[319]

317 By Doctor C - Own work, CC BY-SA 4.0, https://commons.wikimedia.org/w/index.php?curid=48960438

318 Enhancement of the Triple Alpha Process | National Superconducting Cyclotron Laboratory | Michigan State University (msu.edu)

319 Fichier:Reaction-triple-alpha.svg — Wikipedia by Unknown Author is licensed under CC BY-SA The fusion process is not at all simple since there does not exist a stable configuration with an atomic mass of 8 (8Be) that is formed by the fusion of two 4He nuclei. The lifetime of 8Be is a very short 3 × 10-16 seconds. However, if this unstable 8Be nucleus is struck by another 4He nuclei it is possible to form 12C (and a gamma ray). This can occur because the lifetime of 8Be is actually longer than the mean collision or scattering time of helium nuclei at temperatures around 108K which are needed to make this fusion process proceed. In 1952 Edwin Salpeter proposed this process as the way to form 12C.Not long

Once ^{12}C has been formed, it is possible with temperatures around 6 × 108 K to continue forming heavier nuclei by the combination of two ^{12}C nuclei to make 16O, ^{20}Ne, ^{24}Mg and with temperatures around 109K the combination of two 16O nuclei can make ^{28}Si, ^{31}P, ^{32}S.[320]

Professor Weinberg explains the importance of the helium beryllium isotope fusion.

> A beryllium isotope having the minuscule half-life of 0.0000000000000001 seconds must find and absorb a helium nucleus in that split of time before decaying. This occurs only because of a totally unexpected, exquisitely precise, energy match between the two nuclei. If this did not occur there would be none of the heavier elements. No Carbon, no Nitrogen, no life. Our universe would be composed of hydrogen and helium. [321]
>
> Furthermore, the mysteries are enhanced when we consider how surprising it is that the laws of nature and the initial conditions of the universe should allow for the existence of sentient beings who could observe it. Life, as we know it, would be impossible if any one of several physical quantities had slightly different values. The best known of these

afterwards Fred Hoyle suggested that the fusion of 8Be and 4He would be greatly enhanced if the 12C nucleus possessed an energy level near to the combined energies of the two nuclei involved. The subsequent reaction would then be a resonant reaction. This resonant energy level was found by experiments at the Kellogg Radiation Laboratory at CalTech. The triple alpha process will occur in red giant stars that have left the main sequence (and have consumed their core hydrogen) and have core temperatures of 108K and higher. Once 12C has been formed it is possible with temperatures around 6 × 108 K to continue forming heavier nuclei by the combination of two 12C nuclei to make 16O, 20Ne, 24Mg and with temperatures around 109K the combination of two 16O nuclei can make 28Si, 31P, 31S and 32S.

320 Triple Alpha Process | COSMOS (swin.edu.au)

321 The Fine Tuning of the Universe | Gerald Schroeder There is an essential step in the chain of nuclear reactions that build up heavy elements in stars. In this step, two helium nuclei join together to form the unstable nucleus of beryllium 8, which sometimes before fissioning absorbs another helium nucleus, forming carbon 12 in this excited state. The carbon 12 nucleus then emits a photon and decays into the stable state of lowest energy. In subsequent nuclear reactions carbon is built up into oxygen and nitrogen and the other heavy elements necessary for life. But the capture of helium by beryllium 8 is a resonant process, whose reaction rate is a sharply peaked function of the energies of the nuclei involved. If the energy of the excited state of carbon 12 were just a little higher, the rate of its formation would be much less, so that almost all the beryllium 8 nuclei would fission into helium nuclei before carbon could be formed. The universe would then consist almost entirely of hydrogen and helium, without the ingredients for life. Weinberg: Life in the Quantum Universe (ucla.edu)

quantities is the energy of one of the excited states of the Carbon 12 nucleus.[322]

The nuclear fusion process (CNO Cycle) occurs very marginally in the Sun, but is the dominant fusion pathway in stars 1.5 times more massive, than our Sun. This process also fuses four protons into a Helium nucleus, by using Carbon (C), Nitrogen (N) and Oxygen (O) nuclei as catalysts. [323]

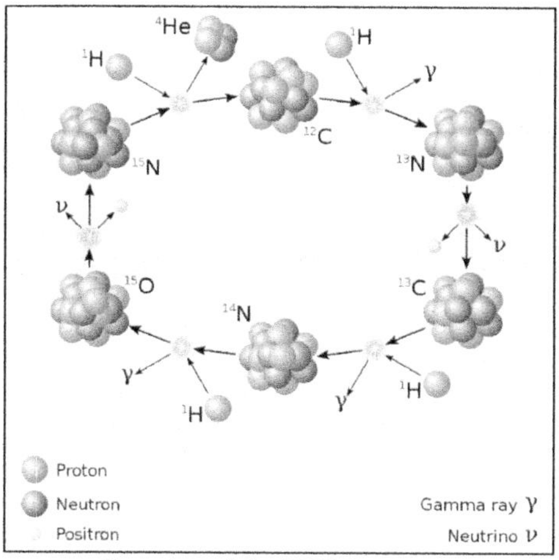

Overview of the CNO-I process.

There are several alternative CNO pathways that can lead to Helium-4 production. This process produces only 0.8% of the Sun's total energy output. Like p-p chain reaction, the CNO cycle has several alternative paths, but the dominant one, occurring in the Sun, is primarily CNO-I. Thus, the line of demarcation is drawn at Iron which is the most stable nucleus in nature. In addition to being the line drawn in the sand by stars, it is also the line drawn during the radioactive decay series from Uranium to Iron.[324]

---

322 Weinberg: Life in the Quantum Universe (ucla.edu)

323 (1) New Message! (yourwiseanswers.com)

324 Radioactive decay occurs when an unstable (radioactive) isotope transforms to a more stable isotope, generally by emitting a subatomic particle such as an alpha or beta particle http://www-eng.lbl.gov/~shuman/NEXT/MATERIALS&COMPONENTS/Background_measurement/natural-decay-series.pdf

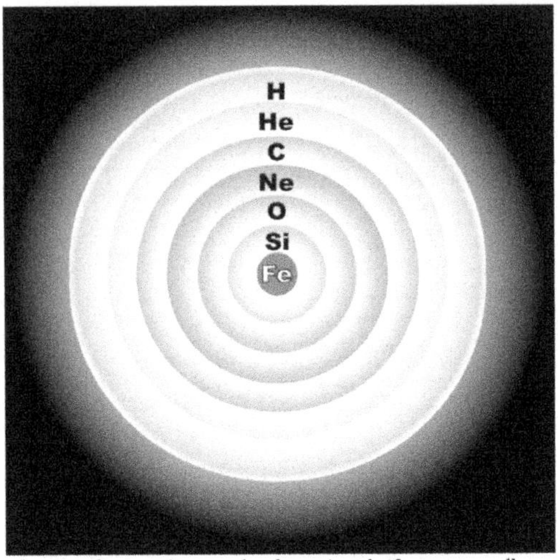

The onion-like layers of a massive, evolved star just before core collapse (not to scale)
This diagram shows a simplified (not to scale) cross-section of a massive, evolved star (with a mass greater than eight times the Sun.) Where the pressure and temperature permit, concentric shells of Hydrogen (H), Helium (He), Carbon (C), Neon/Magnesium (Ne), Oxygen (O) and Silicon (Si) plasma are burning inside the star. The resulting fusion by-products rain down upon the next lower layer, building up the shell below. As a result of Silicon fusion, an inert core of Iron (Fe) plasma is steadily building up at the center. Once this core reaches the Chandrasekhar mass, the Iron can no longer sustain its own mass and it undergoes a collapse. This can result in a supernova explosion.[325]

The following is a chart of the elements produced in a star by nuclear fusion. [326]

Elements Produced by Autonomous Nuclear Fusion[327]

Atomic Number	Chemical Symbol
1	H
2	He
3	Li
4	Be

325 By User:Rursus - R. J. Hall, CC BY 2.5, https://commons.wikimedia.org/w/index.php?curid=2565862
326 [Macdougall: Nature's Clock, p.31]
327 http://www.astrophysicsspectator.com/tables/Abundances.html

5	B
6	C
7	N
8	O
9	F
10	Ne
11	Na
12	Mg
13	Al
14	Si
15	P
16	S
17	Cl
18	Ar
19	K
20	Ca
21	Sc
22	Ti
23	V
24	Cr
25	Mn

During stellar nucleosynthesis, an element is formed by the reaction of Hydrogen (H), Carbon (C), Nitrogen (N), Oxygen (O), Phosphorous (P), Sulfur (S), Chlorine (Cl), Sodium from Natrium (Na), Magnesium (Mg), Potassium from Kalium (K). These elements are what powers our solar system and other stars.[328] This means that the elements of the building blocks of life, i.e., Carbon, Hydrogen, Oxygen, Nitrogen, Phosphorus and Sulfur (CHONPS), etc. can all be produced within a star without external intervention of forces such as gravity required to fuse them. The four most abundant elements in the universe are: Hydrogen, Helium, Oxygen and Carbon. Both animals and plants have a high percentage of these elements.

---

328 The End Product Of Stellar Fusion: Iron – BosCoin

Percentage of Elements in Plants and Animals

Element	Mass in plants	Mass in animals	Biological uses
Carbon	12%	19%	Found in carbohydrates, lipids, nucleic acids, and proteins.
Hydrogen	10%	10%	Found in carbohydrates, lipids, nucleic acids, and proteins.
Nitrogen	1%	4%	Found in nucleic acids, proteins and some lipids
Oxygen	77%	63%	Found in carbohydrates, lipids, nucleic acids, and proteins.
Phosphorus	<1%	<1%	Found in lipids and nucleic acids.
Sulfur	<1%	<1%	Found in proteins.

CHONPS in the human body: Four elements compose about 96% of the weight, and major minerals and minor minerals[329]

All living organisms, without exception, need water to survive and grow, making it one of the most important substances on Earth. In complex organisms it is used to dissolve vitamins and mineral nutrients. It is then used to transport these substances, as well as hormones, antibodies, oxygen and other substances around and out of the body.[330]

329 CHON - Wikipedia

330 Biogeochemical Cycle - Definition and Examples | Biology Dictionary

Elements beyond Iron are formed by a different process and gravity takes control. The star can no longer sustain the innate fusion process that established equilibrium with gravity and in larger stars, gravity begins to fuse Iron into heavier elements. Iron fusion absorbs energy and causes the core to collapse. Each element is not only different in its composition of electrons, protons, neutrons, etc. but emit their own signatures within the electromagnetic spectrum called "spectral or absorption lines." Absorption lines allow scientists to discern the composition of stars vast distances away.[331] In this regard, they talk to us not only telling us how far away they are but what they are composed of.

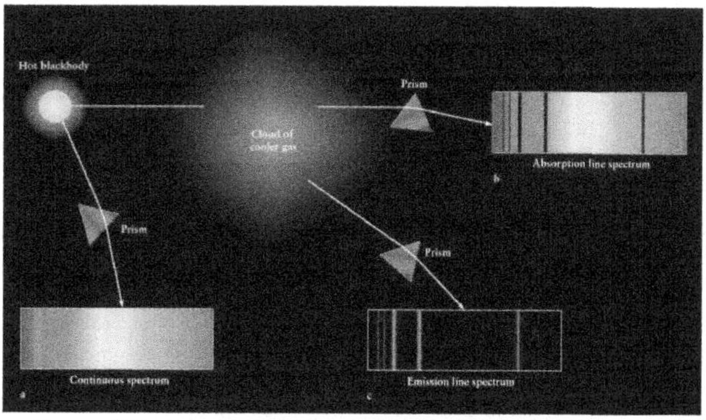

Absorption and Emission Lines of a given element: The thickness of the line is determined by the amount of the element present

When larger stars explode; they not only form the elements beyond Iron but provide the impetus for other stars to be born and may even become centers of galaxies. And so, we are once again faced with an axiom of the universe, that death sustains life and life sustains death. And we now know that we are all made of stardust blowing, like pieces of sand, in a cosmic wind. Not only are we composed of elements made in distant suns, but we use the energy stored in other life forms or energy stored millions of years ago in fossil fuels created during the Age of the Dinosaurs to sustain our local equilibrium. In this regard, we are like the Sun, and we also radiate energy. To keep our lives going, we must take in about two thousand calories per day and we give

---

331 The absorption lines in the Sun and stars can be identified with individual chemical elements or molecular compounds by comparing their positions in the spectrum (their wavelengths) with those observed from pure sources in the laboratory. Some absorptions are very weak, just shallow dips in the spectrum, whereas others are completely black. The "strength" of an absorption line -- the amount of energy removed from the spectrum -- depends on the amount of the particular chemical element in the star causing the line and on the efficiency of absorption. http://stars.astro.illinois.edu/sow/spectra.html

off as much radiation as a 100-watt light bulb per person. And at death, the light goes out.

The Periodic Table is a chart that describes the known elements (atoms). Most of the elements were produced in the Sun, however, a few were man made. First constructed by Dmitry I. Mendeleyev in the mid-19th century, the table has been of enormous value in the development of chemistry. In a stroke of brilliance, Mendeleyev left openings for elements that had not yet been discovered and he even went so far as to predict the properties of those missing elements. Although the table was first constructed using atomic weight and not number, Mendeleyev's overall concept forms the basis of the chart we used today.[332] However, it was not recognized until the second decade of the 20th century that the order of elements should reflect the atomic number, which is equal to the positive electrical charge of the atomic nuclei.[333]

The Russian chemist Dmitri Mendeleev in the 1890s

The table is read from top left to bottom right with the elements arranged by. their atomic numbers. As you move across and down the table, the atomic number

332 Mendeleev didn't conjure up the first periodic table on his own. At least six people invented it working independently, but Mendeleev became the most important character in the story of the periodic table. [Kean: The Disappearing Spoon, p.21]

333 Periodic table | Definition, Elements, Groups, Charges, Trends, & Facts | Britannica

(number of protons) and weight (protons and neutrons) increases. Each horizontal row on the periodic table is called a period. There are seven periods on the periodic table and eighteen columns. Each element has its tile or cell in the periodic table. The exact information given for the element varies, but there is always the atomic number, the symbol for the element (denoted by one or two letters), and the atomic weight.[334]

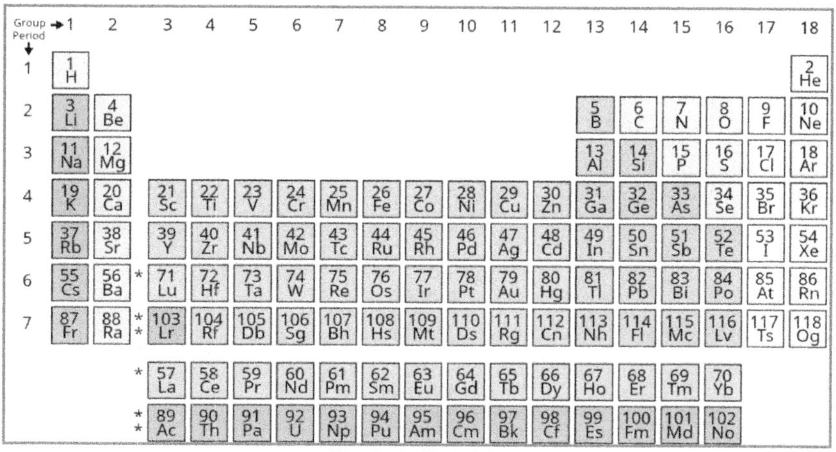

The Periodic Table

The table is also arranged in groups according to the number of electrons in the outer shell. The elements in each group have similar chemical properties due to the similarity in their valence electron configurations.[335] For example, the noble gases (column 18) outer's shell all are full which makes it difficult to bond with other elements. Based on selected physical properties, elements are classified into metals and nonmetals. Metals are located on the left side of the periodic table, and nonmetals are located on the right. Therefore, the Periodic Table is a succinct way of identifying the property of an element and its relationship to other elements. There are two rows of elements found below the main body of the table. They are placed there because there wasn't room to put them where they should go.

Elements very rarely appear in nature alone but as part of molecules. Atoms such as Helium "normally" stand alone because the completeness of their outer shell

334 How Is the Periodic Table Organized Today? (thoughtco.com)

335 The arrangement of the periodic table | Learning Lab (rmit.edu.au)

inhibits the electrical bonding of other atoms to form molecules. In this regard, they are the bachelors of the atomic world. However, implicit in the structure of most atoms is the ability to bond with other atoms electrically if they have a complementary set of electrons necessary to complete the outer shell. Take the molecule of water, H_2O, for example. Oxygen needs two electrons to complete its outer shell which is provided by two Hydrogen atoms. Although Helium contains two electrons in its outer shell; Oxygen could not bond with the Helium because the outer shell of the Helium atom is complete. Whereas the outer shell of the two Hydrogen atoms is not complete making bonding possible. Without such a relationship, water could not exist. In this regard, the Periodic Table serves as a roadmap of how atoms bond to form molecules.

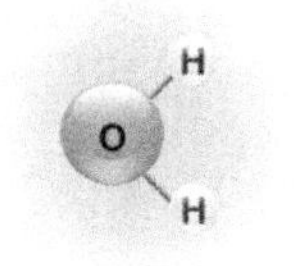

Water

It is as if these elements, cooked in stars and enclosed in amino acids and other molecules were produced with life in mind. Thus, stars are the factories of life, and without them, the molecules required for proteins and DNA would not be possible. However, once this goal is met, gravity steps in to make the gold found in planets such as our own. It is only in larger stars that heavy elements such as gold are made. Therefore, our cosmos is dependent on one source of finite energy and one rule, the *Conservation of Energy*, to generate all the known elements and all subsequent laws. The speed of light of a photon is a consequence of the conversion of mass into energy. It is as if entropy asks the star, how much mass do you have and based on that answer it knows how and when the star will die. Thus, the eulogy of a star is written in the language of its mass, and its mass consists of the residual energy left over after the cosmos was created by the Big Bang sequence. This created a system of procreation and death that would be sustainable for at least trillions of years or until all the energy in the cosmos is exhausted.

Amino Acids, Proteins and DNA

It is also very important to note the interaction between Hydrogen, Oxygen, Nitrogen and Carbon and their ability to bond with each other. This is essential to creating molecules and long chains of amino acids and proteins that lie at the core of our genes. [336] All the information for encoding all those amino acids and proteins is contained within your genome! There are about 20 amino acids, and they link together in molecular chains called polypeptides, which are the building blocks of proteins. In the structure, the R represents a side chain specific to each amino acid.

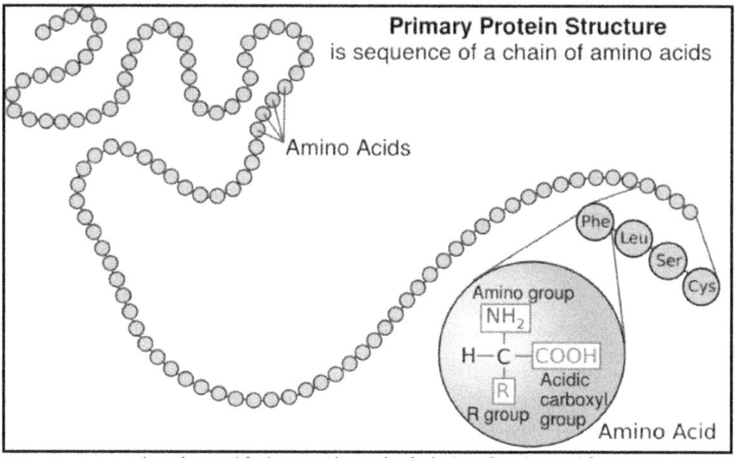

A polypeptide is an unbranched chain of amino acid

---

336 Amino acids are organic molecules that, when linked together with other amino acids, form a protein. Amino acids are essential to life because the proteins they form are involved in virtually all cell functions. Some proteins function as enzymes, some as antibodies, while others provide structural support. Although there are hundreds of amino acids found in nature, proteins are constructed from a set of 20 amino acids. Amino Acids: Structure, Classification and Function (thoughtco.com)
A peptide formed by the union of an indefinite (usually large) number of amino acids by peptide links (-NH-CO-). Polypeptide | definition of polypeptide by Medical dictionary (thefreedictionary.com)
The functional or R groups of amino acids are the groups that define the chemistry of proteins. They are the basis on which amino acids are classified, and according to the Miller-Urey experiment, possibly the start of all organic life on Earth. Not only is "What are amino acid functional groups?" a stimulating question to ask, but the answer to that question comes awfully close to revealing the true answers behind life, the universe, and... well...everything. https://aminoco.com/blogs/amino-acids/function-amino-acid-groups

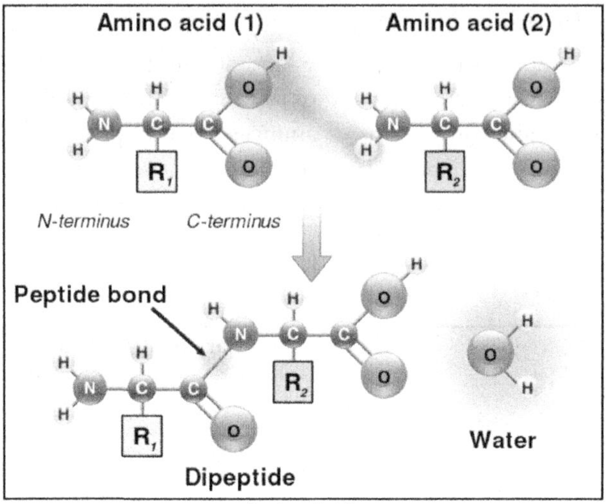

Amino Acids

On Earth, the amino acids and proteins of life are all "left-handed" and cannot be exchanged for their right-handed lookalike. Meanwhile, all sugars molecules are "right-handed." The opposite hands for both amino acids and sugars exist in the universe, but they just aren't utilized by any known biological life form. (Some bacteria can actually convert right-handed amino acids into the left-handed version, but they can't use the right-handed ones as is.) In other words, both sugars and amino acids on Earth are homochiral: one-handed and mixing the two does not appear to work.[337] There's nothing to prevent life from developing from right-handed amino acids, it just happened that on our planet the left-handed amino acids were what was available. Scientists do not know why amino acids are left-handed and sugars right-handed. However, it is certainly feasible that it could have been the other way around. One prevalent theory that is being explored is that meteorites could have delivered left-handed amino acids to Earth.[338]

---

337 https://www.smithsonianmag.com/space/must-all-molecules-life-be-left-handed-or-right-handed-180959956/

338 Life's 'Left-Handed' Amino Acids Still A Puzzle (forbes.com)
Meteorites then might have served as the cosmic Mayflowers that transported molecular seeds to Earth. In 1969, the Murchison meteorite that fell in Australia contained dozens of different amino acids—the building blocks of life. The origin of life on Earth, explained | University of Chicago News (uchicago.edu)
The OSIRIS-REx mission (Origins, Spectral Interpretation, Resource Identification, and Security Regolith Explorer). The primary objective of the mission is to return at least 60 g of "pristine" material

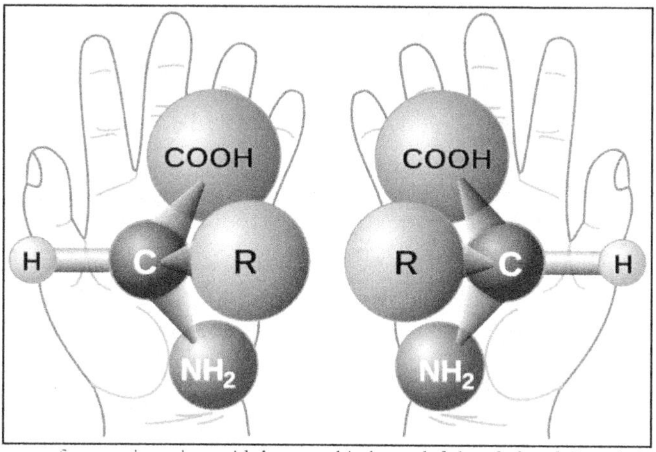

Two enantiomers of a generic amino acid that are chiral, one left-handed and the other right-handed

Proteins are made up of one or more long, folded chains of amino acids (polypeptides) and the structure in which the different amino acids are arranged helps to determine the role of the protein. Protein-encoding deoxyribonucleic acid (DNA) specify the sequences of amino acids in each protein. In turn, proteins are responsible for coordinating nearly every function of the cell. Both protein-encoding genes and the proteins that are their gene products are absolutely essential to life as we know it. They are critical to most of the work done by cells and are required for the structure, function and regulation of the body's tissues and organs.[339]

The DNA *Double Helix Structure* has two strands. Each strand is made up of a polynucleotide chain consisting of a Deoxyribose Pentose Sugar attached to Phosphate group with Nitrogenous bases (Adenine, Guanine, Cytosine, Thymine) being projected from the strand. Adenine always pairs with thymine, and cytosine always pairs with guanine as shown below. [340]

from an asteroid (101955) Bennu, which is spectrally similar to organic-rich CI or CM meteorites NASA scientists will analyze the OSIRIS-REx sample to determine Bennu's chemical composition. They hope to find water ice and organic molecules such as amino acids — both precursors to life.

339 Protein (genome.gov) https://www.hou.usra.edu/meetings/abscicon2017/pdf/3138.pdf

340 DNA | Definition, Discovery, Function, Bases, Facts, & Structure | Britannica Gregor Mendel, known as the "father of modern genetics," was born in Austria in 1822. A monk, Mendel discovered the basic principles of heredity through experiments in his monastery's garden. His experiments showed that the inheritance of certain traits in pea plants follows particular patterns, subsequently becoming the foundation of modern genetics and leading to the study of heredity. Gregor Mendel - Life, Experiments &

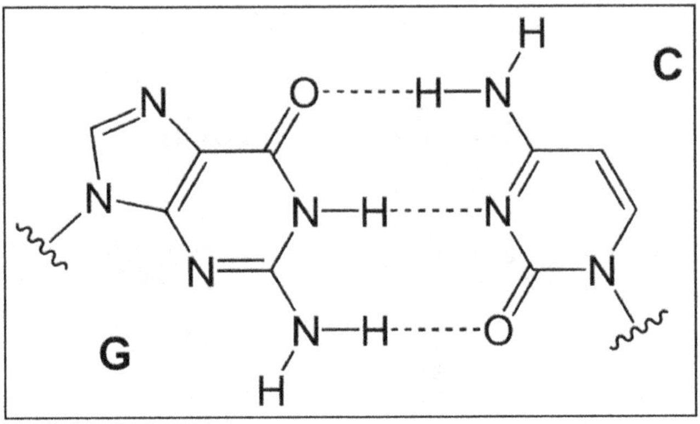

A GC base pair with three hydrogen bonds. Bottom, an AT base pair with two hydrogen bonds. Non-covalent hydrogen bonds between the pairs are shown as dashed lines

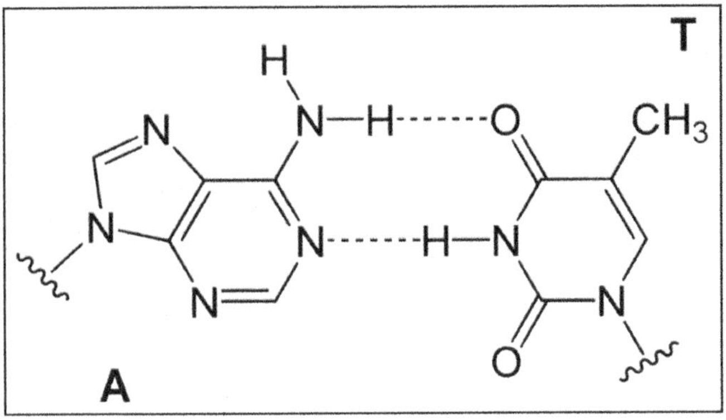

One strand is attached to the other strand forming a twisted ladder-like helical structure. We refer to "genetic code" as the term used to define how the A, C, G, and Ts are strung together and how ribosomes (the cellular machinery responsible for making proteins) read the mRNA sequence and translates each three-base codon into its corresponding amino acid in a growing protein chain. A DNA strand averages about two nanometers (or two billionths of a meter). In comparison, a human hair is about 40,000 times wider than a DNA strand. [341]

Facts (biography.com)

341 DNA | Definition, Discovery, Function, Bases, Facts, & Structure | Britannica

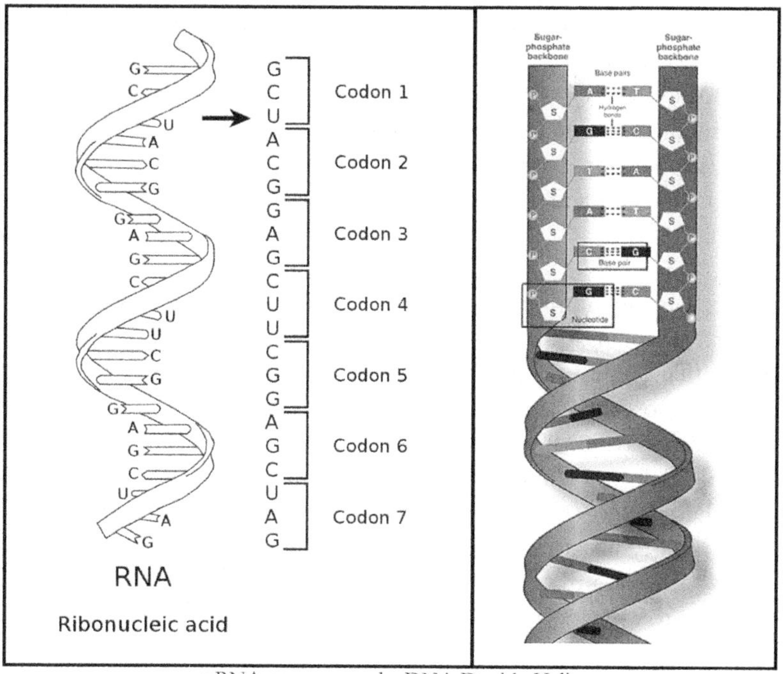

mRNA sequence and a DNA Double Helix

The two strands are held together by hydrogen bonds between the bases, with adenine forming a base pair with thymine, and cytosine forming a base pair with guanine. DNA's unique structure enables the molecule to copy itself during cell division.[342] It replicates by separating into two single strands, each of which serves as a template for a new strand. The new strands are copied by the same principle of hydrogen-bond pairing between bases that exists in the double helix. Two new double-stranded molecules of DNA are produced, each containing one of the original strands and one new strand. Replication is the key to the stable inheritance of genetic traits.[343]

---

342 When a cell prepares to divide, the DNA helix splits down the middle and becomes two single strands. These single strands serve as templates for building two new, double-stranded DNA molecules - each a replica of the original DNA molecule. In this process, an A base is added wherever there is a T, a C where there is a G, and so on until all the bases once again have partners. Deoxyribonucleic Acid (DNA) Fact Sheet (genome.gov)

343 DNA | Definition, Discovery, Function, Bases, Facts, & Structure | Britannica

When proteins are being made, the double helix unwinds to allow a single strand of DNA to serve as a template. This template strand is then transcribed into mRNA, which is a molecule that conveys vital instructions to the cell's protein-making machinery. [344] Within a cell, DNA is organized into chromosomes. Chromosomes are thread-like structures located inside the nucleus of animal and plant cells.[345] Each chromosome is made of protein and a single molecule of (DNA). Passed from parents to offspring, DNA contains the specific instructions that make each type of living creature unique.

Chromosomes vary in number and shape. Most bacteria have one or two circular chromosomes. Humans, along with other animals and plants, have linear chromosomes that are arranged in pairs within the nucleus of the cell. Humans have 23 pairs of chromosomes, for a total of 46 chromosomes. In fact, each species of plants and animals have a specific number of chromosomes. For example, a fruit fly, has four pairs of chromosomes, while a rice plant has 12 and a dog, 39. The only animal cells that do not contain pairs of chromosomes are reproductive cells, or gametes, which carry just one copy of each chromosome producing one pair of sex chromosomes (XX for a male or XY for a female). [346] Thus, we went from chemicals made in the stars to cells in the body in one fell swoop as if the stars had life in mind.

Succinctly put, the Big Bang sequence did not have enough time to burn all its energy so it did the next best thing and stored it in mass that could. If all the energy of the cosmos was consumed during the Big Bang sequence, life as we know it could not exist. Everything in the cosmos, from the macro to the micro, is predicated on a series of events that happen in a relatively short time in the history of the cosmos. Many scientists contend that because the universe is designed to support life that it must indeed exists elsewhere and that we are not the only winners in the cosmic lottery.

344 Deoxyribonucleic Acid (DNA) Fact Sheet (genome.gov)

345 https://onlyzoology.com/bonds-in-dna-double-helix-structure/

346 https://www.genome.gov/about-genomics/fact-sheets/Chromosomes-Fact-Sheet

PARALLEL AND MULTIVERSES

In 1915, Einstein printed the general relativity theorem. And that was a breaking point in the history of cosmological evolution. Since then, scientists have created many universe models by using the general relativity equations. Some of them were plausible and others sounded like sci-fi. [347]

The possibilities of other worlds, multiverses, and dimensions postulated by some scientists continue to intrigue us. Time has erased superstition as the medium of understanding and placed it within the hands of science. It is through science that the horizon of our cosmos comes into sharper focus as we search with greater enthusiasm to discern our place, not only in our cosmos, but within what is plausible and possible. Mathematics is a universal language, and it is through the eyes of math that we can visualize dimensions unseen. We are not the inventors of mathematics, but discoverers of its laws imbued in the nature of things and taught to us by Nature. Using math as a tool, we are able to use physics, chemistry, and biology to help us understand the nature of our reality. Everything that science understands about our way of life has been revealed to us through mathematics. The French mathematicians of the early eighteenth century, Galois, Fourier, Laplace, etc., had no idea of how their mathematics would be used in the coming centuries.

Feynman and Gödel

Science has always followed the mathematical breadcrumbs through the wilderness of the unknown. The concept of extra dimensions is a mathematical construct. It is paramount to understanding how dwellers of the quantum world interact and lie at the core of string theory. Science often reveals that what is true in the micro world is also true in the macro world. Therefore, the understanding of extra dimensions may also shed light on how denizens of the cosmic world interact as our cosmic world may in fact be a microcosm of a greater macro existence.

But how do we define a dimension? Is it a world of bounded infinities, a world among worlds, and a universe among universes? If so, how do these worlds interact and how

347 https://medium.com/predict/extraor"dinary-universes-g%C3%B6dels-rotating-universe-6b44938b2a5

do we define the line of demarcation between worlds and universes some large, some small, of which our species and our world may set an equal distance apart bound by infinity in both directions? Perhaps, each dimension is bounded by its infinities, and you can only enter the next dimension after all the possibilities of the current dimension have been exhausted. This may be the basis for Feynman's summation over histories, i.e., that a photon takes all possible paths. Not only does it take all possible paths, but it must also take all the paths simultaneously. [348] However, these paths make take place outside of our dimension.

Richard Phillips Feynman (1918–1988)

348 This approach, also known as 'sum over histories', describes an averaging of all the paths a particle can possibly take from its starting point to its destination. According to Richard Feynman, particles must be viewed as travelling from one location to another along every possible path and along all possible paths between them. For example, if you shine light on a barrier with two slits in it with a photographic plate behind to see which electrons make it through which slit, the experiment will confirm that each electron that makes it through to the screen actually goes through both slits. Not only that but the same electron must be passing simultaneously through both slits Feynman's sum over paths - overview - ELWYN'S BIG BANG PAGE (weebly.com)

The following explains:

> The electron, according to Feynman, simultaneously 'sniffs' out every possible path connecting its starting location with its final destination", and you add up the probabilities of all conceivable paths. The probability that the electron (or photon - any particle) arrives at any chosen point is therefore built up from the combined effect of every possible way of getting there, rather like working out the probabilities of a person who enters a subway station from one entrance or turnstile exiting via another specific location when numerous routes and choices are available[349].
>
> According to Heisenberg: The path of the electron comes into existence only when we observe it." When we measure it, we always find the electron localized as a particle. We may say that our measurement reduces the electron wave to the particle state. Whenever we measure it, a quantum object appears at some single place, as a particle. The probability distribution simply identifies that place (or those places) where it is likely to be found when we do measure it—no more than that. When we are not measuring it, the quantum object spreads and exists in more than one place at the same time, in the same way that a wave or cloud does—no less than that.[350]

Between measurements in its wave state, we have no idea where the electron or the photon resides.[351] Perhaps, the actual movement relative to the photon or electron takes place outside our dimension (our spacetime) and we observe the effects of that action in our dimension(spacetime). The reason that we cannot determine its location is because we are missing the coordinate supplied by the other dimension as it moves or jumps to our spacetime. By chance, the photon travels at the speed of light because it is the limit within our dimension and hence the closest thing to infinity the latter being the boundary of all dimensions. Likewise, it is the medium of spacetime that

349 Feynman's sum over paths - overview - ELWYN'S BIG BANG PAGE (weebly.com)

350 [Goswami: The Self-Aware Universe p.39]

351 Posed by the Danish physicist Niels Bohr, this Copenhagen interpretation says that all quantum particles don't exist in one state or the other, but in all of its possible states at once. The sum total of possible states of a quantum object is called its wave function. The state of an object existing in all of its possible states at once is called its superposition. Do Parallel Universes Really Exist? | HowStuffWorks

limits the photon to the speed of light. In actuality, the photon may travel at superluminal speeds relative to the dimension from which it emanates. This is a necessary condition for it to take all the possible paths within our dimension simultaneously.

Feasibly, our entire dimension, which we call spacetime, is also traveling at the speed of light, or perhaps even greater constant speed, and the aforementioned conditions form our version of reality. We must be forever mindful of Einstein's *Theory of Special Relativity* which states that an object moving at a constant speed for all intents and purposes is at rest relative to other bodies traveling at that speed. Therefore, what appears to us to be at rest is in a constant state of motion relative to something outside of spacetime. In this regard, we are all chained members in Plato's Cave within our dimension and subject to the laws of its reality which in turn brought it into being. The action itself, in accordance with *Gödel's Incompleteness Theorem,* lies outside of our reality.[352] To quote Gödel:

> It's indeed the case that if the laws of physics are formulated in a formal system (S)which includes the concepts and axioms of arithmetic as well as physical notions such as time, space, mass, charge, velocity, etc., and if (S) is consistent then there are propositions of higher arithmetic which are undecidable by (S). But this tells us nothing about the specifically physical laws encapsulated in (S), which could conceivably be complete as such.[353]

Furthermore, Gödel argued that an axiom statement that is taken to be true can never truly be proven or complete within its own system. In any given system, at least one axiom must be false or unproven.[354]

---

352 The theorem states that in any reasonable mathematical system there will always be true statements that cannot be proved. The result was a huge shock to the mathematical community, where the prevailing view was an unshakeable optimism about the power and reach of their subject. It had been assumed that math was "complete", meaning that all mathematical statements are either provable or refutable. The 25-year-old Gödel demonstrated this was incorrect by constructing a true statement that was not provable. Maths, he announced, has its limits. Can you solve it? Gödel's incompleteness theorem | Mathematics | The Guardian

353 'The Fabric of the Cosmos' | Freeman Dyson | The New York Review of Books (nybooks.com)

354Despite his successful career, Kurt Gödel's mental health never seemed to stabilize. Towards the end of his life, Gödel's paranoia only grew. He became very suspicious of all food and was convinced that someone was trying to poison him. He refused to eat any meals that had not first been tasted by his wife.

From *Gödel's Incompleteness Theorem,* we can infer the following:

A closed system (S) could be described as a group of things(objects) that are dependent on each other and linked by some common entity. In such a system you can draw a circle around what is inside the system and what is not or outside the system. All closed systems depend on something outside of the system and cannot be explained inside without suggesting something outside that is believed to be true or to proceed elements within the system. No system alone can completely prove itself to be true. There must be something outside any circle that is invisible and unknown to us. You can always draw a bigger circle but there will still be something outside the circle. Future events are predictable but often uncertain.[355]

Kurt Friedrich Gödel (1906 – 1978)

However, when she became ill in 1977 and had to be hospitalized for six months, Gödel simply refused to eat anything at all, effectively starving himself to death. He died on Jan. 14, 1978, of malnutrition, weighing only 65 pounds. The Tragic Life Of Kurt Gödel, The Renowned Mathematician Who Starved Himself To Death (allthatsinteresting.com)

355 https://evo2.org/incompleteness/

Multiverses and Alternative Realities

The possibility of multiverses or parallel universes is consistent with *Gödel's Incompleteness Theorem.* The following explains:

> One of the many implications of recent cosmological observations is that the concept of parallel universes is no mere metaphor. Space appears to be infinite in size. If so, then somewhere out there, everything that is possible becomes real, no matter how improbable it is. Beyond the range of our telescopes are other regions of space that are identical to ours. Those regions are a type of parallel universe. Scientists can even calculate how distant these universes are, on average! And that is fairly solid physics. When cosmologists consider theories that are less well established, they conclude that other universes can have entirely different properties and laws of physics. The presence of those universes would explain various strange aspects of our own. It could even answer fundamental questions about the nature of time and the comprehensibility of the physical world.[356]

In 1954, Physicist Hugh Everett came up with the idea of parallel universes. He theorized the following:

> A measurement taken of a quantum object causes an actual split in the universe. The universe is literally duplicated, splitting into one universe for each possible outcome from the measurement. For example, say an object's wave function is both a particle and a wave. When a physicist measures the particle, there are two possible outcomes: It will either be measured as a particle or a wave.[357]
>
> When a physicist measures the object, the universe splits into two distinct universes to accommodate each of the possible outcomes. So, a scientist in one universe finds that the object has been measured in wave form. The same scientist in the other universe measures the object as a particle. This also explains how one particle can be measured in more than one state.[358]

---

356 file:///multiverse_sciam.pdf

357 Do Parallel Universes Really Exist? | HowStuffWorks

358 Do Parallel Universes Really Exist? | HowStuffWorks

Like galaxies and black holes that once remained outside the boundaries of conventional physics, the concept of multiverses remains within the theoretical realm until concrete evidence of their existence is confirmed. However, in every case hitherto, our estimation of the size of the cosmos has been short by orders of magnitude. Therefore, the idea of no other universes outside our own may be just as naive as the once held belief that everything was part of the Milky Way. However, the possibilities are endless. Parallel universes, in which our physical laws prevail, may be as plentiful as galaxies. Likewise, there could be a large–perhaps infinite–number of universes, with the fundamental parameters of physics varying from universe to universe. The "brane" multiverse theory suggests that our universe is just one of an infinite number of universes that are floating in a higher-dimensional space.[359], Perhaps the Big Bang represents the moment our universe emerged from another multiverse, after the arrow of time had already been established.

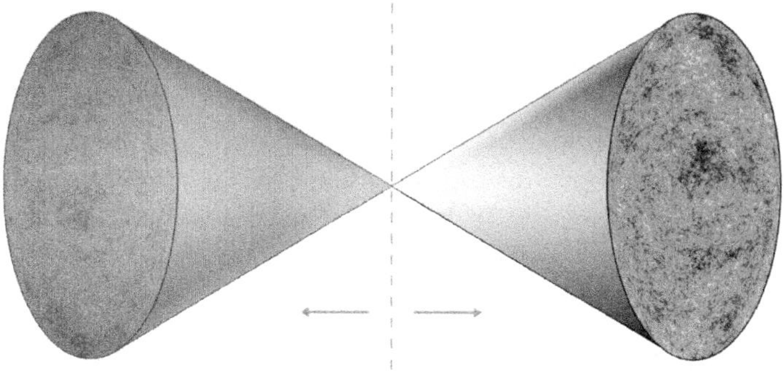

Concept of a twin universe, with the beginning (of time) in the middle[360]

Everett extended the phenomenon of the behavior of the photon and the electron to parallel universes. While in this universe, you may choose path A, an alternate you will choose path B in a parallel universe. [361] Parallel universes, born around the same time as ours, may have started with similar entropy to ours but perhaps with a different

359 The Brane universe exists on a higher dimension where vast multitudes of membranes in higher dimensions collide to form 3-dimensional universes such as ours. This theory is used within String Theories that use higher dimensions to explain the structure of reality. The Multiverse Theory Explained (With Examples) and Whether it's Possible - History-Computer

360 By Eduemoni - Own work, CC BY-SA 4.0, https://commons.wikimedia.org/w/index.php?curid=86585553

361 What are Parallel Universes? (with pictures) (allthescience.org)

arrow of time. Some may live in a world where the Nazis won World War II, or where the dinosaurs survived, or where things fall up instead of down or the past comes before the future. Multiverses may exist, where every possible world exists, along with all the infinite versions of you.[362] Perhaps when we dream that something is happening, it really is happening in an altered state at a different time and place.

Moreover, there could be just as many universes with similar, but different, paradigms that render our laws incomplete or even useless. Just because our laws are predictable and consistent may only mean that they apply to our version of reality and are not necessarily applicable to other realities nor are they absolute. Within our universe, we remain inside the bubble of our existence and a universal truth that may not be absolute. One that circumscribes our reality. Just because spacetimes seems to be the fabric of our universe does not constitute that it is the holy grail of the cosmos, nor does it preclude the existence of extra dimensions within our universe and beyond. The following explains:

> It's not just quantum mechanics that leads to the inescapable conclusion that our universe is just a speck in a vast sea of universes. Different branches of physics spin off very different kinds of multiverses. Our current theory of how the universe came to be predicts an infinite expanse of other universes. If the Big Bang started with a period known as inflation – in which space itself expanded much faster than the speed of light so that there would be a multitude of other universes a lot like ours but causally disconnected from it – this could also potentially be the case with different arrangements of matter. Another level of multiverse arises from the theory of eternal inflation, when the space between universes continues to expand, and a limitless number of new 'bubble' universes with very different properties continue to form.[363]

Reality can take on many faces. Take, for example, the virtual reality of a video game. It could be viewed as a two-dimensional projection of our three-dimensional world, one less dimension than our own. Furthermore, it could be viewed as a system, (S), constructed from an outside higher three-dimensional world. All its movements,

362 New Scientist: The Universe Next Door: A Journey through 55 Alternative Realities, Parallel Worlds and Possible Futures

363 [New Scientists: The Quantum World, pp. 55,56]

elements and actions are dictated by mathematical constructs well understood in the higher dimension. Yet, the actions within the two-dimensional space are not consequences of things generated within system, (S), itself and the objects in system, (S), have no idea of their source or how they are being manipulated by an outside source.

The elements and actions within the game "seem" real and it is only our knowledge of the outside dimension that we know that they are not "real" in our sense of the word. At some point, Artificial Intelligence from outside may be introduced into the video game in which the characters "think" that they are real and may even have the capacity to generate ideas and phenomena independent of the outside system. If we extend this analogy to our world, we can easily see the similarities. Take for example a bee. A bee does not know anything about Euclidian geometry, yet it constructs beehives that are consistent with the laws of Euclidian geometry. How did the bees know that although hives could be built using squares and triangles that the hexagon is the most efficient form?

Another example is a beaver. Beavers do not walk around with slide rules on their hips, calculators in their little hands/feet or blueprints of their next dam. Nor do they walk around with sweatshirts from Drexel University, Rensselaer Polytechnic Institute, MIT or their particular alma mater. Frankly, they don't give a damn about such things. Yet they construct dams that are just as efficient as those that graduated from such outstanding engineering institutions. How is it that these living things know how to do such things without innate knowledge of such laws?

Furthermore, realities that have different reference frames could develop different laws. For instance, a goldfish in a fishbowl could develop laws that are predictable and consistent within its reference frame but completely erroneous within ours because we are not bounded by the distortions of the fishbowl itself or his sense of reality. We can impact the fundamental laws of his fishbowl by performing actions outside that conflict with the laws of his fishbowl. All this leads us to an interesting question: Are we objects in someone else's video game or fish in someone else's fishbowl or beavers building dams without a blueprint?

Like the beavers, we are intuitively equipped with the tools to build without knowledge of where those tools came from or the science that created them. Yet the beaver's dam is not only driven by his need to store food (bark), but by the greater ecological need of which other living things, e.g., fish need to live within system S. The only difference between us and the beaver is that scientists are smart enough to substitute a variable, X, known as the *Anthropomorphic Principle.* Yet at the same time, these same scientists fail to recognize that the reality of the video game also conforms to the *Anthropomorphic Principle* although we "created" it. Just as the beaver is ignorant of his place in the ecological system, we are just as ignorant of our place within the cosmos.[364]

Similarly, experiments and laws within our reference frame may not be understood within our reference frame as natural phenomena and therefore appear to be supernatural. However, to another reference frame, our view and even our laws may seem myopic or void of the essence of the phenomena observed. The biggest trick of science is to separate the illusion from reality. Therefore, when the double slit experiment is performed, the photon takes all possible paths and perhaps what we observe is the reflection, shadow or trace within our dimension of the action taken in a higher dimension by the photon.

Beyond Our Cosmic Horizon

Our cosmic horizon is limited to what we observe in spacetime. However, if we adhere to the laws of inflation and follow them to their logical conclusion; superluminal speeds are not only possible but may be the norm. Time and space may have existed before Planck time, but we just have no idea of what it was like because it lies outside of our cosmic horizon. Before the formation of baryonic matter, we essentially had a quantum physics laboratory in the sky as the laws of our cosmos came into fruition. Inflation of the universe at the Hubble rate opens Pandora's box beyond the Big Bang to reveal the possibility of infinite universes beyond our cosmic horizon. The following explains:

> Inflation tells us that the period of time before the Big Bang was extremely cold, almost at absolute zero, and it was empty of everything but empty space, and that empty space carried energy that stretched the

364 The anthropic principle definition is the belief that scientists may use human life as a starting point to infer predicted properties of the universe that are compatible with the creation of human life if we consider human life as a given state of the universe. Anthropic Principle - Definition, Types, Applications and FAQs (vedantu.com)

universe out to this enormous size and into the initial state before the Big Bang.

During inflation, space expands so fast that portions of the universe are swept outside the cosmic horizon as objects recede from each other at faster than the speed of light; the horizon contains less and less. Inflation and its relationship to the cosmic horizon help to explain a number of properties of the early universe. The presence of horizons in the universe presents an interesting problem for science because they represent a limit to our knowledge. We cannot, even in principle, see beyond the edge of the observable universe.[365]

Using the James Webb telescope, we can see back to the beginning of the Big Bang. The Hubble sphere, however, is much larger than the 13.7 light-years of the Big Bang, that being its smallest possible radius. Notwithstanding, scientists estimate the radius to be at least four times that. Perhaps more astonishing is the number of potential universes given by the inflationary model. The following explains:

> Inflation already ensured that certain traces of the universe's initial conditions would be forever out of reach. Those conditions would put even branes to shame, in terms of defying perceptions. If inflation can pop one quantum universe into existence, why not many? In fact, according to quantum theory, it should. It would, if inflation actually happened. In that case, our inflationary bubble would be one of an ensemble of 10^{500} inflationary bubbles, each its own universe. That's 100,000,000,000,000,000,000,000,000,000,000,000,000,000,000,000,000,00 0,000,000,000,000,000,000,000,000,000,000,000,000,000,000,000,000,000, 000,000,000,000,000,000,000,000,000,000,000,000,000,000,000,000,000,00 0,000,000,000,000,000,000,000,000,000,000,000,000,000,000,000,000,000, 000,000,000,000,000,000,000,000,000,000,000,000,000,000,000,000,000,00 0,000,000,000,000,000,000,000,000,000,000,000,000,000,000,000"" universes. Our universe would just happen to be the one with a value of lambda suitable for the existence of creatures that can contemplate their hyper-Copernican existence."[366]

365 What came before the Big Bang? UB physicist's new popular science book explains one leading theory - UBNow: News and views for UB faculty and staff - University at Buffalo

> Even if there are 10^{500} universes (but not, perhaps, if there are an infinite number of them), they could have been providentially created by the almighty God with a purpose we cannot fathom. Why not? It has even been suggested (by Paul Davies) that multiverse explanations are reminiscent of divine explanations and unintentionally reintroduce a transcendent creator.[367]

It has been said that when you eliminate the impossible what you are left with is the possible regardless of how bizarre it may seem. The possibility of other universes with different special dimensions and physical laws pushes the envelope even further beyond our comprehension. Moreover, since mass is energy in another form, mass may reflect partial components from the other dimension or universes that manifests within our spacetime.[368] As far as information is concerned, it can only travel at the speed of light within our spacetime but can move at superluminal speeds relative to other dimensions not restricted to our laws of space, time and mass. In this regard, a photon could be a transcendent influence operating outside of spacetime but has influences within spacetime.[369] It may be a Cosmic Messenger yet to be understood and there is more to the Electromagnetic spectrum than meets the eye. This hidden information conveyed may be viewed as *Consciousness* from which all the necessary elements and conditions required to conserve and dissipate energy within our dimension emerges over time, from Planck time to the end of our existence. If the proton can excite subatomic particles at the quantum level, then what else can it excite?

366 [Richard Panek: The 4 Percent Universe: Dark Matter, Dark Energy, and the Race to Discover the Rest of Reality pp.236, 237]

367 Cosmology and Theology (Stanford Encyclopedia of Philosophy)

368 For more information on this subject, please refer to my paper entitled: The Music of the String.

369 [Goswami: The Visionary Window p.37].

Part II

MY SCIENTIFIC ESSAYS

INTRODUCTION TO MY SCIENTIFIC ESSAYS

This section is a collection of my scientific essays. I have always had a love for science in general and physics in particular. String theory has always fascinated me along with the wonders of the cosmos. Therefore, in my scientific essays, I try to address the quantum and the cosmic worlds. In fact, the two are inextricably linked and you cannot understand the cosmic world without understanding the quantum world. I often try to combine some of my philosophical beliefs with my scientific knowledge as a means of thinking outside of the box. I truly believe that science is the search for the spiritual through knowledge of the physical and I try to reflect this belief in many of my scientific essays.

Some of the essays in this section may be considered outside of mainstream science. They constitute thought-experiments that I have conducted throughout the years. The line of demarcation between physics and metaphysics has always been blurred, with some scientists refusing to recognize the latter as a viable alternative to what is observed in the physical world. Yet, there remain psychic phenomena that cannot be explained within the known scientific realm such as How is it that a photon is conscious of being observed and acts in accordance with the experiment being conducted? When we expect it to be a particle it is a particle and when we expect it to be a wave it is a wave. It has also been said that math, and thus science, by its own nature is incomplete according to *Gödel's Incompleteness Theorem*. Therefore, something outside of science must act as an impetus to what is observed and later ascribed to the physical laws. This is a philosophical barrier that I have tried to explore in my writings.

Yet one cannot deny that in the modern era, our scientific knowledge has grown exponentially. Over the last several decades, the beauty and complexity of the cosmos have unfolded before our eyes. We now know that space and time are relativistic, and that mass and energy are two sides of the same coin. Yet there is so much more to be understood. Throughout the 20th century, more questions have been raised than ever before as the mask of the cosmos is slowly lifted by the hands of science. As we stand on the shoulders of giants, we peer into worlds unknown in search of our identity and the nature of our existence.

THE NATURE OF OUR EXISTENCE

It has always been my belief that there is a mysterious *Consciousness* to the cosmos far beyond the grasp our mortal minds. Perhaps, it is part of the entropic nature of the universe that we live in. Science has shown that when we gaze into the abyss of the quantum world, the abyss looks at us in ways that can only be described as supernatural. We see a world within a world with denizens far beyond our current comprehension. The following essays represent my feeble attempt to understand this microworld and its nexus to the macroworld. However, we must always keep in mind that as we revolve around the sphere of the unknown, the view of the micro and micro worlds change as our knowledge continues to evolve.

EXTRA DIMENSIONS

THE RELATIONSHIP BETWEEN ENERGY AND DIMENSIONS

BEYOND OUR REALITY

COSMIC STREAMS OF CONSCIOUSNESS

THE CONSCIOUS NATURE OF THE STRING

THE MUSIC OF THE COSMOS

THE MUSIC OF THE STRING

TOWARD A HIGHER CONSCIOUSNESS

THE DOGONS AND SIRIUS B

EXTRA DIMENSIONS

Imagine a vast sheet of paper on which straight Lines, Triangles, Squares, Pentagons, Hexagons, and other figures, instead of remaining fixed in their places, move freely about, on or in the surface, but without the power of rising above or sinking below it, very much like shadows—only hard with luminous edges—and you will then have a pretty correct notion of my country and countrymen

Edwin A. Abbott, Flatland: A Romance of Many Dimensions

The concept of extra dimensions was brilliantly put forth in the fictional novel *Flatland* written by Edwin Abbott in 1884. The book is still widely read and often serves as an introduction to the discussion of extra dimensions. Flatland is a world of two dimensions and when a three-dimensional sphere passes through Flatland, it appears as a series of circles at the point of intersection. Forty years after the book, the possibility that space might actually have four spatial dimensions rather than three was suggested by Theodor Kaluza.[370] A fourth dimension would allow a fourth direction perpendicular to all three therefore, if a fourth dimension could exist; why not five, six or even more? Thus, our three-dimensional universe could be embedded in four space dimensions just as two-dimensional space is embedded in three-dimensional space. Our existence within three dimensions does not preclude the existence of a fourth dimension no more than the existence of a square precludes the existence of a cube.

For instance, if we examine dimensions less than our own, we see a consistent process. Take for example a line of any length, which is an infinite number of points, which will manifest itself in terms of form in our dimension with a finite length. However, its existence is comprised of an infinite number of points independent of its length and since all infinities are equal, all line segments are equal in their composition. Yet the form appears different within the reference frame of our dimension. In this regard, the one-dimensional line is the summation of the histories of all points. Moreover, the current dimension is equal to the summation from minus infinity to plus infinity of $(d-1)$ dimensions where, d, represents the current dimension. Similarly, a square which is one dimension above a line consists of an

This ideal was also expressed in Robert T. Browne's book The Mystery of Space written in 1919

infinite number of lines. and a cube an infinite number of squares each bounded by separate but equal infinities. This may continue *ad infinitum.*

With this in mind, the fourth dimension could be represented by an infinite number of three-dimensional elements. The size of the element in all the aforementioned cases is inconsequential as each is equal to the other relative to its own infinities and size is merely an illusion necessary for reality to exist within the dimension itself. This is because each element within a given dimension is composed of infinite elements of $(d-1)$ dimensions, which in itself, is the summation of histories of dimensions hitherto. The Incompleteness in Gödel's Theorem is due to the nature of infinity since it could never be attained within the dimension itself therefore, it must remain outside the dimension and reflect itself onto the $(d-1)$ dimension, thus giving it a relative form independent of its size. Size is determined by what is required to bring the form into existence, which in itself, is the driving force of all forms in all dimensions. Just as each point has no *Consciousness* of the line, and its purpose or relative position within the line, it is impossible for anything within the existence of $(d-1)$ dimensions to be cognizant of the actions occurring in dimensions higher than its own. It would be in effect, as logical as the reflection or shadow of an image of a man being cognizant of the existence and moreover the actions of the man.

Notice also in this analogy, that the reflection or shadow itself is in fact $(d-1)$ dimensions. Just as we can see the shadow, but it cannot see or fathom us, something outside of our dimension may have the same awareness of us simply because our reference frames and hence our existence are different. With a greater *Consciousness*, they not only see but understand the movements within our dimension and our dependency on their dimension and may even direct our movements. *Gödel's Incompleteness Theorem* maintains that a higher level of *Consciousness* could succinctly explain something that is unintelligible at the level of lower *Consciousness.* Life and death may be a mystery to us but within the higher dimension, they may be understood as a necessary and inevitable consequence of an entropic universe and moreover how such phenomena relate to their dimension and possibly to their existence. Furthermore, their existence may depend on the entropy of ours just like the line depends on everything that the point consists of, the square everything that the line consists of, etc...

In this regard, dimensions are not only orthogonal but stacked as well. Spacetime, as we know it, could be the result of several internal dimensions "crushed" or flattened into an external surface that renders itself in three spatial dimensions and one time dimension. These hidden dimensions are "crushed" by their own infinities like points on a line or lines within a square. They are like invisible stitches of threads that bind infinite nothingness into the fabric of the universe. Although a point has no dimension it has essence, and it is this essence that is the beginning and the end of all things large and small. Incidentally, the essence of a line is the summation of the essence that preceded it. Because size is an illusion and a manifestation of our reality. When the extra dimensions were first discovered in the 1970s, it was clear that they must be hidden in some way. After all, we certainly don't see more than three spatial dimensions. In reality, dimensions can be of any size, just as a cube could be any size, and although the size appears to be variable the process that manifested the form is consistent and independent of the dimension itself. Regarding form, Heraclitus (535-475 BCE) may have been right after all not only in his emphasis on form but his assertion that there is nothing permanent but change. All objects contain infinities within their own dimension that are bounded but by definition, not finitude. However, the number of orthogonal vertex points within a dimension is proportional to the dimension itself. [371]

By this I mean, take a point with zero dimensions, it has a vertex number of 1. Next a line with one dimension but with a vertex number of 2, one at each end of the line. Next a square with a dimension number of 2 and an orthogonal vertex number of 4. Next a cube with a dimension number of 3 and an orthogonal vertex number of 8. From the following examples, it could be seen that the number of vertex points is equal to 2 to the power on the dimension or 2^d. This would render the fourth dimension with 16 vertex points, the fifth with 32 and so on. It is therefore the number of vertex points that determines the form of any given orthogonal dimension. It is also my belief that all the spatial dimensions are orthogonal to each other.

Regarding reference frames, each object is bound by the dimension that it uses as a reference. For example, a triangle drawn using a two-dimensional reference obeys the Euclidean laws of that dimension. However, the same triangle drawn against a three-dimensional reference, such as a sphere, obeys an entirely different set of laws in this

[371] This excludes two dimensional and three-dimensional objects such as triangles and pyramids which are not orthogonal at all their vertices.

case Riemannian. Therefore, it is logical to assume that one drawn against a four-dimensional reference would enjoy yet another set of laws. Thinking out of the box, Einstein made much use of Riemannian geometry in his development of his *Theory of General Relativity*.

In each case, we may ask the question: Did the triangle change or did our perception of the triangle change when the two-dimensional plane was wrapped around the three-dimensional sphere like the skin on an onion? In this regard, objects are like water they take the shape of the container which in this case is the dimension. Notice that the inherent form of the triangle has not changed since each one, regardless of the dimension used as a reference point contains an infinite number of points. The line dilates or contracts based on the amount of space provided thus the form is changed. That is to say, the one drawn against the sphere appears to be larger than the one drawn against the flat surface, and because the space is curved the triangle appears stretched. As Einstein's theory predicts length is a function of reference frames.

When the ancients looked at the stars and two stars appeared to be close to each other, they had no way of knowing that two stars were not close to each other because they were missing a third coordinate vital to the true position of the stars. Just because a set of laws apply to a given set of standards does not preclude them from being applicable and more accurate when a different set of standards are applied as demonstrated by Newtonian and Einsteinium physics. We assume that parallel lines never intersect, but that depends on our definition of a "straight" line within three-dimensional space. If a straight line is defined as the shortest distance between two points, then the line in reality may be curved and thus intersect at some point.

Perhaps, we observe the other spatial dimensions above ours as reflections or shadows. However, these extra dimensions are required to give the form that we see in each dimension. Using the analogy of Plato's Cave, the shadows are representative of the reflection of d dimensions onto a $(d-1)$ reference frame. They exhibit the same form and movement within the same time dimension. Thus, the time dimension is constant and coeval within all spatial dimensions as time applies to us. The shadow itself is the summation of all other dimensions that give it its form. In theory, the shadow cannot discern any reflection equal to or above its own. Therefore, a two-dimensional object would reflect the same as a three-dimensional object onto a two-dimensional plane albeit the sphere passing through Flatland. And a four-dimensional object would appear as a three-dimensional object within a three-dimensional sphere. However, a one-dimensional object is not reflected onto a two-dimensional plane, instead, it is absorbed.

From our three-dimensional perch, we cast our eyes upon the laws of Nature that govern our reality. Such laws have been rendered to be both true and malleable but not necessarily absolute. From the Clockwork Universe to the Unified Theory, we seek to understand our existence through a better understanding of the laws of Nature. However, from this perch, we can see but so far, and imagine things such as dark matter and dark energy hitherto unseen.[372] We realize that there is simply not enough visible mass to make the universe function. Furthermore, we do not know if this dark matter is baryonic (composed of protons, neutrons electrons, etc., the usual suspects) or non-baryonic which is a totally different class yet to be understood. Now, this non-baryonic matter may represent a different storage mechanism for energy not based on mass at all, and moreover, render Einstein's *Theory of Relativity* inapplicable.

We tend to define spacetime and our reality in terms of mass, which is only 4 % of what we postulate, is out there. The other 96 % remains in the darkness, which implies that our knowledge base is very incomplete to what could be known. Succinctly put, Mass is equal to Energy divided by the speed of light squared, but what if the energy was divided by something else, and what would we call it? $E= mc^2$ is only based on 4% of what is out there hardly universal in the grandeur sense. When we express this formula in terms of the ratio of Energy to Mass, $E/m=c^2$, we find that it is constant because c is constant. However, from this simple formula, we may postulate others. For example: Is dark matter equal to dark energy divided by another universal constant that we are unaware of? And if so, does this interaction produce a different kind of gravity, and a different set of fundamental forces hitherto unseen but universally felt?

Furthermore, we imagine Higgs fields and vibrating strings to explain unknowns within the baryonic sphere. And just as we know that two lines segments that appear equal are dependent on their reference frames, and that time and space are also

[372] It is estimated that only 4 percent of the known universe is luminous or composed of baryonic matter [Wrinkles in Time, p. 164].

likewise dependent, we search for an answer to a question that may not emanate from our dimension; and thus, renders itself unsolvable in accordance with *Gödel's Incompleteness Theorem.* Our knowledge like the universe may seem unbounded, but, in reality, it is bounded by the nature of its existence and the horizon of a dimension outside of our own. We appear as dwarfs in a giant cosmos, standing on the shoulders of a science so conceived trying to bring conclusions to questions and greater understanding that we hope we can achieve. But what are humans after all, but animals in the vast wilderness of the cosmos searching for breadcrumbs to lead to a greater understanding of reality. We use science to collect these breadcrumbs and religion to believe that we know Who spread them. We know what our brains consists of, but we do not know what imagination consist of, or even if it is something that lies inside or outside of our brains. We are thus bounded by the myriad thoughts and ideas that it can generate and understand.

However, there is one law throughout the cosmos that could never be broken and that is the *Conservation of Energy* or sometimes referred to as *the First Law of Thermodynamics.*[373] Simply put, the law states: that energy may neither be created nor destroyed. Therefore, the sum of all the energies in the system is a constant. In this case, the system is the cosmos. The laws and forces of our cosmos must not only adhere but also conform to the *Conservation of Energy.* With this in mind, we can view the *Conservation of Energy* as the ultimate universal constant. In this regard, mass could be viewed as the byproduct of two constants, the initial energy of the cosmos and the speed of light. Thus, the conservation of the initial energy of the cosmos and the speed of light being constant and the limit are two "Holy Grails" of our universe. Both restraints could possibly emanate from outside of our dimension yet act within our dimension. Although the photon, which travels at the speed of light, is an artifact of nuclear reactions within our dimension, its speed may be determined outside of our dimension or by the fabric of spacetime resulting from the *Conservation of Energy.*

Both energy and the speed of light could be products of a higher dimension reflected onto our dimension producing what we refer to as mass. However, mass, as we know it, may not even exist within higher dimensions. In such dimensions, unlike ours, the

[373] I am not ignoring the conservation of momentum; I just view this as a corollary to the Conservation of Energy because of its dependency on mass. In his book Just Six Numbers, Sir Martin Rees identified six fundamental numbers that govern our universe. However, I contend that these six numbers are dependent on the Conservation of Energy and the initial Energy of our universe.

speed of light is not necessarily the limit. In fact, it may be a dimension of pure and pristine energy, in which the conversion of energy into mass is not only unnecessary but inhibiting. Our dimension may only be a reservoir to dump its energy into for reasons unbeknownst to us. When we say that the speed of light is the limit, it is because any rest mass that moves at the speed of light becomes infinite; what exactly do we mean? We mean that it is too large to exist within our dimension, not that such phenomenon is a prerequisite for leaving our dimension. If we define the boundaries as I previously suggested, i.e., that infinity describes the boundary of dimensions, then infinite mass means outside of our dimension or in a dimension that has no rest mass i.e., pure energy.

As mentioned previously, mass is equal to the energy divided by the speed of light squared. In this regard, Einstein's equation reveals that mass is consistent with other phenomena within our dimension and obeys the inverse square law in a similar manner as gravity. Energy must be conserved, and the speed of light must also remain constant. Perhaps, all universal constants and the nature of our forces emanate as a result of constraints placed on our dimension from outside of our dimension. The inverse square law may be an indication of such restraints as it is inherent to three-dimensional space. I also believe that universal constants are necessary to dictate the forms and hence the laws of our dimension. This is because the amount of energy dumped into our dimension from the outside must be finite in one sense and relativistic or bounded in another.

Perhaps, anything that emanates from within our dimension cannot alter the essence of something that comes from outside of our dimension, it can only transform it. For example, when mass is transformed from one element to another, it must give off a corresponding energy equal to the delta in mass, i.e., Hydrogen atoms forming Helium or in the case of radioactive material, neutrons changing into protons, electrons and neutrinos. Force is normally the catalyst of mass transformation; and force is nothing more that the product of mass and acceleration *(F=ma)* which in turn warps the spacetime around it. Although, gravity is the weakest force and although it can be used to produce or even bend light by altering the path that it takes, it cannot alter its speed.

THE RELATIONSHIP BETWEEN ENERGY AND DIMENSIONS

When Sir Fred Hoyle was researching how carbon came to be, in the "blast-furnaces" of the stars, his calculations indicated that it is very difficult to explain how the stars generated the necessary quantity of carbon upon which life on Earth depends. Hoyle found that there were numerous "fortunate" one-time occurrences which seemed to indicate that purposeful "adjustments" had been made in the laws of physics and chemistry in order to produce the necessary carbon[374]

In this essay, I will put forth an idea that has been floating around in my head for some time. I have struggled with the *First Law of Thermodynamics,* mainly that matter cannot be created. Perhaps a better explanation is that it cannot be created in our spacetime or dimension. If we view our universe as a closed system, ***(S)***, and since within this closed system energy cannot be created then it had to be created by some means outside of our universe. Scientists postulate that holes could be punched in spacetime through black holes that possibly link us to unknown universe(s) possibly with a completely different origin and physics that we will never be able to prove. These points of numerous singularities may represent gateways to other dimensions that feed ours.

374 The Fine Tuning of the Universe | Gerald Schroeder Sir Fred Hoyle was an English astronomer and cosmologist, primarily remembered today for his contribution to the theory of stellar nucleosynthesis, and his often-controversial stance on other cosmological and scientific matters, such as his rejection of the Big Bang theory in favor of a steady state universe and the panspermia theory of the origin of life on earth. He is considered one of the most creative and provocative astrophysicists of the second half of the 20th Century. It was also this work that caused Hoyle, an atheist until that time, to begin to believe in the guiding hand of a god (what would later be called "intelligent design" or "fine tuning"), when he considered the statistical improbability of the large amount of carbon in the universe, carbon which makes possible carbon-based lifeforms such as humans.
https://www.physicsoftheuniverse.com/scientists_hoyle.html

The Cosmic Egg?

The most prevalent theory is the cosmic egg, which I have a problem with because there is no discussion of a cosmic chicken or other cosmic eggs. Furthermore, a nexus to other possibilities beyond the cosmic egg hypothesis is seldom discussed. Moreover, the cosmic egg is often viewed as part of system *(S)* which is a violation of *Gödel's Incompleteness Theorem.* The Big Bang Theory is often presented as follows: Once upon a time there was this cosmic egg that exploded spreading its contents throughout the known cosmos[375]. When Fred Hoyle coined the phrase "Big Bang," he did it sarcastically and did not view it as the holy grail of astronomy. However, most scientists accept it as part of the *Anthropomorphic Principle.* Only time will tell whether the Big Bang Theory will prevail or be supplanted by yet another theory of our beginning. It may, however, prove to be the biggest mumpsimus in scientific history. Yet, one thing is for certain. There appears to be a certain precision to the universe. Professor Weinberg quantified the tuning to be one part in 10^{120}.[376]

Possibly, the cosmic egg paradigm focuses too much on our cosmos and its vastness and baryonic matter. As a result, serious consideration of a more pervasive vastness and the possibility of other universes and dimensions beyond our own with autonomous cosmic origins, best exemplified by the variation of stars in our cosmos, are often excluded. Although all stars adhere to the basic fundamentals, they also have their own uniqueness. Conceivably, such uniqueness could be extended to our universe and like the stars in our universe they could be one among many universes. We must be forever mindful that "the absence of evidence is not evidence of absence." We must not forget that it was less than a century ago that the concept of galaxies was foreign to us.

---

375 Quantum theory suggests that the universe could have begun as a `random fluctuation of the vacuum'- literally coming into being from nothing. In this theory, equal amounts of positive and negative energies appeared, which underwent expansion and transformations that led to the primordial explosion of a compact ball of matter, the `cosmic egg.' This explosion has resulted in the expanding universe we know today. https://www.csmonitor.com/1988/0616/2quant4.html

376 According to growing numbers of scientists, the laws and constants of nature are so "finely tuned," and so many "coincidences" have occurred to allow for the possibility of life, the universe must have come into existence through intentional planning and intelligence. In fact, this "fine-tuning" is so pronounced, and the "coincidences" are so numerous, many scientists have come to espouse The Anthropic Principle, which contends that the universe was brought into existence intentionally for the sake of producing mankind. Even those who do not accept The Anthropic Principle admit to the "fine-tuning" and conclude that the universe is "too contrived" to be a chance event. The Fine Tuning of the Universe | Gerald Schroeder

THE NATURE OF OUR EXISTENCE

Perhaps the best way to understand how our cosmos and others could have emerged from pure energy and how it is maintained by the *First and Second Laws of Thermodynamics* is best explained by a thought experiment. Imagine, if you will, that our cosmos is a box, a cosmic box, with an open top. Each side of the box represents what we view as infinity from within the box in a manner similar to an event horizon within a black hole. Since black holes warp spacetime in such a way that it is infinitely curved, such points may represent points of singularities that are tied to the sides of the cosmic box and the warping is actually caused by the bending of spacetime by the higher dimension giving us the illusion of infinite gravity, infinite pressure, infinite density and infinite curvature of spacetime at the point of singularity.[377] Notice, in this case, a singularity represents a nexus point where infinities from various entities converge and are basely "crushed" to a point of zero dimensions. And like my analogy earlier of a line being the summation of an infinite number of points forming a new dimension, a singularity could represent such a point at the cosmic level.

It is also at this point that Einstein's *Theory of General Relativity* breaks down in the same manner that Special Relativity breaks down due to infinite mass when the velocity approaches the speed of light. In both cases, the effect of the warping by the higher dimension appears as infinite relativistic mass or infinite acceleration or perhaps more important is that the rest mass is traveling at or above the speed of light. Nonetheless, such speed may be a prerequisite for the penetration of the cosmic walls of infinity that confines our cosmos. In this regard, the speed of light represents the escape velocity of our universe with the associated appearance of infinite mass as in the case of the event horizon of a black hole. Infinite in this case does not indicate "gigantic" mass but something that is beyond the limits or outside of our universe or cannot be contained within our universe. In other words, infinity is the event horizon of our universe. The same is true for superluminal speeds. The radiation of a black hole could indicate the superluminal nature of things that leave our universe. The infinite gravitational force and the subsequent warping of light could represent the phenomena as the boundaries of infinity are approached.

---

377 By definition, a black hole is a region where matter collapses to infinite density, and where, as a result, the curvature of spacetime is extreme. Moreover, the intense gravitational field of the black hole prevents any light or other electromagnetic radiation from escaping. But where lies the "point of no return" at which any matter or energy is doomed to disappear from the visible universe? http://archive.ncsa.illinois.edu/Cyberia/NumRel/BlackHoleAnat.html

Using my analogy of our cosmic box being surrounded by walls of infinity in all directions means that not only is it warped at the point of singularity of the black hole, but the entire cosmos is also warped by infinity. The appearance of infinite mass just may be an illusion and an indication of the transference of all the mass back into its original state of pristine energy as it existed before Planck time. This illusion may occur because we do not truly understand the intrinsic nature of mass/energy as it is manifested within our cosmic box since Einstein's formula is limited by the premise that nothing can travel faster than the speed of light.

However, by supposition, everything inside the cosmic box is finite and superluminal. Similarly, outside of our cosmic box, there could exist other cosmic boxes orthogonal to us also warped by the higher dimension with adjacent walls of infinity. Therefore, any object that falls into a black hole, i.e., within the influence of the event horizon surrounding the point of singularity will be lost to our universe forever and perhaps land in an adjacent universe or a higher dimension.[378] If this is true, then a black hole could be a link to adjacent universe(s) with an autonomous spacetime and independent physical laws. Thus, black holes represent objects and phenomena that are not part of our universe but have effects on our universe and moreover may provide the glue that holds our universe together. The following explains:

> Physicists are still unsure what happens to things that meet their demise within a black hole. The contemplation of what happens to stars and other objects that get pulled into them has led to some scientists postulating that our own universe could have emerged from the other side of a black hole in a parallel universe and that there are vast numbers of such universes that have been spawned by black holes.[379]

As part of this proposed scenario, the sizes of these cosmic boxes are determined by their initial energy which in turn may render different physical laws and a different fabric of spacetime. The nature of "forms" in these peripheral cosmic boxes may also be different than our own yet linked by the fundamental law of energy conservation and the dissipation of its energy. The speed of light may even propagate through their spacetime at a different rate. Within our cosmic box, we can observe and understands all phenomena of dimensions less than our own to have

378 [Kaufmann: Black Holes and Warped Spacetime, p.87]

379 The Multiverse Theory Explained (With Examples) and Whether it's Possible - History-Computer

finite values although, in reality, they are composed of an infinite number of relative infinities bounded within our dimension which is contained within a more absolute infinity (everything inside and outside of the box).

By relative, I mean, a finite object traversed infinite times. For example, we know that the Earth is finite and can even gaze down on it from above. However, to an ant that is asked to travel every inch until it finds the end, the Earth appears infinite because it could never stop even if it arrived at the same point, it could go around an infinite number of times. This is because the Earth could be viewed as a two-dimensional plane wrapped around a three-dimensional sphere. In other words, *(d-1)* dimension shaped by *d* dimension in this regard, *d* dimension can only exist as a summation of an infinite amount of *(d-1)* dimensions. That is to say that: $d = \sum (d-1) + (d-2) + (d-3) \ldots + (d-n)$.

Succinctly put, we are the summation of all the infinities of dimensions less than our own and our dimension warps those dimensions thus bending their form to create our version of reality. This includes all things within our spacetime including space and time. Possibly, the appearances of dilations and compressions albeit, sizes and shapes are due to the restrictions of the box which, in this case, is based on the *Conservation of Energy* and the amount of energy initially placed in our box. Furthermore, the higher dimension could be composed of an infinite number of cosmic boxes just like ours in form but not in content as the amount of energy inside each cosmic box may differ.

The reflections or projections as the Physicist David Bohm (1917-1992) are the results of actions taken place outside our dimension. This he defined as the difference between *Implicate Order* and *Explicate Order.* [380] To put it succinctly, we draw and create with lines boxes and cubes, the dimension above us may draw and create with cubes made of our dimension to form a tesseract in their dimension as familiar to them as a cube is to us.[381] Thus, our dimension is to the higher dimension as a

[380] David Bohm was a theoretical physicist, who contributed new ideas to quantum theory and neuropsychology including Bohm diffusion and holonomic brain theory. He worked at Berkeley Radiation Laboratory during the Manhattan Project and was well acquainted with J. Robert Oppenheimer and Albert Einstein. Bohm introduced concepts of *Implicate Order* and *Explicate Order*. Bohm defined explicate order as the order of the physical world. He defined implicate order as the source of explicate order, and as an underlying whole that physical form constantly unfolds out from and enfolds back into. http://everythingforever.com/Bohm.htm

[381] The painting *"Crucifixion (Corpus Hypercubus)"*, by Salvador Dalí, 1954, featured at the

point is to a line, a line to a square, or a square to a cube in our dimension. Moreover, their dimension warps ours in a similar manner as a light beam is warped by the gravity of a star according to *General Relativity*. I also believe that the higher dimension uses gravity to warp our dimension of spacetime around itself.

The laws and rules of one dimension do not necessarily apply to a higher dimension best exemplified by the difference between a triangle drawn on a two-dimensional surface verse that same triangle drawn on a three-dimensional surface. What if for the sake of argument, this higher dimension is not restricted to the *Law of Conservation of Energy*, but another law? For any set of laws contained within a system (S), there must be a law or set of laws outside of that system that brought that system into existence.

Let us suppose that in the higher dimension, energy is constantly created by a means unbeknownst to us, but it cannot be dissipated or consumed instead it must flow out of its dimension. In other words, the *First and Second laws of Thermodynamics* do not apply to the higher dimension. To support this point, we evoke the laws of relativity. As we approach the speed of light, time slows down. Let's take this to the limit and assume that time is zero. Since time is the measurement of Entropy, then Entropy must also be zero. And since Entropy is the measurement of energy dissipated or available for use, must also be zero. Such a world, outside of the boundaries of our reality, does not preclude the creation or flow of energy. It only precludes the "use" of energy and that is an important distinction. This energy appears in our dimension as mc^2 to be transformed into mass which lies at the core of our existence. However, relative to its own dimension, the speed of light is not the limit and mass, as we know it, does not exist and is only a phenomenon of our three-dimensional space.

Let us further assume that to be within the laws that govern its dimension, it creates "cosmic boxes" or universes of which *Conservation of Energy* and the dissipation of energy are fundamental laws. Now, these cosmic boxes are filled from the outside with energy, in the same manner that we could pour water into an open box. The energy, like the water in a box, is finite and the amount of energy is dependent on the size of the box. Now this box has a hole in it from which the water, or in our case, the energy must flow but under no circumstances can any more energy

Metropolitan Museum of Art in New York, depicts the crucified Jesus upon a tesseract which is representative of the fourth dimension.

be poured into the box until all the energy is out because this would violate the *First Law of Thermodynamics.*

Now, the hole in our cosmic box represents *the Second Law of Thermodynamics.* It is from this flow that the energy vector is established as we move from an ordered to a disordered system. All the subsequent vectors, entropy, time, the flow of heat, and etc. flow in one direction because no additional energy can enter into the box until all of the energy is exhausted. In this regard, both dimensions share a symbiotic relationship in a similar manner that plants and animals share a symbiotic relationship in their exchange of Oxygen and Carbon Dioxide. In nature, things derive their existence from mutual dependency and in isolation, they only represent potential and not kinetic energy which requires motion from one complementary pole to another or mutual attraction in the case of gravity.

Conservation of Energy

In such a scenario, one dimension creates energy, and the other dissipates and consumes it under the direction of the *First and Second Laws of Thermodynamics.* The lesser dimension is governed by the dissipation of energy and the higher by the creation of energy. Both dimensions are inextricably linked to each other and are parts of a much greater whole yet to be understood. The simple fact that energy does exist in our dimension and cannot be created within our dimension, and dimensions less than our own are part of our dimension or the closed system, *(S),* strongly suggests that it had to have been created in a higher dimension according to *Gödel Incompleteness Theorem.*

Since the amount of energy created and the rate of its flow out of the higher dimension is unknown, there could be myriad or even infinite cosmic boxes like ours required to maintain the rate of flow of energy out of the higher dimension. Each would be its own separate universe bounded by the conservation and dissipation of energy with perhaps a unique Big Bang sequence. Perhaps these unseen cosmic boxes are separated by their various versions of reality. Therefore, they may not be aware of the larger phenomena of their existence or that we or the higher dimension exist. Like us, their consciousness, thoughts, etc., may be reflections of the higher dimension and we all act as shadows of the higher dimension.

For all we know, extra-terrestrial life may or may not be organically based, with a different set of amino acids, DNA structure, etc. In this cosmic scheme, the existence of sentient beings such as humans are key to shaping the reality of the planet. In such as system, everything within the cosmic box moves toward the dissipation or consumption of energy and is bounded by the *Conservation of Energy*. This is the essence of our cosmic journey, the search and dissipation of energy, which links us to our planet, our planet to its star, the star to the galaxy and the galaxy to our universe and moreover possibly our universe to the multiverse.

In the proposed scenario, before the Big Bang sequence of events, the cosmic box was being filled and there was no energy dissipation within the box[382]. Perhaps, the Big Bang started not with the explosion of a cosmic egg, but rather like dropping a pebble into an ocean of cosmic energy of which everything is a ripple of waves caused by the flow of energy that the wave created. Consequently, it is not matter being exploded out of some cosmic cannon, but waves drifting across a cosmic ocean. Although energy provided the building blocks for life, Perhaps, these actions and thus the *Consciousness* of the cosmos shaped the energy from the higher dimension. Conceivably, this is the mechanism manifested throughout our dimension that brings order to a universe that is inherently disordered. In this regard, we are puppets that move to the rhythm of the strings of a higher dimension and history is a record of the movement of such *Consciousness*.

[382] I believe in the Big Bang sequence but not necessarily the explosion of a cosmic egg I always had a problem with the cosmic chicken and the lack of a connection to a greater whole.

BEYOND OUR REALITY

Imagination will often carry us to worlds that never were. But without it we go nowhere

Carl Sagan

In recent years, scientists have looked for alternative paradigms that would explain the expansion of the universe. Could the universe be static, existing for all time, and also expanding? If we view every piece of mass made by the aforementioned process as bubbles of energy floating on a cosmic ocean, similar to ice floating on water, then a different paradigm emerges from the so-called "loaf of raisin bread" or "balloon with dots" that are commonly used to explain the expansion of the universe. It was a long-established belief that galaxies were spread uniformly after the Big Bang like pellets shot from a shotgun. However, investigations into the redshifts of galaxies revealed that galaxies resemble "suds in a kitchen sink" and there appear to be large sections of space void of stars throughout the universe. The following explains:

> According to the vacuum fluctuation models, our universe, along with these other universes, were generated by quantum fluctuations in a pre-existing superspace. Imaginatively, one can think of this pre-existing superspace as a infinitely extending ocean full of soap, and each universe generated out of this superspace as a soap-bubble which spontaneously forms on the ocean.[383]

Perhaps, the biggest mistake was to believe that all the energy is contained only in the "floating suds" of energy known as stars and galaxies. Such energy suds or bubbles are central to the dissipation of energy within our cosmos. Maybe the bigger question is what are these suds floating on? It could be that they are floating on non-baryonic matter that permeates space of which Einstein's *Theory of Relativity* is inapplicable. This non-baryonic matter (which could be synonymous with dark energy/matter) is undetected by us but sensitive to the movement of the wave aspect of particles large and small. Thus, we are as blind to it and its relationship to baryonic matter and our

---

[383] The Fine-Tuning Design Argument | Discovery Institute

physical laws as our predecessors were blind to magnetic fields or the laws of Electromagnetism before the discoveries of Michael Faraday and James Clerk Maxwell.

It is understood that the universe does not consist of motionless aether required to transmit light waves. Maybe it is a mistake to assume that space is void just as it was a mistake to assume that the universe was uniform. It is, therefore, better to think of space as not being empty but uniformly constructed, by that I mean with the same basic properties or fabric. Perhaps, instead of a cosmic egg, there was a cosmic sperm that fertilized space to create baryonic matter. This space was not benign nor empty but contain non-baryonic energy such as dark energy and are part of our spacetime.

Continuing with my analogy, energy bubbles could form in the box as a result of energy flowing out of the cosmic box in the form of heat dissipation or cosmic friction as the temperature of the cosmos cooled like a patch of ice forming on a lake. In this regard, it is consistent with the laws of our universe. Therefore, energy bubbles in their nascent form of Hydrogen and Helium condensed under their own gravity forming stars, galaxies, etc., as long as the energy flows out of the cosmic box. All the suds of energy appear to be isolated bubbles, some in groups. However, they are connected by an unseen cosmic ocean of energy. Any change or addition of any suds, ripples through the cosmos like a wave on water.

The wave particle duality is important in this regard as the wave is the conduit that connects the part to the whole. Therefore, it is possible that when you observe a particle as a wave you are observing "how" the energy suds float on the cosmic ocean of energy, like ice blocks floating on an ocean thus generating waves associated with its movement like a gigantic cosmic boson. Movement of any part is felt by the cosmic ocean, and each part reacts not to the suds of energy but to the movement caused by the suds of energy which in turn is a result of the movement of energy throughout the cosmic box. Space, time and the speed of light are references used in our dimension as a means to describe the movement of waves manifested as particles or mass within our cosmic box.

Regarding the particle wave duality, particles could trace out an interference pattern over time because they reveal a little piece of their history of histories as they traverse through the vast cosmic ocean. Furthermore, the *Consciousness* that controls the actions of the photon is conscious of the nature of the experiment and the environment thereby rendering an answer suitable to the environment and the experiment performed. Although these experiments seem to defy what we view as reality, i.e., two objects being in the same place at the same time, perhaps they are nothing more than reflections of a greater reality hitherto unknown and presently misunderstood. Additional associated phenomena such as the *Heisenberg Uncertainty Principle* which states: The more precisely the position is determined, the less precisely the momentum or velocity is known in this instant, and vice versa may fall into the same category. Our understanding of the aforementioned phenomena is incomplete because our knowledge of the cosmic ocean or things such as a Higgs field is not understood. The Higgs field may be the stuff that the suds are actually floating on. It may be similar to a magnetic field with monopoles dispersed throughout the universe to create a field that essentially holds the fabric of spacetime together.

Although the Big Bang model answers and resolves many questions, some questions still remain. One of the major questions often asked about the Big Bang is: How did the universe get so lumpy? The Big Bang concept assumes that since galaxies are moving away from each other at one time they must have been all together. The lumps may have an alternate explanation. To invoke the ocean analogy again, if you were to see patches of ice floating on the ocean would you necessarily assume that they were all once together? Or would you assume that they possibly froze at different times and different locations and are drifting away from each other due to the current of the ocean? Scientists and even geniuses are not immune to inserting gaps into hypothesis when they have no clear explanation. This is perhaps best demonstrated by Einstein's Cosmological Constant which he called his "greatest blunder."

It is the nature of the spacetime medium that needs to be better understood. Regarding the expansion of the universe, it is possible that the "dots on the balloon" or "the raisins in the loaf" are not expanding in the manner described. Instead, the suds of energy are drifting on the energy or nonbaryonic (exotic) matter away from each other and toward the infinite boundaries of our dimension in all directions through the spacetime medium. Although Hubble's expansion equation tells us that the farther away a galaxy is the faster it moves and we attribute such phenomena to the Big Bang explosion, we could be as wrong as in the case of the uniformity of galaxies. Such moments may also be explained using a different expansion model. It may be the result of the summation of energy, i.e., the summation of waves that are synchronized and thus constitute a cosmic wind blowing across the cosmic ocean like tides moving across spacetime. In this regard, any movement in any part of spacetime affects the whole of spacetime like a pebble being dropped in a lake with the wave generating cosmic harmonics. The phenomena attributed to the Hubble expansion may be due to a cosmic current resulting from the dark energy which causes the universe not only to expand but to accelerate and moreover is part of the entropic puzzle that will eventually lead to the destruction of the cosmos trillions and trillions of years from now.

Yet there is still another possibility that needs to be explored. Imagine that spacetime is a flat two-dimensional place and everything in it is a result of the intersection between this two-dimensional place and a three-dimensional place albeit the three-dimension sphere intersecting the two-dimensional Flatland. In this case, we could predict the effects that the sphere has on Flatland by observing it over time. This is because the sphere is symmetrical. However, if the same phenomena happen with asymmetrical or random objects moving through Flatland it would be difficult to predict the effect.

For example, let's substitute the symmetrical sphere with a tree with branches and three-dimensional leaves. As this tree passes through flatland, it would appear that the leaves are not connected and are therefore isolated. However, in reality, they are connected by the branches and the tree trunk that we cannot visualize. Each one of these leaves could be galaxies and the gravity that hold them together is local to each leaf but not to the tree itself. In this regard, it is really the tree that holds everything together. And what we observe as entropy in spacetime is really related to the time that this asymmetrical object takes to pass through spacetime and the local symmetry of the object itself.

The leaves (galaxies) appear to be moving away from us because the branches that they are attached to, intersects spacetime at a different angle as it traverses through spacetime giving us the perception that they are expanding. In this regard, dark energy/matter is the tree outside of our dimension and entropy is the movement of that tree through our spacetime. The space that we see and the objects/phenomena that we observe are because our system (S) is bounded by spacetime. Consequently, there could be other spacetimes above and below ours that experience the same phenomena as the dimension one level above theirs; (the tree) passes through their spacetime. In this paradigm, our beginning started not with a big bang, but with the creation of the spacetime dimension. Therefore, we can describe the Big Bang not as the explosion of a cosmic egg, but as an intersection between dimensions at Planck time, 10^{-43} seconds. Incidentally, 10^{-43} seconds may be the rate at which one dimension travels through another.

Although I disagree with the cosmic egg as the genesis of our cosmos, I do believe that certain aspects of the Big Bang sequence of events are correct mainly the expansion of the universe. The start of expansion/inflation is the conception of the universe as we know it. Before Planck time, the universe was extremely hot, above 10^{32} degrees Kelvin. It was not until the universe cooled to 10^{13} degrees Kelvin that bubbles of energy began to form.[384] All the matter in the universe was created in the first few minutes of the universe. This cooling is a result of the *Second law of Thermodynamics.* In a universe poured like water; and then cooled to form patches of ice, it could be smooth with lumps of crystallized ice acting (like the suds in a kitchen sink) mentioned hitherto distributed throughout it. Before Planck time 10^{-43} secs, time and space as we know it did not exist. At 10^{-33} secs, we have the unification of forces and the emergence of matter. At 10^{-10} secs, we have quarks combining to form protons and neutrons and all forms of baryonic matter. Within the first three minutes, nuclear reaction and the synthesis of elements such as Helium and Lithium emerged. It takes at least one year for stars and galaxies, i.e., the stuff that we are made of, to form.

In both the ocean and the kitchen sink analogy, perhaps it is the Higgs field that provides the buoyancy which allows the objects to float. And like the water in the ocean or the kitchen sink, it is within the Higgs field that most of the energy is contained. Rather than our universe starting with a bang, it started with a variable flow that over time has stabilized with the rest of the universe. The cooling of the universe could be the result of the slowing down of the energy flow or vice versa. In this regard, the known forces may act like "energy rivers" that flow into the cosmic ocean. We so often forget that heat is a sign of the flow of energy. The universe may have been hotter in the beginning because the rate of energy flow was greater.

384 [Quarks and the Cosmos, p. 153] Most

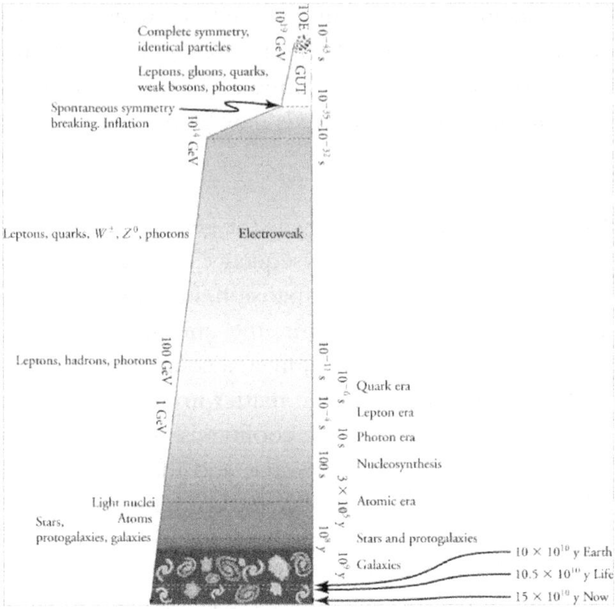

The start of the universe, beginning with the Big Bang[385]

We may not understand the relationship between baryonic matter to the nonbaryonic (exotic) matter and the potential effect they have on each another.[386] For example, when a piece of Iron is placed in a magnetic field, we do not see the field, but the Iron reacts to the field due to the intrinsic properties of Iron and the field. For years, the phenomenon was observed, and no one thought to link the symbiotic relationship of magnetism with electricity. It was not fully understood until Maxwell produced his equations which also linked them to the speed of light. This could also be the case, when baryonic matter "floats" on a nonbaryonic cosmic ocean causing a symbiotic relationship between the two. In this particular case, the nonbaryonic matter acts like a field and we know of its existence because of the effect that the field has on matter within the universe.

385 23.3 The Unification of Forces - Physics | OpenStax

386 All the matter that we ordinarily deal with in everyday life is baryonic matter. Baryonic matter is composed of atoms which have a nucleus that is positively charged surrounded by an electron or electrons that are negatively charged. Baryonic matter is also called "ordinary matter" or "normal matter." The particles within the nucleus of an atom of baryonic matter are protons and neutrons. Both these particles are categorized as baryons. Thus, the term "baryonic matter." Baryons are subatomic particles that are composed of three quarks held together by the strong force. baryonic matter - Quantum Physics Lady

Since we do not know the relationship between baryonic matter to non-baryonic, we cannot begin to fathom such a relationship within the confines of our physical world. It would be like describing the effects of a magnetic field on a piece of Iron without the knowledge of Maxwell's equations. However, when we think of the space between the energy bubbles, we often think of it as void and even benign. When in fact, the space between the energy bubbles may contain significantly more active energy in the form of fields or non-baryonic matter than within the energy bubbles. This may explain why there is not enough mass to hold a galaxy together hence, the notion of dark energy. As in music, it is perhaps the space between baryonic matter along with the resonant frequencies within that holds everything together.

Perhaps this is better explained by using an analogy between water and alcohol. At temperatures between the freezing point and boiling points of water and alcohol, the alcohol and the water are both liquids. However, as the temperature begins to drop, the water begins to crystalize, but the alcohol does not. Let us, for the sake of this analogy, say that the alcohol represents energy in another form (non-baryonic matter or dark energy) that was also present since the beginning of our universe. However, together they may represent energy in its pristine form. Moreover, energy in its crystallized form (baryonic) could float on the other, like suds on water or ice cubes floating on alcohol.

If we think about *(d-1)* dimensions as being the final layer wrapped around a *d* dimension, then the surface of the Earth resembles the skin of an onion. From this analogy, we see how the observer of *(d-1)* dimensions thinks that it is the only dimension. Each time a particle/wave goes around a circle of an extra dimension the wave pattern of every loop or path gets added as part of its sum of histories. When it has traversed an infinite number of times and resonance is achieved via alignment of the waves, the particle/wave could exist in the extra dimension because all paths within its dimension have been exhausted.

On the other hand, photons have no mass and no acceleration and relative to each other could travel at superluminal speeds outside the spacetime continuum but maintain a constant speed within it. This gives the photon the ability to exhaust every possible path within spacetime simultaneously in accordance with Feynman's histories over histories and to communicate superluminally with each other. However, all these things that act or react within the spacetime continuum could ultimately be traced to its most fundamental law of energy conservation. It is hard for humans to fathom any event or phenomenon in our cosmos that does not have this fundamental law at its core, from the formation of stars and galaxies to the operation of minute denizens of the quantum world. *Conservation of Energy* defines the boundaries of our cosmos and the nature of our forces.

In the case of Feynman's histories, when you add up the contributions from an infinite number of histories the answer is likewise infinite with the electron having infinite charge and mass. Perhaps, this is the result of the electron leaving our dimension and then returning through the walls of infinity that bound us. [387]Once again, the walls of infinity must be breached before the higher dimension could be reached. Because the final layer or skin is one dimension less from the vantage point of the object moving through a higher dimension (independent of time and space and with a thickness or depth unknown), it would appear to be still, or better yet, to have infinite mass. [388]Thus, as we approach the speed of light the rest mass appears to become infinite according to Einstein's *Theory of Relativity*.

We should not, however, view infinite mass in isolation but rather examine the wave aspect at the point of infinity. Mass may appear to be infinite just before it goes to zero and at this point, the wave aspect is completely dominant. However, it could be that the "appearance" of infinite mass is a phenomenon of breaching three-dimensional space at its boundaries. Therefore, a photon reduces it speed as it traversed through our spacetime like a stone reduces its speed once it hits the water. I suggest that when the double slit experiment is performed, and we see a photon sometimes as a wave and sometimes as a particle, we are witnessing the photon as it splashes back into spacetime or the reflection of the photon's movements relative to spacetime. I also contend that the photon uses the outside dimension as a shortcut between points within our dimension. The speed of light being constant will always give you an accurate measurement of the distance between two points within spacetime and should not be considered an indication of the paths that the photon took to arrive from one point in spacetime to the other. Furthermore, outside of spacetime, a photon may not be a particle, but just a wave. The only difference is that in the case of the photon, its speed remains constant at one foot per nanosecond throughout the medium of three-dimensional spacetime.

---

387 [Hawking: The Grand Design, p. 107]

388 [Hopper: Dark Energy, pp.129-131].

Maybe this is best illustrated by the following example. If we consider spacetime as a flat surface wrapped around a higher dimension of a sphere with points *A* and *B* representing two points on the surface of the sphere. Although the surface of spacetime appears flat, it is actually curved by the higher dimensions which manifests itself as gravity within our dimension. (see diagram below).

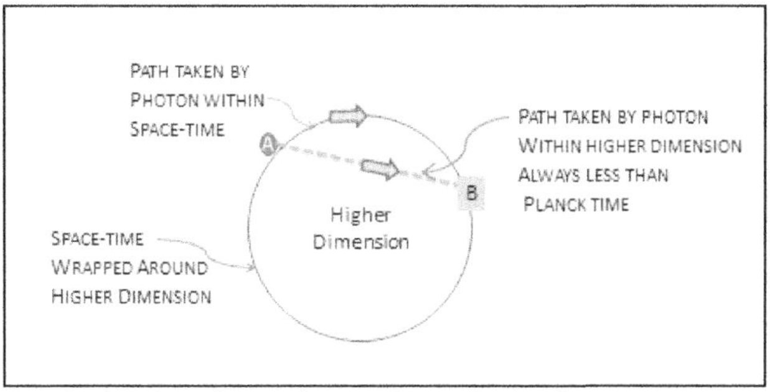

Path taken by a photon in multiple dimensions

Any photon traveling within spacetime must appear to travel along the curve and its distance is measured by how long the photon takes to traverse from point *A* to point *B*. However, if the photon actually takes the shortcut through the higher dimension, in which the time and distance is always the same, let's just say zero or at least less than Planck time, it will essentially appear at points *A* and *B* simultaneously regardless of the distance that is observed to travel within spacetime. In fact, it could be that everything that happens within the higher dimension is less than one Planck time and shorter than one Planck-length relative to our own.[389]

---

[389] The Planck length is the scale at which classical ideas about gravity and spacetime cease to be valid, and quantum effects dominate. This is the 'quantum of length', the smallest measurement of length with any meaning. And roughly equal to 1.6 x 10^{-35} m or about 10^{-20} times the size of a proton. The Planck time is the time it would take a photon travelling at the speed of light to across a distance equal to the Planck length. This is the 'quantum of time', the smallest measurement of time that has any meaning, and is equal to 10^{-43} seconds. No smaller division of time has any meaning. Within the framework of the laws of physics as we understand them today, we can say only that the universe came into existence when it already had an age of 10^{-43} seconds.
http://www.physlink.com/Education/AskExperts/ae281.cfm

Within these ascribed limits, Quantum Mechanics breaks down and we have no mechanism for piercing such barriers. Since the laws of the higher dimension do not subscribe to ours, we cannot observe the relativistic path or time within the higher dimension but only its reflection which is the curved path within spacetime.

However, relative to the higher dimension the photon is "always" at both slits. Therefore, events that are happening in what we perceive as the "present" have already passed and what we are viewing during the experiment is essentially a video replay within spacetime akin to watching a two-dimensional movie when the action occurred in three dimensions much earlier. Regarding the infinite mass when approaching the speed of light predicted by Einstein's equation it could be that when mass leaves our dimension it appears to us as infinite and when it comes into our dimension from the outside, it appears to be zero as in the case of the massless photon and messenger particles or perhaps pristine energy poured into the cosmic box.

$$m = \frac{m_0}{\sqrt{1 - \frac{v^2}{c^2}}}$$

When v=c, m is infinite

Nevertheless, there are no free lunches and at some point, the cosmic box with all the energy dissipated, its entropy returned to zero and order once again will be restored. However, with this order, the order achieved by local equilibrium systems disappears and life or the universe as we know it cannot exist. Because of the relationship between time and entropy, at this point time cannot exist. The "Big Crunch" may in fact be viewed as a resurrection from which the fall of our universe is sprung anew by the filling of the cosmic box repeatedly. In this regard, perhaps the analogy of an hourglass is more appropriate with the top representing the initial energy and the bottom the amount of energy dissipated or consume.

As you can see in this analogy, the direction of the flow of energy, and entropy are fixed as they were in the analogy that I used previously to describe the cosmic box. Order is only restored when all the energy is on one side or the other. At the current rate of energy dissipation, our universe will still be around literally, billions and billions and billions and billions of times longer than the current age of our universe or our sun.[390] Not only would the lives of our Sun and the Earth lie in the distant past, but our galaxy would have merged with our sister galaxy Andromeda. Thus, the cosmos will rage on without us as it did before us. Moreover, as I have mentioned before, our cosmos may only be one branch of a cosmic tree with its branches extending like tentacles deep into the abyss of universes and dimensions hitherto unseen. In this regard, humankind is an infinitesimal speck, with dreams of immortality perhaps achieved by other beings in other places and spaces unbeknownst to us.

Although our planet, our sun, and our universe will meet their demise, the existence of other dimensions and universes will continue perhaps forever. Since, mass, time, space, entropy, etc. may not exist; then three of our known forces may only exist unified as they were before Planck time (as in the case of gravity and electromagnetism demonstrated by Kulza). Therefore, there is no independent gravity, and no independent strong or weak nuclear forces because there are no atoms. The only force that could possibly independently exist is the electromagnetic force which is what Kaluza/Klein predicted would be the effect that a spatial fourth dimension would have on our dimension.[391] Perhaps other dimensions and universes could be described as non-baryonic (dark) invisible to the eye, immune to entropy, time and space; the realm of the immortal bounded by what we call forever.

Furthermore, it is perhaps no coincidence that the massless photon comprised of such energy could exist in both dimensions. Therefore, it would be the perfect conduit for information to be transferred from one dimension to the other and there may be more to the Electromagnetic spectrum than we currently believe. This could further imply that all actions in the higher dimension are instantaneous and thus always present at all points within the higher dimension, thus representing a point of singularity. Moreover, as Einstein has said: matter/energy warps our space into the fourth dimension. In such a dimension our past, present and future are always part of their continual present. Even in our dimension, if things could travel at the speed of

---

390 [Krauss: Atom p.280]

391 In 1919, Theodor Kaluza attracted the attention of the physical community since 1921 with his unified field theory of gravitation and electromagnetism in five dimensions. The first intimation of hidden dimensions began with the work of the theoretical physicist Theodor Kaluza. In a 1921 paper Kaluza showed that, by adding an extra dimension to the equations of Einstein's theory of general relativity, he could obtain an extra equation that seemed to predict the existence of light. That looked promising. But where, then, was this extra dimension? The Swedish physicist Oskar Klein offered an answer in 1926. Perhaps the fifth dimension was curled up into an unimaginably small distance: about a billion-trillion-trillionth of a centimeter. Parallel Universes: The Brane Multiverse | Made in Atlantis

light, they have no time differential relative to their reference frame. In such a scenario, $E= mc^2$ is only relevant to our dimension.

Yet there is another formula for energy that is just as important as the one made famous by Einstein which is $E= h\nu$. However, if we equate the two together, we have, $h\nu=mc^2$ with a wave on one side and a particle on the other both multiplied by universal constants. The speed of light is incredibly large, 299,792,458 meters/second and Planck's constant is incredibly small, $6.62606957 \times 10^{-34}$ joule second. Dividing both sides by h, we see that ν is equal to mass multiplied by some cosmic constant c^2/h. Thus, the relationship between mass and frequency becomes clear. Although these formulas for energy were equated before to give a relationship between particles and waves, I would like to emphasize here that the formula may give us further insight into what happens when mass is infinite at the boundaries of the cosmic box. Quite often when scientists and mathematicians encounter infinities, they tend to "renormalize" them to make sense of them within the realm of finite numbers without stretching their imagination into what the math is saying and what infinity may truly represent in a larger scheme.

Life, as we know it, uses mass as its reference frame in defining and framing events within spacetime. However, this definition of life would have no meaning in a massless higher dimension not restrained by spacetime. Nevertheless, if we define life based on *Consciousness*, i.e., an awareness of something independent of mass, then such a definition could apply to both dimensions. Perhaps unlike Edwin Abbott's *Flatland*, where beings are distinguished by the number of sides they possess, beings in a higher dimension may be comprised of energy in their pristine form and may also be distinguished by the amount of energy that they possess or can create. To use an analogy of light, they may glow at different colors in proportion to their energy production or wavelengths. The amount of energy they create may be viewed as a measure of their enlightenment, status or connection to something or some dimension greater than their own which likewise is dependent on its dimension in a way unfathomable to us. Since this energy cannot be dissipated, once born they are forever immortal. For all we know, this could be the realm of souls and we humans with mass innately seek such immortality as part of our *Consciousness.*

In this regard, the relationship between our dimension to the higher dimension is analogous to the positive (+) and negative (-) of a battery with energy flowing between them but in one direction or heat flowing from a hotter element to a cooler one. Such intrinsic properties are inherent throughout nature and present on all levels. The unidirectional nature may also apply to reality as they see us perhaps as goldfish in a fishbowl with a myopic view of reality and is therefore unable to fathom the reality of those outside that gaze upon them. In this regard, the flow of information or *Consciousness* may also be unidirectional flowing from a higher state of *Consciousness* to a lower state. Furthermore, energy may be a prerequisite of the closed system, ***(S)***, for the *Consciousness* to flow.

COSMIC STREAMS OF CONSCIOUSNESS

With all your science can you tell how it is, and whence it is, that light comes into the soul?

Henry David Thoreau

The self-aware universe in which *Consciousness* creates the material world has been proposed by scientists and luminaries for ages.. Although most of it is conjecture or hyperbole, it is only by stretching the limits of our imagination that a greater truth is revealed. I believe that science has its place and so does philosophy and religion in the constant search for the truth. Furthermore, a philosophical or religious perspective could provoke profound insights into the nature of our reality henceforth confirm by science. Quite often in science, it is the philosophical or religious thought that sows the seeds later reaped by science. For instance, Einstein imagined riding on a light beam before he envisioned his *Theory of Relativity* and the early astronomers, Newton, in particular, were driven by religious convictions.

In many respects *Consciousness* is like light, the latter often used as a metaphor for *Consciousness. Consciousness* illuminates your reality, and it is what makes the realm of the possible perceptible. The following explains:

> One theory is that *Consciousness* is a field of energy that permeates the universe like the force of gravity does, in various degrees of intensity. And the complexity of the human brain has evolved to the point of being able to respond to that universal field of *Consciousness* like a receiver, a receptor. In the same way a radio picks up radio signals. And much like the flower that opens by the sunlight coming to it from millions of miles away. In this analogy, *Consciousness* is like gravity that saturates the entire universe, and so touches our minds too. Mythologist Joseph Campbell stated that *Consciousness* is like the light of a lightbulb. And if you smash the lightbulb the light will go out. But you will not

have destroyed the source of electricity that enabled the lightbulb to shine. It may be discovered someday that *Consciousness*, like gravity and electromagnetism, is a basic, irreducible quality of the universe.[392]

Perhaps this is the road that the physical, as we know it, diverges and connects to the metaphysical world or what Plotinus (204-270 AD) would call the *Divine Mind.*[393] It is a road from which one becomes many, and many become one, a road that takes infinity to nothingness in ways unfathomable to us, linking all dimensions however great or however small into a *Consciousness* of oneness. Maybe this represents a sort of cosmic singularity in which all dimensions converge. Perhaps in humans, such universal *Consciousness* is experienced by their souls and felt by their spirits and may have also been the mechanism that brought such genius to the minds of Einstein, Srinivasa Ramanujan, and others.[394]

To Plotinus, all matter is nothing more than the manifestation of *Consciousness* or in his words the *Divine Mind.*[395] Moreover, he believed that the soul is a conduit to the *Divine Mind* and hence the nexus to a greater *Consciousness* that connects one to the many and the micro to the macro.[396] In this regard, the soul is the mechanism from which *Consciousness* flows from a higher or spiritual state to a lower or physical state. Using Plotinus' model of the *Divine Mind,* the realm of the soul and the physical world, we could suggest that the fourth dimension is the realm of souls that connects us to the *Divine Mind* and that our three-dimensional realm is only the abode of the physical. Everything to be observed in the universe implies *Consciousness. Consciousness* creates

---

392 Consciousness as Light and Space - SAND (scienceandnonduality.com) Joseph Campbell was an American author and teacher best known for his work in the field of comparative mythology. He was born in New York City in 1904, and from early childhood he became interested in mythology. He loved to read books about American Indian cultures, and frequently visited the American Museum of Natural History in New York, where he was fascinated by the museum's collection of totem poles. https://www.goodreads.com/author/show/20105.Joseph_Campbell

393 Plotinus is generally regarded as the founder of Neoplatonism. He is one of the most influential philosophers in antiquity after Plato and Aristotle. http://plato.stanford.edu/entries/plotinus/

394 When he was 15 years old, Ramanujan obtained a copy of George Shoobridge Carr's Synopsis of Elementary Results in Pure and Applied Mathematics, 2 vol. (1880–86). This collection of some 6,000 theorems (none of the material was newer than 1860) aroused his genius. Having verified the results in Carr's book, Ramanujan went beyond it, developing his own theorems and ideas.http://www.britannica.com/EBchecked/topic/490500/Srinivasa-Ramanujan

395 There are some that would refer to this as Intelligent Design which seems to be emerging in some circles as a compromise between science and religion.

396 [Abhayananda: History of Mystics, pp.145- 147]

our reality and makes it knowable. In this regard, the soul is the wave, and the body is the particle.

Plotinus (204-270 CE)

One may therefore ask: Do the angels sing in a key and live in an abode one dimension higher? Do they sing their divine song in a subconscious voice that we hear but do not understand? And since four-dimensional space bends three-dimensional space; perhaps, it is the soul that shapes our bodies using energy as a conduit. Such paradigms are core to the thinking of many Eastern religions. For example, Kundalini Shakti, (coiled and dormant cosmic energy) is the belief that every individual is a manifestation of cosmic energy and that our universe is a result of the same *Consciousness,* ever revealing itself in various forms.[397] According to some Eastern

[397] Kundalini Shakti is the divine spiritual power within every human being. Known by many names in many lands, this sacred inner presence is the divine within, the sacred light and love that illumines each of us and guides us to union with the One, the ultimate Reality beyond all phenomena. http://kundalinicare.com/?page_id=77Shakti is pulsating, vibrating, throbbing creating power or force of the manifest. These vibrations create the universe, the Big Bang, and all names and forms in manifested Creation. Everything that has ever existed is through Shakti. These subtle vibrations of Shakti are represented as mantras (sounds) and movement (dance), and when a series of sounds and physical movements, are infused with Shakti, we transform our physical vehicle into a potent tool that will awaken the conscious mind and reveal knowledge and wisdom which leads to higher experience.

principles of the universe, all that exists in the universe must also exist in the individual body because it is believed that all in the cosmos is built on the same plane [398] Therefore, we can only see what is outside by looking deep within.

However, within our current state of *Consciousness*, we may not have the ability to look deep within to see the outside and we may have to leave the body, or the mind may have to evolve to obtain such skills. The following explains:

> The brain operates in time and space, having linear thoughts that are the end point of a selective filtering process. So, whatever is outside time and space is inconceivable, and unfiltered reality would probably blow the brain's circuits, or simply be blanked out.[399]

The concept of a stream of *Consciousness*, emanating from a central source such as Plotinus' *Divine Mind,* or *Kundalini Shakti* leads to a greater question regarding the nature of the reality perceived. For example, the reality of a tree or an animal with a lower *Consciousness* regarding its surrounding and its place within its environment differs vastly from the reality of humans. More to the point, the DNA difference between humans and chimps is very small, yet the *Consciousness* between the two is great. What is it about the human brain that makes it much more cognitive than that of a chimp when the DNA variance between the two is so small? We understand the process of evolution and we understand the mechanisms of DNA, but what we do not understand is why and how such processes came into existence. Equally as important, we do not understand the motivation behind life in the universe. Perhaps this points to an additional Law of Thermodynamics based on Plotinus' *Divine Mind,* and maybe a culmination of previous Laws of Thermodynamics but not quantifiable by science. It may simply state that: *Consciousness* flows from a higher state to a lower state and like energy cannot be created or destroyed.

Without such *Consciousness,* the physical world would be as useless as a computer sitting on a desk without software. Everything must be a part of a greater whole and a greater *Consciousness.* This includes our planet which has a *Consciousness* of its own and

---

http://kundalinicare.com/?page_id=77

398 [Grof: Ancient Wisdom and Modern Science, pp.115,116]

399[Deepak Chopra, Roger Penrose, Brandon Carter: How Consciousness Became the Universe: Quantum Physics, Cosmology, Relativity, Evolution, Neuroscience, Parallel Universes, p.476]

through such *Consciousness*, often referred to as *Gaia*, it has maintained an environment suitable for life for millions of years.[400] However, such *Consciousness* must be linked to a greater cosmic *Consciousness*. For example, due to the dynamics of nuclear fusion, the luminosity of the Sun has changed over the life of our planet. Yet, the temperature of the Sun has remained constant for millions if not billions of years despite such changes in the luminosity of the Sun. Why? Because the environment of the Earth has changed to adjust to changes in the luminosity of the Sun. More precisely, the transformation in the chemical makeup of the Earth's atmosphere was directly proportional to the increased output of the Sun. This was a result of plant life that changed the Earth's atmosphere from Carbon Dioxide to Oxygen and the Carbon once in the atmosphere went into the development of other organic life forms that produced Carbon Dioxide required to sustain plant life. This sequentially produced more Oxygen which consecutively produced the ozone layer which in turn blocked the dangerous ultraviolet rays.[401]

Similarly, the complexity of living cells arises out of vast numbers of interacting chemical processes forming interconnected feedback systems that cannot be described merely in terms of those individual components or processes and their interactions.[402] Take, for example, the molecule that produced chlorophyll which allowed plants to make their own food. What was the cognitive *Consciousness* that brought such a system into being thereby setting the stage for all other forms of life from unicellular organisms to human beings? Moreover, what was the *Consciousness* that allowed the chemicals to be cooked inside of stars and then dispersed throughout the universe?

All of this is germane to the nature of our existence; and is part of a cosmic self-regulating feedback system of which life on this planet is only a small part. In this regard, we are symbiotic pieces of a much grander puzzle in which no piece truly understands its place within the context of the whole. Nature seems to do what is necessary; such is the nature of Nature. Furthermore, humans may be an insignificant and infinitesimal part of the cosmos, but it is through their *Consciousness* that humans realize that the cosmos exists and perhaps more important: yearns to understand it. However, according to *Gödel's Incompleteness Theorem*, since our *Consciousness* is the only

400 [Dyson: Imagined Worlds, P169]

401 [Davies: The Cosmic Blueprint, p. 132]

402 [Venter: Life at the Speed of Light, p.17]

tool that we must use to fathom the universe, we can never understand the nature of the universe because we are within it. Stars, however bright cannot illuminate or release us from the ignorance by which Nature bounds us. Humankind will never conquer Nature because it is the nature of Nature to conquer Humankind.

The double slit experiment is an example of such arcane *Consciousness*. First conducted by Thomas Young, it is one of the most famous experiments in the history of physics. It demonstrates that light does not always obey our laws, behaving at times as a particle and as a wave without rhyme or reason. More shocking is that the experiment reveals that observation of the light influences its behavior. For instance, when scientists placed detectors at each slit to determine which slit each photon passed through, the interference pattern disappeared. Something about the act of measuring which slit is used removed the wave element completely. At this point, the photons acted exactly as we'd expect a particle to behave. The very uncertainty in position is related, somehow, to the manifestation of wave effects. This suggests that observing the photons alone changes our reality. It implies a level of *Consciousness* beyond our comprehension.[403] As Gödel has demonstrated, there are things that are true that can never be proven and perhaps the answer lies not within our dimension but in one a dimension higher.

403 Double slit experiment: how consciousness determines our reality (lawofheartcoherence.com)

THE CONSCIOUS NATURE OF THE STRING

The supreme paradox of all thought is the attempt to discover something that thought cannot think

Kierkegaard

There are many arguments for and against string theory. Some have gone as far as not to call it a science at all and have let it dwell between the netherworld of science and philosophy. It has been said that string theory is as much of a mathematical construct as any put forth in science. It has also been said that math, and thus science, by its own nature is incomplete according to *Gödel's Incompleteness Theorem.*[404] Therefore, something outside of science must act as an impetus to what is observed and later ascribed to the physical laws. But just what is this thing? Some have suggested that this thing is conscious energy, and it is the essence of the vibration of the string. Philosophers and scientists have long pondered the line of demarcation and the nexus between metaphysics and physics. The dispute is at least as old as Aristotle from which the term "metaphysics" is derived. The advent of quantum mechanics and furthermore the discovery of wave-particle duality has raised the debate to a higher level.[405] At least one renowned physicist has accused adherents of this branch of physics of suggesting that "God did not play dice with the universe" although few physicists today subscribe to this particular view.

In ascribing the vibration of the string to *Consciousness*, i.e., energy with both awareness and purpose, we have crossed over the line to the realm of metaphysics. Science is based on proof, not conjecture; it requires experimentation supported by its laws. However, we are approaching a point where the phenomena observed may not be ascribed to the physical realm alone and the enigmatic void between what we believe to be true and what we observe must be crossed. The seeds of science are often

---

404 Kurt Gödel was an outstanding mathematician and close friend of Albert Einstein; through his math he presented many intriguing questions regarding the nature of time and the universe. Physicists and mathematicians for decades have tried to find flaws in his theories without much success.

405 Louis De Broglie was an early proponent of the wave-particle duality concept. He also favored the philosophical side and argued in addition to science we need a "supplement of the soul".

planted in the fertile soil of philosophy with Nature being our greatest teacher. It speaks to us in a language often misunderstood. Some may look at a tree and see a trunk with its branches; Others look at it and see a string vibrating with its trunk as its fundamental frequency and the branches as harmonics which further serve as fundamentals to subsequent branches. This is the voice of Nature as it unfolds its *Consciousness*. It speaks to us through analogies and paradoxes yet to be understood.

To cite a more relevant example, let us turn to music with its vibrating elements. One cannot deny that we humans are conscious of music, yet we do not know why particular pulse modulated frequencies are pleasing to the ear. However, men of science know that music can be in effect broken down into a group of Fourier series that exhibit the same phenomena or *Consciousness* as the tree previously described. Moreover, it is through the relationships of fundamental frequencies and their subsequent harmonic content that musical harmony is achieved. Yet, our ears are already attuned to discern just what that relationship is and to assign dissonance to those sounds that do not subscribe to those laws. We could reduce music to a set of physical laws; however, we cannot ascertain why we humans are in tune with such laws therefore the system is incomplete within the bounds of physical science. Such phenomenon is replicated repeatedly in Nature to the point where it is idiosyncratic of Nature itself. Nature is a collection of dualisms each being complimentary and complementary, and maybe the nature of all things is based on this premise.

Nothing exemplifies this better than a DNA strand's ability to replicate or the ability of hydrogen to generate all the natural elements.[406] Replication and evolution are the two most prominent leitmotifs of Nature and Nature, in general, is a collection of innumerable systems in which everything succeeds from something else. This is all part of the *Consciousness* of Nature. Although some events in Nature could be attributed to serendipity as an engineer and designer of computers, I know that complex things by their nature are not serendipitous and thus require some type of plan and harmony. Just as a computer is more than a pile of transistors randomly arranged, science has unfolded that the cosmos itself is more than just a group of stars shining brightly for our adornment.[407]

406 Before a cell divides, its DNA is replicated (duplicated). Because the two strands of a DNA molecule have complementary base pairs, the sequence of each strand automatically supplies the information needed to produce its partner. Hence each strand is both complementary and complimentary when operating as a double helix.

407 This is akin to the famous 747 junkyard quote by Fred Hoyle in his book Intelligent Universe he

The music analogy could be extended to string theory. We could analyze strings *ad infinitum* within the physical domain and never arrive at a complete answer. It is like conducting Fourier analysis on music, we could probably develop a computer to write music and even one to listen to it, however, both would probably lack that certain *je ne sais quoi* that we humans have learned to appreciate and still there would be no computer music at all if it were not provided from outside the computer within the mind of the programmer. Perhaps, music resonates with us because it is in tune with our *Consciousness*.[408] So why should string theory be any different? The vibrations of the string cannot be totally explained within the confines of physics; the causation of the effects that we view must be from beyond the realm of the physical, perhaps from another dimension, and what we are observing could be the effect of that dimension on our own.

For example, in our own dimension, we do not attribute a shadow to its own two-dimensional entity, but as the effect of a source of (energy) light shining on a three-dimensional object. If the three-dimensional object runs the shadow runs, if it stops the shadow stops. The person exists in his dimension without the shadow, but the shadow does not exist at all without the person or without the energy required to reflect it onto a two-dimensional surface. To believe so, we would be like the men in Plato's cave who believed that the shadow before them was the only reality.[409]

This analogy could be applied to our current milieu. However, in this case, the flat surface on which the image is projected is the spacetime plane. The energy emanates from outside of the spacetime continuum and its effects are viewed as mc^2 onto the spacetime plane. Thus, mc^2 applies only to the plane, which is the physical realm that we live in, i.e., the third dimension. Accordingly, time and entropy are also subsumed

states: A junkyard contains all the bits and pieces of a Boeing 747, dismembered and in disarray. A whirlwind happens to blow through the yard. What is the chance that after its passage a fully assembled 747, ready to fly, will be found standing there? So small as to be negligible, even if a tornado were to blow through enough junkyards to fill the whole Universe." (p.19)

408 It is also interesting to note that Pythagoras associated consonance tones with proportions of matter, i.e., the mass of a hammer or the length of a string.

409 The metaphor of the cave from Plato's Republic represents a concept that contrasts the way in which we perceive reality. According to Plato, all we perceive are imperfect "reflections" of the ultimate forms, which subsequently represent truth and reality. In this allegory, the men in the cave only see shadows of events performed and not the actual events and thus, interpret the shadows as the only reality.

within this realm. This could explain the directional nature of both. Therefore, the Big Bang could have been a fundamental vibration of *Consciousness* of which everything else is a harmonic. This fundamental vibration could have only come from outside of the physical dimension in accordance with *Gödel's Incompleteness Theorem.* Physicists imagine a string and try to analyze its vibrations but that is as far as it goes. The reason why vibrations appear to be a physical thing could possibly be attributed to our prejudice toward a physical solution to the dilemma.

If we were to think of space and time as this two-dimensional grid warped not by gravity but by *Consciousness* in the form of energy, then mass and the speed of light could be attributes that represent the intersection of this energy with the spacetime plane. In this sense, it would be like the sphere passing through Flatland with mc^2 representing the intersecting circle and perhaps the rate at which it passes through spacetime is equal to the speed of light. Things are not often what they seem. The Moon and the Sun may seem to be the same size when viewed from Earth, but this is more of a celestial coincidence than an actual fact. However, in the esoteric sense, this celestial phenomenon could also be viewed as being in harmony with the fundamental vibration of the Big Bang and the underlying *Consciousness* that governs all things.

There is a reason why some scientists and philosophers view the physical realm in this way. We know that the speed of light is finite and constant within the physical plane. And we can define the physical plane as being the realm of space, time, mass and the speed of light. And why is it that all stars emit photons at this particular speed? It appears to be intrinsic to the Nature of mass/energy conversion from the formation of the first star until today. However, it does not necessarily mean that it is inherent in all energy to mass conversions. The speed of light could be an artifact of our dimension, the ultimate reference to the rate of change within a spacetime universe. It was indeed the constant speed of light that led to the Michelson-Morley experiment and the response by Lorenz and Fitzgerald to their undisputable experimentation.[410] Both are well recognized today as being the antecedent of the relativity concept. The equations developed by Lorenz and Fitzgerald form the basis for space and time dilation and contraction and have established the speed of light as the limit.

410 The constant speed of light was predicted by Maxwell's Equations and proven by the Michelson-Morley experiment.

These assumptions go well beyond what we observe as physical. If we substitute infinity, ∞, for the fastest speed and make the mass zero, the energy at this stage is not zero but indeterminate. If the velocity is infinite, the math tells us that the energy at that point is uncertain.[411] Therefore, a string without mass could vibrate far beyond the speed of light. At this point, relative to this string, or elements within the string, time and space are reduced to a singularity. Another way to view this equation is to deduce that when mass is minimized (zero) velocity is maximized (infinity) and the amount of energy cannot be determined. Thus, a string without mass composed of only energy could vibrate all the time, everywhere and moreover appear static to an outside observer. Therefore, I think the true Nature of *Consciousness* is pervasive throughout the cosmos. Early in the formation of the cosmos, such speeds could have been achieved and perhaps it was the Big Bang that generated the initial velocity (vibrations) within our spacetime. As the energy was converted, to what we call mass, the speed required became finite and constant like a decaying exponential converging on a finite limit within the context of a physical reality.

If we regard, E, as being the energy (in the form of photons) released from the fusion of four hydrogen atoms to form one helium atom, there is a gap in time when the speed could have exceeded the speed of light.[412] All we can really infer from $E = mc^2$ is that in the presence of mass, the amount of energy converted is proportional to its escape velocity squared. And since, c, is squared in Einstein's equation, theoretically, c, could be positive or negative and yield the same result. Furthermore, what are the implications of a plus and a minus c? Are they related to each other in a manner similar to particles and anti-particles? Do time and entropy reverse and the universe contract? Under such conditions, are there other parallel universes that have minus c as their basics?

In fact, there would be no limit to the amount of such alternating (mirror) universes existing like electricity and magnetism one generating the other repeatedly. One of the universes moves toward chaos, the other universe moves toward order, both within

[411] The energy at this point is uncertain as zero times infinity is indeterminate. In addition, all disciples of quantum mechanics acknowledge that uncertainty is a core principle within their field. In fact, in the nascent days of this branch of physics, it was a source of major discussion between its primogenitors Einstein and Bohr.

[412] This period could be up to 500,000 years after the Big Bang. The formation of strings could have taken place as early as 10^{-35} seconds.

the spacetime continuum, thus, a series of "ramifying universes" in which the net change in time does not exist. Again, I point to the complementary and complimentary aspect of Nature. This could possibly explain how the cosmos began from energy without mass, each particle of mass that was generated from energy has its mirror particle in an adjacent three-dimensional universe with an opposite time reference and entropy. The mass in each universe could be nothing more than the replication of strings as energy passes through spacetime. In this regard, the string could be the "DNA" of our reality. One thing for certain, we do not understand how our universe works, there simply isn't enough "visible "mass in it to hold everything together. I think it was the philosopher Immanuel Kant that first attributed the squares that are so prominent in many of our equations to the fact that we live in a three-dimensional world. Moreover, he contended that if we lived in a four-dimensional world the relationships would be cubed.

We may go even further to suggest that gravitational attraction is not due to mass but to the vibration or construction of strings, the mass being an indication of the density of strings (localized energy) within a given area. These strings will naturally want to coalesce (perhaps due to resonance) hence, the attractive nature of the gravitational field. In fact, all fields could be based on this premise. However, in a magnetic or an electrical field polarization plays a factor perhaps stimulated by the frequency of interactions between photons and electrons. This may have more to do with the arrangement (orientation) of strings as opposed to the density (quantity) of strings. Similar statements may be made about nuclear forces and thus within the string all the known forces could be unified. In any event, the speed of light may act as a resonance frequency that reacts or perhaps stimulates vibrations within the string. Or on the other hand, generate harmonics that vibrate in a higher dimension.

Nature is very idiosyncratic from the simple to the sublime. The wave movement experienced on the beach is analogous to the movement of waves in general and light in particular both creating interference patterns along the way. We could also cite other phenomena such as the Doppler Effect and wave resonance. Hence, the understanding of *Consciousness* is essential to understanding the nature of the cosmos. The Earth must have some degree of *Consciousness* to know how to conduct photosynthesis or to control the sequence of events generating life, not to mention the myriad ways in which every form of life is dependent upon its existence and equilibrium. It is, after all, the one body that connects and shares a symbiotic relationship between our macro and micro worlds.

And why do stars shine? Perhaps it is because sunlight is a conduit of conscious energy that is necessary to maintain states of equilibrium in an entropic physical universe. Moreover, photons could not only be viewed as energy, but as packets of *Consciousness* detailing vital information about their origins while providing the required energy to sustain equilibrium. The speed of light could be constant because it acts as a "master key" at the subatomic level opening doors to everything within the physical realm. We understand light in terms of its Electromagnetic spectrum. However, we may be blind to the *Consciousness* of light which may be just as intrinsic.

Complexity in Nature seems to be the culmination of many simple events and processes each when separated is not as complex as the whole. Each entity is sentient of its corresponding element or its adjacent layer but not conscious of the whole. This is true of the human in the cosmos and the cell in the human, etc. Just as we have learned that the set of numbers contains both rational and irrational numbers, the essence of the string may have an imaginary element that denotes its *Consciousness.* This may also be ascribed to every particle and every object in our physical universe. To illustrate my point, I cite the experimentation conducted on photons. How is it that a photon is conscious of being observed and acts according to the experiment being conducted? When we expect it to be a particle it is a particle and when we expect it to be a wave it is a wave. [413]

413 The Uncertainty Principle says that we can neither detect nor measure fundamental particles without altering them.

414 In quantum physics, entangled particles remain connected so that actions performed on one affect the other, even when separated by great distances. The phenomenon so riled Albert Einstein, he called it "spooky action at a distance." The rules of quantum physics state that an unobserved photon exists in all possible states simultaneously but, when observed or measured, exhibits only one state. Spin is depicted here as an axis of rotation, but actual particles do not rotate. Entanglement occurs when a pair of particles, such as photons, interact physically. A laser beam fired through a certain type of crystal can cause individual photons to be split into pairs of entangled photons. The photons can be separated by a large distance, hundreds of miles or even more. When observed, Photon A takes on an up-spin state. Entangled Photon B, though now far away, takes up a state relative to that of Photon A (in this case, a down-spin state). The transfer of state between Photon A and Photon B takes place at a speed of at least 10,000 times the speed of light, possibly even instantaneously, regardless of distance. A proposed experiment would send one photon of the entangled pair to the orbiting International Space Station, a distance of around 310 miles (500 kilometers). This would be the largest distance that has been experimentally tested. How Quantum Entanglement Works (Infographic) | Live Science

Moreover, how is it that two photons are conscious and react to each other when such a reaction must take place at a superluminal speed?[414] These are observable and repeatable phenomena in which the only thing that changes during the course of the experiment is *Consciousness* resulting from the act of observation, which may take place outside of the physical domain and our dimension. The act of observation and measurement changes the nature of the system and the perspective of the photon in an invisible dimension in which it also operates. In the life of a photon, it is always a wave and a particle and "just because we choose not to look at the Moon doesn't mean that it is not always there"[415].

The mathematics of string theory predicts the presence of other dimensions and using what we know about dimensions less than our own we could generate concepts of higher dimensions. Examination of dimensions reveals that higher dimensions are constructed by stacking up or folding elements of the current dimension. Thus, a square is nothing more than a stacking of points and folding of lines and a cube is a stacking of lines and a folding of squares. Using such logic, a four-dimensional object is nothing more than the stacking and folding of three-dimensional objects. To extend the analogy further, it would be quite easy for us to inject or remove objects from a world that exists in two dimensions (Flatland), we just simply "come in" through a dimension outside of the two-dimensional reference.

Thus, it would be possible to go beyond the limits of that realm and alter its environment and our perspective of such a world from the outside would be vastly different from those confined to two dimensions. Similarly, a four-dimensional view of our world always would see time not as a single instance but as a spectrum from past to future unrestricted by the linear and directional motion of time or entropy. Correspondingly, conventional spatial laws are altered and mass as we know it does not exist in a dimension composed only of energy as our universe may have been before the Big Bang. [416] However, energy could exist in both dimensions and adjacent universes.

---

415 Quote by Albert Einstein: "I think that a particle must have a separate reality independent of the measurements. That is an electron has spin, location and so forth even when it is not being measured. I like to think that the moon is there even if I am not looking at it."

416 Early in the 20th century Theodor Kaluza suggested a fourth dimension of space. When he solved Einstein's equations for general relativity using four dimensions, he discovered that Maxwell's equations for electromagnetism emerged spontaneously.

Furthermore, the concept of having only one three-dimensional world from a four-dimensional perspective would be just as infeasible as having only a single square from a three-dimensional perspective. Thus, the possibility of other three-dimensional worlds folded beyond the realm of spacetime. Black holes may represent the points where three-dimensional universes are stacked or folded. The immense density of the black hole may be nothing more than the effects of the bonding of such universes. string theory, or M theory to be more precise, predicts as many as eleven dimensions and the underlying *Consciousness* of all these dimensions may lie in the highest dimension predicted.

In summation, I would like to cite an arcane concept that is as old as any form of mental consciousness expressing such beliefs. The ancient Egyptians depicted missing *Consciousness* in their *Eye of Horus,* which was comprised of fractions, i.e., the eyeball was 1/4 the eyebrow 1/8, etc. However, the eye itself, the sum of all its elements was 1/64 short of being complete and they believed that this element came from outside. This esoteric representation of the finite realm of numbers along its trek toward infinity and incompleteness is perhaps *Gödel's Theory of Incompleteness* depicted in another form as does Zeno's Arrow.[417] In any event, maybe Einstein's theory of relativity is incomplete and that he replaced one demon with another.

Hence, the belief that the speed of light is the ultimate speed maybe our modern day "luminiferous aether." And our prejudices toward only physical explanations and our three-dimensional world may be akin to the ignorance attributed to the greatest minds before the scientific revolution of relativity. The common impediment of that era was the belief that both time and space were absolute when in fact they are not. Similarly, the combined wisdom of Aristotle, Galileo and Newton could not discern galaxies from stars nor fathom the intricacies of Quantum Mechanics. Moreover, the legendary Ptolemy misled men of intellect for over 1,500 years with a celestial model that proved in the end to be erroneous. This in no way should detract from the genius and accomplishments of Einstein. Nonetheless, it has

---

417 Zeno states that for motion to occur, an object must change the position which it occupies. He gives an example of an arrow in flight. He states that in any one (duration-less) instant of time, the arrow is neither moving to where it is, nor to where it is not. It cannot move to where it is not, because no time elapses for it to move there; it cannot move to where it is, because it is already there. In other words, at every instant of time there is no motion occurring. If everything is motionless at every instant, and time is entirely composed of instants, then motion is impossible. logic - Zeno's arrow paradox - Philosophy Stack Exchange

been over a century since his *Theory of Relativity* and it is now time to rethink the fundamental question of the relationship of matter and energy, and to view matter and energy within the greater context of higher dimensions and *Consciousness* that may not ascribe to his laws. Quite often when a scientific paradox presents itself, the solution lies in discerning what phenomena we assume to be true that is actually false. For example, that time and space are not relativistic or that all the stars lie in the Milky Way galaxy. Once this dilemma is resolved, new doors of knowledge are opened, and a greater understanding of our reality is revealed. After all, Nature is our greatest teacher.

THE MUSIC OF THE COSMOS

Music is the pleasure the human mind experiences from counting without being aware that it is counting

Gottfried Leibniz

In the next two essays, I would like to discuss music and its relationship to the cosmos and string theory. In the second essay, I will make hypothetical analogies between vibrations of strings at the subatomic level and vibrations of notes in music. Maybe, music is much too precious to exist solely for our pleasure. Therefore, it must sing a sentient chorus that lies at the core of our reality. It has a *Consciousness* of its own, that makes us conscious of the inherent and natural beauty of its resonance and makes us want to become one with it. string theory was developed to explain the symphony at the smallest atomic level. [418]For centuries, the elusive atom envisioned by the Greeks, was assumed to be the smallest of all things. However, advancements made during the 20th century proved that the atom is composed of much smaller subatomic particles, mainly electrons, protons and neutrons all sharing a specific relationship to each other. Moreover, various atoms share relationships that constitute the bonding of molecules. Further investigation led to the subdivision of protons and neutrons into quarks and yet other subatomic particles were revealed.[419] The vibration of strings may indeed be the music of the cosmos with stars, planets and creatures dancing to its majestic rhythm.

Perhaps one of the greatest epiphanies was that light was both a particle and a wave. This subsequently led to the discovery that electrons also exhibit wave-like characteristics. string theory is, in some ways, the culmination of work in Mathematics

---

418 M-theory and string theory are both attempts to provide a single theory that combines the physics of the observable macro-world with that of the micro-world of quantum physics. string theory and M-theory postulate other dimensions beyond the three that we interact and observe, namely height, width, and length. Greater support for additional dimensions has arisen from the fact that the mathematics of quantum theory has been found to work better within models of reality comprised of 10 or more spatial dimensions in addition to one or more time dimensions. The Multiverse Theory Explained (With Examples) and Whether it's Possible - History-Computer

419 A quark is an elementary particle with an electric charge equal to one-third or two-thirds that of the electron. The name "quark" was taken by Murray Gell-Mann from the book "Finnegan's Wake" by James Joyce. The line "Three quarks for Muster Mark..." appears in the fanciful book. Gell-Mann received the 1969 Nobel Prize for his work in classifying elementary particles. http://hyperphysics.phy-astr.gsu.edu/hbase/particles/quark.html

and Physics that explores the possibility that all particles are composed of strings vibrating to produce the phenomena that we ultimately deem as mass. The latest iteration includes as many as ten spatial dimensions and stands to shatter the rules of reality that we have embraced hitherto.[420] In addition to the subatomic world, some scientists contend that string theory contains a fundamental object called a brane, which can have many dimensions. In some *braneworld* scenarios, our universe is actually "stuck" inside of a 3-dimensional brane (called a 3-brane)[421]. In this regard, string theory represents the latest quantum leap into the unknown.

The analogy of a string vibrating to form a fundamental particle is a mathematical construct being embrace by physicists such as Leonard Susskind, Brian Greene, and John Schwarz. Scientific verification has not yet been achieved; hence string heory is not in the class of *Special or General Relativity*. However, a couple of decades ago, black holes were also constructs and the recent verification of planets outside of our solar system orbiting around distant suns was just a theory as well. My primary reason for making an analogy between string and music theories is to expand upon the original construct keeping in mind that Nature tends to repeat motifs as previous parallels drawn between the solar system and the atom, the Doppler Effect in light and sound, ocean waves and waves of light, etc., demonstrates.[422] Therefore, music could be more than meets the ear and subliminally tell us about the rhythms of our universe. Hence, understanding music theory and how our current musical system works may teach us about the harmony of the cosmos.

---

420 One of the most remarkable predictions of string theory is that spacetime has ten dimensions! At first sight, this may be seen as a reason to dismiss the theory altogether, as we obviously have only three dimensions of space and one of time. However, if we assume that six of these dimensions are curled up very tightly, then we may never be aware of their existence. Furthermore, having these so-called compact dimensions is very beneficial if string theory is to describe a Theory of Everything. The idea is that degrees of freedom like the electric charge of an electron will then arise simply as motion in the extra compact directions! The principle that compact dimensions may lead to unifying theories is not new, but dates from the 1920's, since the theory of Kaluza and Klein. In a sense, string theory is the ultimate Kaluza-Klein theory. http://www.damtp.cam.ac.uk/research/gr/public/qg_ss.html

421 The Basics of String Theory (thoughtco.com)

A braneworld is a theory in which the four space–time dimensions of the universe that are apparent make up a surface, called the brane, in a higher-dimensional space–time, called the bulk. One of the appealing features of the brane world is that it can explain why gravity is much weaker than the other forces, with the nongravitational forces being localized to the brane and the gravitational force not limited to the brane. Braneworld - Oxford Reference

422 The red shift in light is actually caused by the gradual loss of light energy over great distances.

The Fundamentals of Music

As legend has it, Pythagoras heard hammers pounding and noticed that some of the sounds were harmonious and others were not. Being a person of science and philosophy, Pythagoras set out to find the source of this phenomenon. It had long been known that some sounds produced by hammering could, in some cases, be consonant. It was also well known that this phenomenon could also be produced by strings of various lengths and thicknesses and with wind instruments. What Pythagoras did was quantify the relationships that caused such phenomena. He started out by plucking a string of a given length, tension and thickness; then he listened to its sound as he continued to pluck. Of course, it was harmonious with itself.

Next, he cut the string in half and plucked it along with the first string, the two were also harmonious. He continued, cutting the string in thirds, and the sound was also harmonious with the other strings. Next, he cut the string into fourths, and not to his surprise; they all were in harmony with each other. What Pythagoras discovered were the four basic notes, or vibrations of frequencies that were most harmonious with each other. Moreover, they were associated with each other by whole numbers. The distance between notes is referred to as intervals, and for that matter, when we speak to each other we speak in intervals. Sounds that share the same interval relationship are harmonious with each other, and that is why some words and phrases rhyme. The plucking of the first string created what is called unison and it is also referred to as the tonic, the second string the octave, the third string the dominant and the fourth string the subdominant. By cutting strings into these various whole number proportions, Pythagoras increased the vibrations of the strings in corresponding proportions. That is to say, the string with half the length vibrates twice as fast, a third the length three times as fast and so forth.

The ratios are as follows:

String number:	1	2	3	4
String period of oscillation	1/1	2/1	3/2	4/3

The third harmonic occurs in the second octave and the fourth harmonic in the third octave.

Although he had his eureka moment, Pythagoras was by no means finished. After all, you could make some music with these four basic notes, and to be honest, people around the world had been making music with them for centuries if not millenniums before Pythagoras. However, making music wasn't the goal, identifying musical relationships was. In music, it is ultimately the relationships of notes (frequencies) to each other that counts, and what Pythagoras discovered were those notes that have the strongest relationship to each other.

What Pythagoras did next is perhaps the most ingenious part. He made a string with a thickness that gave the sound of the previous string that was divided into one-third. He had noticed that the string that was divided in half was essentially the same sound only at a higher pitch. However, the string that was divided into one-third was unique. This is perhaps the most essential aspect of the dominant tone; it is the simplest way to generate new tones. He continued by dividing this new string into one third. He plucked it and a new sound was heard that was also harmonious with the other sounds. This process continued until he found a relationship that was not harmonious, and he stopped. What he had was a scale, a collection of eight notes all harmonious with each other. This became the C major scale which corresponds to the white keys on the piano. The following is the scale:

C D E F G A B C

It is also known as the diatonic scale. The last C is an octave above the first one. Now it is important to note, that plucking a string with one-half the length of the original string was also significant because it defined the range of notes that could be contained without repeating a fundamental tone. With this formula in place, additional octaves were later developed with the keys on the piano essentially representing the range of notes of most instruments from about 30 Hz to 4100 Hz, about eight octaves. Scientists would refer to this as bandwidth. However, notes have additional harmonics associated with them, and since our hearing range extends to 20,000 Hz, we can hear these harmonics. The instruments that we are all familiar with today have a specific range and instruments are distinguished from each other by their range and their harmonic content. By harmonic content, I mean, although two instruments may play the exact note, they vary in how that note propagates through the air, thus a Middle C, 261 Hz, on the piano and one on the guitar will vibrate at the same fundamental frequency, but each will have different combinations(harmonics) of other frequencies that gives it its *timbre*.[423]

During the late Middle Ages and Early Renaissance, musicians, composers, and instrument makers discovered that although interesting, Pythagoras did not go far enough in his exploration of scales. Although the note $F^{\#}$ that was produced by

423 See the Appendix for a table of frequencies.

Pythagoras' successive adding of perfect fifths was not harmonious with the unison or tonic it was harmonious with its fifth G. This was the second great epiphany in music theory although, like Pythagoras, no one knew the source of the phenomena. As a result, of their experimentation, another scale based on G was born which substituted $F^{\#}$ for F. The process of adding fifths continued and other notes and scales were developed leading to our current system known as the chromatic scale. The additional five notes correspond to the black keys on the piano referred to as sharps and flats. It is most ironic that the fifth turned out to be the perfect slice of the musical pie because, unlike the octave, it produced a harmonic frequency that was not generated by C. The chromatic scale is divided into semi-tones and the relationship to each note is equal to the twelfth root of two or approximately 1.06. Hence $C^{\#}$ is 1.06 times C.

The following is the chromatic scale:

C $C^{\#}$ D E E^{b} F $F^{\#}$ G A^{b} A B^{b} B C

In physical terms, an interval is the ratio between two frequencies. For example, any two notes an octave apart have a frequency ratio of 2:1. This means that successive increments of pitch by the same interval results in an exponential increase of frequency, even though the human ear perceives this as a linear increase in pitch. For this reason, intervals are often measured in cents, a unit derived from the logarithm of the frequency ratio.

MUSICAL INTERVALS

Interval	Ratio to fundamental
Octave	2:1
Perfect fifth	3:2
Perfect fourth	4:3
Major third	5:4
Minor third	6:5
Major sixth	5:3
Minor sixth	8:5

	C and ...	D and ...	E and ...	F and ...	G and ...	A and ...	B and ...
C	P1						
D	M2	P1					
E	M3	M2	P1				
F	P4	m3	m2	P1			
G	P5	P4	m3	M2	P1		
A	M6	P5	P4	M3	M2	P1	
B	M7	M6	P5	A4 (TT)	M3	M2	P1
c	P8	m7	m6	P5	P4	m3	m2
d		P8	m7	M6	P5	P4	m3
e			P8	M7	M6	P5	P4
f				P8	m7	m6	d5 (TT)
g					P8	m7	m6
a						P8	m7
b							P8

Interval (music) - Wikipedia

Note that the black keys on the piano, the sharps. and the flats, form a pentatonic scale. Therefore, it could be said that the chromatic scale is a combination of two scales, the diatonic (white keys) and the pentatonic (black keys).

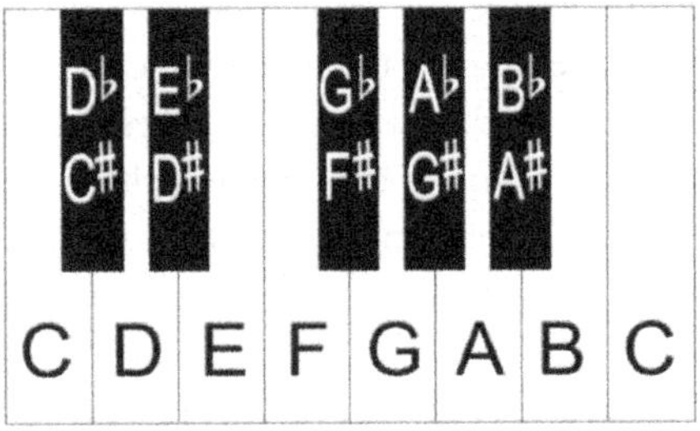

One Octave of a Piano Keyboard

From the musical scale, the relationship between music and mathematics can easily be discerned. A careful examination of the piano keyboard reveals a relationship with the Fibonacci series. First, a black key is always between two white keys, the ratio of 1 to 2 Fibonacci ratios. The first group has two black keys and three white keys, which are Fibonacci ratios, the second group has 3 black keys and 5 white keys, also Fibonacci ratios, in the third group there are 5 black keys and 8 white keys which are Fibonacci ratios and finally in the total set; there are 8 white keys and 13 total keys which are also Fibonacci ratios. Additional links between Fibonacci numbers and sound have also been identified.

For example, the ear uses the Fibonacci series to discern the overtone series effectively performing a Fourier transform on the incoming sound. The first 25 numbers of the Fibonacci series are 1, **1, 2, 3, 5, 8, 13**, 21, 34, 55, 89, 144, 233, 377, 610, 987, 1597, 2584, 4184, 6765, with each new term being the sum of the previous two terms.[424] An interesting fact about the Fibonacci series is that the sum of any consecutive group of ten Fibonacci numbers is divisible by eleven and the answer is always equal to the seventh term in the series of ten. Moreover, the Fibonacci series is related to the Golden Mean and converges on the number 1.618 with perturbations fluctuating above and below the 1.618-line *ad infinitum* without repeating as shown in the following diagram. The vertical scale represents the ratio between two consecutive Fibonacci numbers. The horizontal scale represents the number scale.

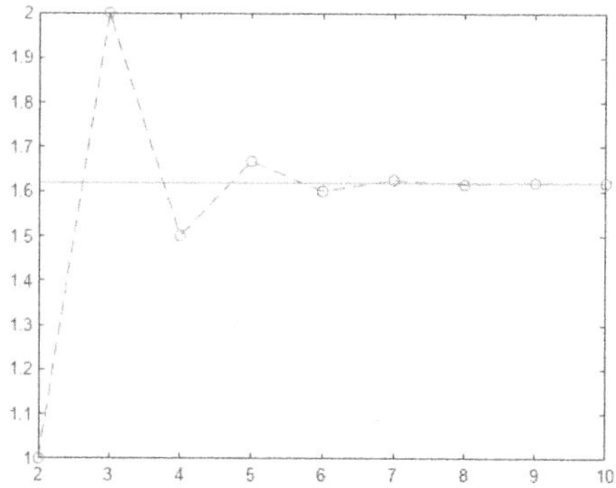

Plot of the Fibonacci series as it converges on the number 1.618.
A similar plot can be made for Its reciprocal .618 [425]

We can also view the above graph as a wave oscillating throughout time and converging on a known state of equilibrium be it 1.618 or its reciprocal .618. In addition to the Fibonacci series another naturally occurring series, the binary series, is also integral to music as harmonic and rhythmic changes normally occur on binary intervals, e.g., 2, 4, 8, 16, 32 beats and every octave occurs at a binary (factor of 2) interval and the frequency of the note in the previous scale is doubled. Throughout nature, we repeatedly find that these two series that tend to define the rhythm of Nature's song. In this regard, music is singing the song of mathematics that our brains readily comprehend and soothes the savage beasts that lie within all of us.

With the final musical structure in place, relationships between the eight notes in a diatonic scale and the twelve notes in the chromatic scale were developed with each note having a musical key associated with it. Each one of the notes in the

424 The Numbers in bold relate to the Piano Keyboard.

425 Fibonacci numbers - MATLAB fibonacci (mathworks.com)

diatonic scale is representative of partials or overtones of the key signature. As mentioned previously, the major tones in any diatonic collection are the tonic, the octave, the dominant, and the subdominant. This is also true of the keys. That is to say, that the key a fifth above (dominant) and a fifth below (subdominant) are harmonically closer to the tonic and their respective scales differ by only one note. For example, the musical keys closest to the key of C, are the key of F and the key of G. These relationships are the most aesthetically pleasing to the ear, and it is as if they are ingrained into the human psychic.

There are countless songs written around this basic relationship and it is common in the music of all cultures. They are as common to us as the primary colors, which by the way are frequencies as well with each color assigned to a specific frequency (wavelength). Using the aforementioned process, twelve scales were identified based on twelve notes. Each note in the Pythagorean scale had its own family of notes or scale and in addition to the seven different notes in the Pythagorean scale, five additional notes were found. With the addition of these notes, the cycle was now contiguous and complete with each note being a fifth apart from the previous note which circumscribed a circle henceforth known as the *Cycle of Fifths.*

The *Cycle of Fifths* forms the roadmap for the harmonic structure of music. It is like a clock with C, at twelve o'clock and its most dissonant note or scale opposite it at six o'clock. This is true of any key.

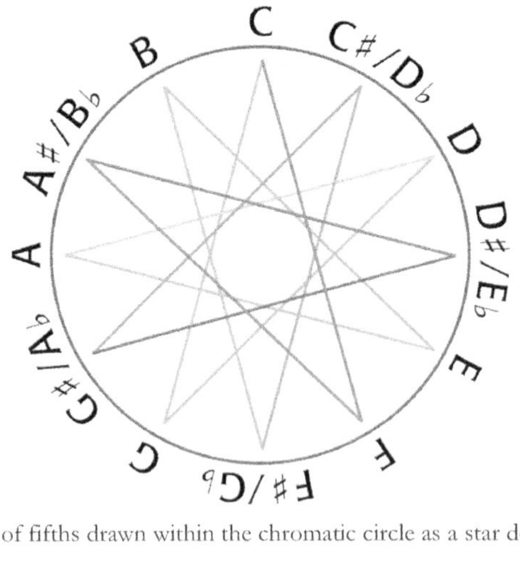

The circle of fifths drawn within the chromatic circle as a star dodecagram

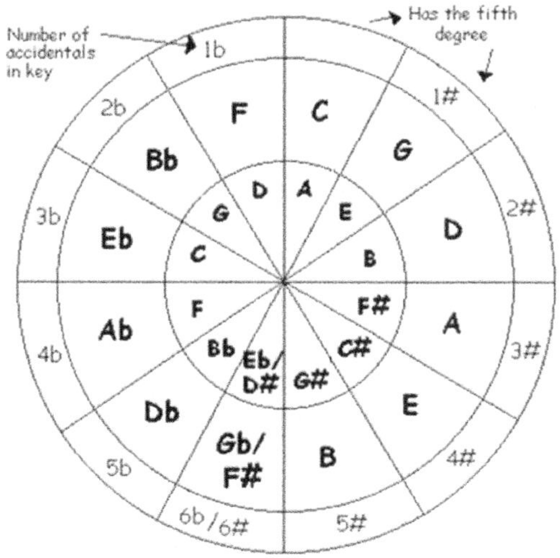

The Cycle of Fifths

The outer circle denotes the number of sharps or flats in each key, the next outer circle denotes the Major key, and the inner circle represents its Relative Minor key which has the same key signature, but its starting point is shifted relative to its major key. For instance, for the Major key of C, instead of starting on C, you would start on A which would yield the A Minor scale. Therefore, by adding five additional notes to the original scale, we progressed from one Major key and one Minor key to twelve Major keys and twelve Minor keys. The Major and Minor keys are like night and day. Typically, the Major keys have a happier sound, and the Minor keys have a sadder sound. Moreover, as Nature would have it, the musical cycle was finally complete. As seen from the chart, keys and scales that are harmonically close to each other are adjacent in the clockwise and counterclockwise directions as shown below.

The beauty of the *Cycle of Fifths* lies in its simplicity and its symmetry. It is important to note, that harmony is maintained not by the individual notes or scales, but by the relationships of notes and scales to each other. This dictates which notes or keys are valid and the migration paths (modulation) from one key to the other. Each additional note added to form a successive key is also a fifth apart from each other. The subdominant of any key and the dominant are in many ways the mirror image of each other, one is a fifth above the tonic and the other is a fifth below the tonic. In terms of vibrations or frequencies, the ratio is *3/2* in the clockwise direction. It is also important to note that the overall ratio between the diatonic scale and the chromatic scale in terms of the number of notes in each scale is *12/8* or *3/2*. Thus, the ratio of *3/2* seems to be integral to the rules of harmony, and as noted previously was the first ratio that Pythagoras observed as truly being different. The rules of harmony must adhere to the relationships delineated by the *Cycle of Fifths* and from such relationships, a musical hierarchy is developed, from notes, to chords, to chord progressions to songs and scores. As the history of music attests, the number of musical permutations from these basic building blocks seems to be inexhaustible.

The Music of the Sphere

Music has been the metaphor used throughout the ages to describe the rhythm of the universe. However, music at its core is based on resonance. Maybe, resonance is the key to understanding the hidden mechanisms of Nature, from the micro to the macro, not only within a dimension but between dimensions. Perhaps the best way to illustrate resonance and the wave phenomena is to use a music analogy. Fourier Analysis since its inception has proven to be a useful tool in describing wave phenomena across a wide spectrum of Engineering and Scientific disciplines ranging from heat to sound. Fourier Analysis of music reveals that it is a series of frequencies (notes) with their associated harmonics vibrating in resonance to produce pleasant sounds.

This relationship which is central to harmony is based on consonance or the alignment of harmonics. It is through resonance that a musical "field" is established which manifests itself as the music being in tune. Frequencies outside of this field interfere and manifest themselves as being dissonant or out of tune. Resonant frequencies are intertwined with spaces or rest to give music its rhythmic structure. These rest are also frequency dependent as they are used to pulse modulate the vibrating frequencies. In music, as in nature, resonance is paramount not only to symmetry but to alignment. Those frequencies that vibrate at multiples of whole numbers of the fundamental frequency are aligned similar to energy levels that are multiples of whole numbers also aligned.

Just as the ear is trained to detect resonances in the air and to discriminate based on alignment and the eye likewise is devised to detect and discriminate wavelengths that perceive colors; the detection, alignment, and responses of various resonances at the cosmic and quantum levels is an integral part of the rhythm and the harmony of the cosmos.[426] Perhaps both are analogs of a more pervasive and fundamental system of resonance. For example, a rainbow may be viewed as a spectral chord with each color representing notes in harmony with each other. Although light travels at a speed, roughly a million times faster than the speed of sound, and the visual bandwidth of light is only about an octave, light and sound are related phenomena as the Doppler effect and the presents of frequencies illustrate. Additionally, when music is played humans seem to have a natural tendency to want to resonate with the music, by clapping, dancing, etc. as the music is played as if we were hardcoded to be a part of the resonance.

Not only are humans hardcoded for resonance but so are the atoms of which we are made. There is no better example than Carbon which serves as a fundamental building block of life as we know it. Carbon is the fourth most abundant atom in our universe after Hydrogen, Helium and Oxygen. However, the Nuclear Fusion of elements required to form Carbon would not be possible without the internal resonance of Carbon. If the resonance level of Carbon had been 4 percent lower there would be no Carbon and therefore no life. And if the resonance level in Oxygen had been a half percent higher, virtually all the Carbon would have been converted to Oxygen. [427]

Moreover, the energy of a photon, from which all the energy required to maintain the rhythm of life on our planet emanates, is directly intertwined with the cosmic rhythm according to the formula: $E = h\nu$; a formula that connects the particle to the wave. It simply states, the higher the frequency, ν, the greater the energy of the photon. Furthermore, without rhythm, a planet does not rotate around a star or on its axis, stars do not rotate around galaxies, etc. Energy requires motion to do work, and the synchronization of motion is part of its efficiency. Since everything is composed of energy, then everything must have its internal resonance as demonstrated by the shattering of the glass when a certain note is sung.

As Kepler once said: "heavenly motions are nothing but a continuous song for several voices perceived by the intellect but not by the ear." Perhaps more to Kepler's point, we may ask: what is the motion of Earth except for a harmonic of the Sun and what is the movement of the Moon, except for a harmonic of the Earth controlled by their

426 This is achieved by performing a Fourier Transform on the sound subdividing it into its various harmonics.

427 [Ferris: The World Treasure of Physics, Astronomy and Mathematics, p.392]

associated masses and distances apart? Additionally, the motions of strings at the quantum level may contain the harmonies that resonate within the universe of the atom and beyond and even throughout the cosmos. Perhaps, the wave aspect of the particle is essential to connecting the energy contained in stars to the energy in the form of fields. In this regard, waves are the conduits of information, and it is from waves that all things in our dimension come into fruition. However, such vibrations are not limited to our dimension, they may also resonate in other dimensions. This lies at the heart of string theory.

Moreover, the music of our cosmos may be orchestrated by a dimension(s) unknown and unseen to us in concert with a greater whole. A plethora of universes may dance to such a beat. For all we know, we may be denizens in a cosmic cave, chained to our ignorance, while viewing shadows projected on the walls of our dimension from a light unseen. Such light may emanate from a higher dimension that warps ours. The reflection of that light appears to as a stream of photons, streaking through our dimension at a prescribed and constant speed, unlocking the door to hidden dimensions with its energy dancing on its beam. It is this rhythm that our atoms dance to and our mass, and therefore our existence, pays homage. It could be the nexus not only to our cosmos, but perhaps to the realm that lies beyond it and our physics. Photons and electrons or any other particle may greet each other as particles; but they dance and exchange their energy as waves.

THE MUSIC OF THE STRING

Sounds never dissipate, they only recreate in another place or time

Earth, Wind and Fire

In this essay, I would like to expand on string theory and to make hypothetical analogies between vibrations of strings at the subatomic level and vibrations of notes in music. The poet John Keats once said: "Heard melodies are sweet, but those unheard are sweeter." String vibrations at the subatomic level could be viewed as unheard melodies playing to the muse of the cosmos. To extend the analogy between music theory and string theory, we will note three important aspects of music. First, the string vibrations are quantitative, i.e., they appear at specific frequencies in whole steps. Second, not all vibrations are valid, some are for specific keys, and some are not valid at all. Third, it is the relationship between frequencies, more specifically, to the tonic or fundamental that is important. The tonic establishes a center of gravity in which all the notes in the scale gravitate toward, especially the dominant. From a mathematical perspective, any complex waveform or sound can be simplified into a series of partial tones resonating at whole number frequencies to the fundamental.

Hence, the octave, the subdominant, and the dominant are strong because they represent the frequencies at, *2x, 3x*, and *4x* the fundamental frequency. Therefore, in the second octave, twice the fundamental frequency and three times the fundamental frequency will appear and in the third octave, four times the fundamental frequency will appear, which, by the way, is the second harmonic of the string that gave us the octave. A similar phenomenon is present for successive octaves. So, when Pythagoras plucked the first string, he didn't generate one frequency, he generated many frequencies all with a whole number related to the tonic. It is also worthwhile to note, that the tonic tone is a fifth above its subdominant. Therefore, the tonic shares the same relationship with its subdominant as its dominant shares with the tonic. It is the symmetrical relationships around the *Cycle of Fifths* in a clockwise or counterclockwise direction that gives music its continual form.

The Relationship Between Music and String Theory

So, what does all this have to do with string theory? Music could be viewed as a closed system *(S)* that is indicative of all closed systems based on the standing wave function. Therefore, string theory may obey the same basic laws hitherto explained. First, the oscillations of strings at the subatomic level are also quantitative and thus, can only occur at specific frequencies. Second, all string oscillations are not valid; any oscillation must have an associate oscillation as its fundamental reference. Third, it is the relationship between oscillating strings to some fundamental oscillation that is important, and it is related to it by a whole number. Such relative oscillations, may account for the creation and actions of subatomic particles which, in turn, form atoms, which in turn, form molecules, etc.

Moreover, each particle has a specific resonance frequency that corresponds to its energy. Therefore, the way that something vibrates determines what it is and perhaps more importantly, its relationship to other things. Just as Pythagoras created new fundamentals by creating new strings that share a relationship to the fundamental pitch previously defined, strings at the subatomic level may act accordingly. Theoretically speaking, this may account for the quantum numbers of electrons: principal, angular, magnetic and spin; and why only energy levels with $2n^2$ electrons are valid and other subatomic phenomena.[428]

The *Cycle of Fifths* provides an excellent example of how a field of resonance works. Just as the notes, chords, etc. are vibrating according to the *Cycle of Fifths,* the strings may be oscillating in accordance with a Higgs-like field. Therefore, only certain oscillations or relationships based on whole numbers of resonance are allowed within this field. In this manner, it operates similar to the *Cycle of Fifths,* it is the arbitrator and the conductor of the strings. As strings propagate through the Higgs-like field, they develop what we observe as mass, but this is merely a reflection of the type of oscillation moving relative to the Higgs-like field.

428The Bohr model was a one-dimensional model that used one quantum number to describe the distribution of electrons in the atom. The only information that was important was the size of the orbit, which was described by the n quantum number. Schrödinger's model allowed the electron to occupy three-dimensional space. It therefore required three coordinates, or three quantum numbers, to describe the orbitals in which electrons can be found. The principal quantum number just indicates the size of the orbital or energy level, the angular momentum quantum number indicates the shape of the orbital, the magnetic quantum number indicates the orientation or position of the orbital, and the spin quantum number indicates the spin of the electron, represented as either +1/2 or -1/2.

There could also be a relationship between frequency and energy according to $E= h\nu$.[429] Thus, the higher the fundamental frequency, the more energy, and the heavier the rest mass of the particle, and conversely the lower frequency the lighter the rest mass of the particle.

Perhaps electrons are smaller in mass because of their charge content and their movement which compensates for their mass, whereas protons and neutrons are heavier because of their stationary position and the force that is required to overcome the repulsion of the electromagnetic force due to two or more positive protons close to each other. This means that their propagation through the Higgs-like Field is much different. [430] Furthermore, protons and neutrons may permeate the Higgs-like field with roughly the same oscillation but with different orientations or spin; with a neutron having the capability of producing a proton, an electron plus a neutrino. Of course, the neutrons are required to help overcome the force of repulsion of adjacent protons and as a result, contribute greatly to the atomic weight of the element without impacting the overall charge. It is the presence of these two elements in the nucleus that is the source of the strong nuclear force.

The Correlation Between Mass, Charge and Multidimensions

The origins of mass and its variance as its approaches the speed of light remains a scientific quagmire. Relativistic mass is expressed by the Lorenz equation as the following:

$$m = \frac{m_0}{\sqrt{1 - \frac{v^2}{c^2}}}$$

Where m_o is the composite waveform of strings passing through the Higgs-like field and v is the overall propagation of the string within the Higgs-like field. As this relates to music, $m_{o,}$ is the note (frequency) and v is the speed of sound or the propagation rate of the Higgs-like field. As seen from the Lorentz equation, the velocity within the spacetime continuum must be less than the speed of light. This is

429 This equation says that the energy of a particle of light (E), called a photon, is proportional to its frequency (v), by a constant factor (h). This means that photons with low frequencies, like radio waves, have lower energies than photons with high frequencies, like x-rays.

430 Particles that interact — or "couple" — with the Higgs field more strongly are granted greater masses. Even the Higgs boson itself gets its mass from its own interaction with the Higgs field. This has been confirmed by watching how Higgs boson particles decay. Higgs boson: The 'god particle' explained | Space

because if the velocity is equal to the speed of light the mass becomes infinite. *Gödel Incompleteness Theorem* proved that there exist questions that while being undecidable within a closed system (such as spacetime) can be seen as actually being true if we jump to the outside. Perhaps, the flaw lies in our interpretation of "infinity" to mean massive when it should be interpreted as unbounded or outside. This is particularly true when we view mass as a wave, with its harmonics extending to infinity. This distinction makes all the difference in the world.

For example, "infinity" may be an indication that the components or partials that constitute the mass and charge lay outside of our dimension or universe and are less than a Planck length. Being inside a Planck length is like being inside an event horizon, in both cases, we have no visibility and we have effectively left our universe. In this regard, the Higgs-like field is very much like the medium of air invisible with some quantified speed of propagation within all spatial dimensions. However, it's less than the speed of light within our three-dimensional spacetime but rendering superluminal speeds within higher dimensions with a size less than a Planck length. Perhaps, this is why our dimension is large and some higher dimensions small. Moreover, photons have no rest mass because of the speed of which they propagate through the Higgs-like field. Hypothetically, photons communicate with each other due to the mutual effect that they have on the Higgs-like field.

The fundamental oscillations of strings are within three spatial dimensions; however, analogous to sine waves in the Fourier series, the other oscillations may occur within other dimensions.[431] In contrast to the harmonics of the Fourier series, each whole number and coefficients may not only represent a harmonic, amplitude and phase but rather the location of the spatial dimension in which the other partials of the string are oscillating with associated amplitude and phase in proportion to the waveform manifested within spacetime. It is a general law of music that two tones sound well together when the ratio of their frequencies can be expressed using small whole numbers, and the smaller the number the better the consonance. Hypothetically, the energy present in each spatial dimension is related to the ratio of its frequency to the fundamental vibration. The spatial dimensions that are closest to the three spatial dimensions could share the greatest relationships in a similar fashion to the keys within the *Cycle of Fifths* from a field perspective. The additional

431 A Fourier Series, with period T, is an infinite sum of sinusoidal functions (cosine and sine), each with a frequency that is an integer multiple of 1/T (the inverse of the fundamental period). Refer to the Appendix for the equations for a Fourier Series.
http://www.thefouriertransform.com/series/fourier.php#series

oscillations, *harmonics,* for lack of another term, transfer to the extra dimension via resonance through the Higgs-like field.

Perhaps the best way to illustrate this phenomenon is to juxtapose the laws that govern the velocity of sound traveling through the medium of air with Einstein's Theory of Relativity traveling through spacetime. The velocity of sound is the conduit of audible frequencies, and the speed of light is the conduit of frequencies within the electromagnetic spectrum. The law governing the velocity of the string is: $v^2 = T/M$ where v is the velocity, T is the Tension of the string and M is the mass of the string. Rearranging the equation *yields* $T = Mv^2$. One can easily see the similarity of this equation to Einstein's Theory of Relativity, $E = mc^2$, where E is equal to the energy and m is equal to the mass and c is the speed of light. Dimensionally they are the same and aligned with the formula for kinetic energy $E_K = \frac{1}{2} mv^2$. As we can see, the velocity through the medium is constant if all perturbations due to temperature and so forth are excluded. Moreover, it increases to the square of the Tension/Mass ratio.

Therefore, if the mass is small and the tension is high, the number of vibrations per second is also extremely high as demonstrated by the strings of the piano. Succinctly put, Einstein's true brilliance was his ability to recognize and extend a natural law of Nature throughout spacetime by using the fastest known velocity for his equation. If by chance a high school algebra student with a love of music and mathematics used a train of thought based on the analog mentioned hitherto, she would have been just as correct in her assumption.

We also know from Fourier analysis that the velocity generates harmonics at whole number multiples of the fundamental frequency, which are a unique set of patterns relative to the velocity so much so that the ear in conjunction with the brain would detect the fundamental even if some of the components are missing as best illustrated by the pipe organ. The question is if the speed of light is traveling through a yet undetected medium akin to the *luminiferous aether* in theory but more sophisticated in scope, i.e., the Higgs-like Field, does a similar phenomenon occur? To expand upon this theory and the juxtaposition of the equations mentioned hitherto, let's propose that this medium expands all dimensions and that harmonics to the fundamental observed in our dimension are present in the other dimension because of the electromagnetic spectrum moving through our dimension at the speed of light. Using the sound analogy, to attain an accurate measurement of the fundamental waveform we would have to go out n number of terms in the Fourier series with each successive term rendering a more accurate waveform.

As in the case of the musical string, all that is needed is a hammer (any one of the plethora of possibilities from the Big Bang) to hit the string to unleash the sound ordained by the tension and the mass of the string. In this regard, the speed of light is just as intrinsic to mass and energy as the velocity is to the tension and mass of the string. By that I mean, the interaction of mass and energy within our dimension in conjunction with the movement of the speed of light through this unseen medium generates or reflects its harmonic content into higher dimensions. Moreover, the laws of universal harmony are just as intrinsic.

Similar to the musical case, each subatomic particle has an associated signature based on its properties and propagation through this unseen field. If we replace the whole number associated with the harmonics with the whole number for the dimension in which it vibrates, then we may be able to guess what is happening within the higher dimension. Let's, for the sake of argument, we attach a sawtooth function to the quark with each whole number representing the dimension in which it vibrates. If we select a fundamental frequency for this quark of 100 Hz. The waveform would look as follows in the time domain:

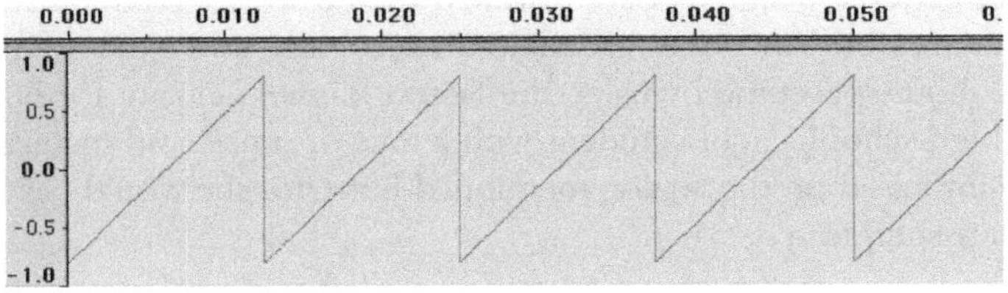

A sawtooth function time domain

Its representation in the frequency (dimension) domain with a fundamental frequency of 100 Hz would be as follows:

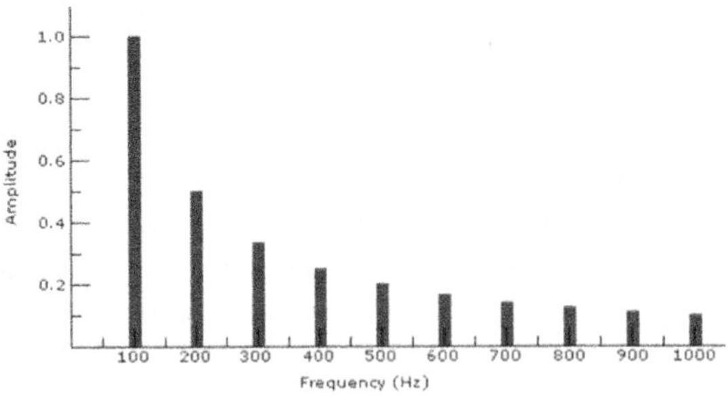

Harmonic frequencies in a sawtooth wave with
a fundamental frequency of 100Hz

As you can see, within the frequency domain the increase in frequency is directly proportional to the harmonic number, which in the case of my hypothesis would render the higher dimension also directly proportional to the type of waveform. In the case of the sawtooth, wave the amplitude is inversely proportional to the dimension. As in all cases, the more harmonics added, the more the waveform approximates what is realized in the time domain, in this case, spacetime. This is the general case, however, depending on the waveform and its harmonic content, accurate approximations could be made with fewer expansions, best illustrated by the difference between a flute that has fewer harmonics versus a trumpet that would require substantially more.

What is key here, and what allows a flute and a trumpet or any other instrument to play together harmoniously, is their relationship to the fundamental. In the case of the flute, most of its harmonic content is contained in the third harmonic and the contribution of other harmonics is insignificant relative to the contribution of the third harmonic. This is true for any note played on the flute. We recognize the different instruments because the timbre is different for each which is determined solely by the harmonic content. We recognize the note as being the same because the fundamental vibration that produced the harmonics is the same although the harmonic content may vary greatly.

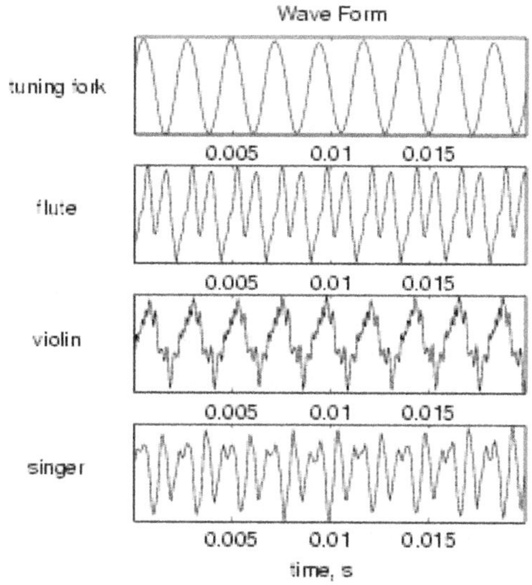

https://th.bing.com/th/id/R.031b971e0dc7b9bb05bede15d6090836?rik=QcFBs6uV%2bq8sow&riu=http%3a%2f%2fohmspeaker.com%2fsite%2fassets%2ffiles%2f2648%2ff_instrumentwaves.gif&ehk=i6s%2fkiOwgHPRtzZXRBrSSu%2bKllpj3gNMI4Sy24AoVVQ%3d&risl=&pid=ImgRaw&r=0

The following explains a possible relationship between harmonics and atomic orbits.

> A musical note from a guitar consists of a whole series of sounds—a spectrum of frequencies. The interesting thing for de Broglie was that the stationary waves along the guitar string make up a discrete frequency spectrum called the harmonics.[432] The lowest-frequency sound is called the first harmonic, which determines the pitch we hear. The higher harmonics—the musical sounds in the note that give it a distinctive quality—have frequencies that are represented as integer multiples of that of the first harmonic. Being stationary is a property of waves in confinement. Such waves are easily set up in a cup of tea. De Broglie asked; Are atomic electrons confined waves? If so, do they produce discrete stationary wave patterns? For example, maybe the lowest atomic orbit is one in which one electron smallest frequency—the first harmonic—and the higher orbits correspond to stationary electron waves of higher harmonics. Graphic representation of a wave[433].

As you can see, De Broglie, the father of the particle/wave duality also used a music analogy to describe wavelike phenomena at the quantum level.[434] However, it is safe to say that within ten terms (dimensions) a good approximation of the waveform present in the time domain (spacetime) could be rendered although the process itself extends to infinity. As in the case of music, each dimension does not necessarily contribute to the function rendered in the time domain (spacetime) it only points to the location where such harmonics could be found as shown by the frequency spectrum for a square wave below and in the case of the sawtooth waveform shown previously:

---

432 The French physicist Louis de Broglie suggested (1924) that particles might have wave properties in addition to particle properties. Three years later the wave nature of electrons was detected experimentally. Objects of everyday experience, however, have a computed wavelength much smaller than that of electrons, so their wave properties have never been detected; familiar objects show only particle behaviour. De Broglie waves play an appreciable role, therefore, only in the realm of subatomic particles. De Broglie wave | Definition & Facts | Britannica

433 [Goswami: The Self-Aware Universe, pp. 32,33]

434 In 1924, Louis de Broglie presented his research thesis, in which he proposed electrons have properties of both waves and particles, like light. He rearranged the terms of the Plank-Einstein relation to apply to all types of matter. The de Broglie hypothesis was verified when matter waves were observed in George Paget Thomson's cathode ray diffraction experiment and the Davisson-Germer experiment, which specifically applied to electrons. Since then, the de Broglie equation has been shown to apply to elementary particles, neutral atoms, and molecules. https://www.thoughtco.com/definition-of-de-broglie-equation-604418

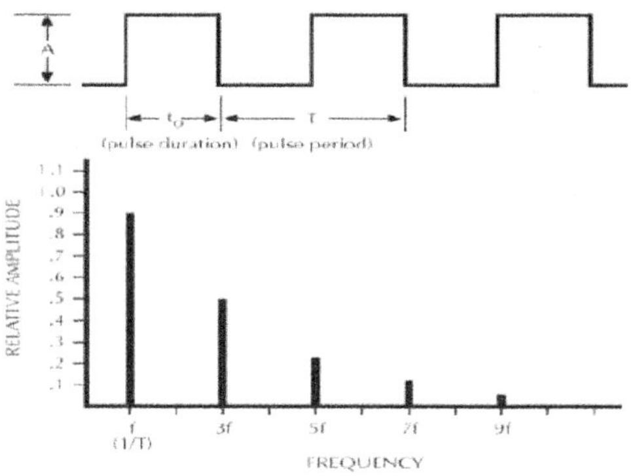

The Fourier series for the square wave above could be written as:

$$F(x) = \sin x + 1/3 \sin 3x + 1/5 \sin 5x + 1/7 \sin 7x + \ldots$$

The harmonic content that constitutes the square wave is shown in the following example with each additional wave giving a greater approximation of the wave.

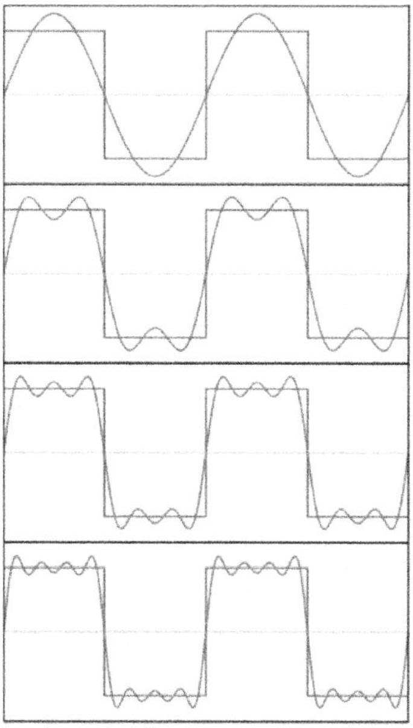

The first four partial sums of the Fourier series for a square wave. As more harmonics are added, the partial sums converge to (become more and more like) the square wave.

Although the two aforementioned waveforms vary in harmonic content, they are harmonious with each other because their fundamental vibration is the same for a square wave at 100 Hz. As you can see, symmetry in the time domain also yields symmetry in the frequency domain, and even in cases where symmetry is not apparent in the time domain it is readily apparent in the frequency domain as the true nature of the waveform is always apparent in the frequency domain. Moreover, those functions that are symmetrical and periodic in the time domain, i.e., even and odd functions, always eliminate either the Sine or Cosine terms in the frequency domain as in the case of the above example. We can determine from the above figure that the square wave has similar symmetry to a sine wave, and this is why only the sine terms appear. The Fourier coefficients provide a measure of the amplitude and phase of the frequencies present within a_m and b_n coefficients.

Perhaps, in this case, they provide an indication of the mass and charge components realized within the dimension of the fundamental vibration, in this case, our spacetime. If the Fourier coefficients are related to the mass and charge, then both would likewise extend to infinity, and it is only in the limit of some finite value ***n*** that they manifest themselves in the real world. Thus, they could be fairly approximated by performing a Fourier transform on the waveform; although the actual components of both mass and charge are infinite rendering new light on the Lorentz equation and the paradox it presents. That is to say, the "infinity" yielded by the Lorentz equation represents the mass as the partials and their coefficients approach infinity in the same manner that Zeno's Arrow approaches a wall. However, if positive whole values greater than the speed of light are inserted into the Lorenz equation, the rest mass relative to the relativistic mass becomes imaginary.[435]

Perhaps, the speed of light traveling through spacetime produces harmonics in other dimensions therefore each dimension has a speed of *Dc* where *D* represents the whole number of the next dimension. By assigning one to three-dimensional space, the fourth dimension would have a speed of *2c,* the fifth *3c,* etc., *ad infinitum* although ten spatial dimensions or more may be required to attain an accurate representation of the phenomenon.[436]

The extra dimensions could be wrapped up in Calabi-Yau space.[437] This could be true if the component of the mass is very small and the energy high relative to our

435 Perhaps a Laplace Transform is in order.

436 This would make the fifth dimension a fifth away. I am aware that some will think that I am mixing apples with oranges here, but intuitively I feel it is correct.

dimension as the speed of light is dependent upon these variables. Such linear increases in the speed of light may not be germane except to prove that the speed of light is not the same in all reference frames and to help describe the differences in propagation of the Electromagnetic field as a function of the dimension. All we can truly say about the term "infinity" is that it lies outside of the realm of finite numbers and each Calabi-Yau shape may be infinite relative to our reality. Even if we say that it goes on forever and forever; that forever is still related to our time base. In a separate reality, Calabi-Yau space may be a never-never land where forever and forever has no meaning relative to itself.

A 2D slice of a 6D Calabi–Yau quintic manifold[438]

Conceivably, within each spatial dimension, time travels faster and distances are extremely short to the extent that relative to us, time and distance do not exist. How soon we forget that time is the measurement of events taking place within a reference frame and without the motion of events time has no significance. This is only possible because each higher dimension is smaller than a Planck length and shorter than a Planck time as they were before 10^{-43} secs of our universe.[439] Perhaps,

437 In 1985, the Calabi-Yau manifolds (created for other purposes years earlier by mathematicians Eugenio Calabi and Shing-Tung Yau) were used by Edward Witten, Philip Candelas, Gary Horowitz, and Andrew Strominger to compactify the extra six space dimensions in just the right way. These manifolds not only preserved the handedness of the particles, but they also preserved supersymmetry just enough to replicate certain aspects of the Standard Model. String Theory and Calabi-Yau Manifolds - dummies

438 By Andrew J. Hanson - Ticket#2014010910010981, CC BY-SA 3.0, https://commons.wikimedia.org/w/index.php?curid=30579072

these universes are so small because they existed before our universe was created and relative to us cannot be created or destroyed. In this regard, we can think of our spacetime as the fruit that blooms on the branch of higher dimensions which connects our universe to the hidden celestial tree. Moreover, there may be other branches that bare similar fruit. To put it succinctly, there is no such thing as spacetime within higher dimensions. Possibly, the only force in existence in the higher dimension is the electromagnetic force. Forces such as gravity and the manifestation of subatomic particles are reflections or echoes of that force in the form of waveforms that resonate throughout higher dimensions. Perhaps in this regard, we should think of our dimension as a celestial sphere that radiates its energy at all points orthogonal to its surface into dimensions hitherto unknown and unseen. Since Calabi-Yau shapes are attached to each point, propagation of waves takes place within Calabi-Yau shapes per their resonance.[440]

Depending on the complexity of the waveform, more than 6 Calabi-Yau shapes may be required. To this end, each dimension of the Calabi-Yau space contains a part of the mass charge per the structure of the waveform. Similar to the case of the flute, there may be waveforms that have most of their mass and charge contained within one Calabi-Yau shape. It is within the Calabi-Yau shapes that the "relationships" are defined similar to the musical scale in which only certain frequencies are valid and perhaps it is a result of those relationships that everything in our dimension is created and maintained.

Regarding the Higgs boson, in music theory, the messenger particle is the relationship of frequencies delineated by the *Cycle of Fifths.* Without such relationships, music as we know it could not exist. Hypothetically, the Higgs boson is the result of relative motion within the Higgs field resulting from resonance. Hence, the Higgs boson must maintain some type of relationship or resonance with the oscillation of the strings that

---

439 The Planck length is the scale at which classical ideas about gravity and spacetime cease to be valid, and quantum effects dominate. This is the 'quantum of length', the smallest measurement of length with any meaning. And roughly equal to 1.6 x 10-35 m or about 10-20 times the size of a proton. The Planck time is the time it would take a photon travelling at the speed of light to across a distance equal to the Planck length. This is the 'quantum of time', the smallest measurement of time that has any meaning, and is equal to 10-43 seconds. No smaller division of time has any meaning. Within the framework of the laws of physics as we understand them today, we can say only that the universe came into existence when it already had an age of 10-43 seconds.
http://www.physlink.com/Education/AskExperts/ae281.cfm

440 A Calabi-Yau shape or space is the extra special dimensions required by string theory can be curled up consistent with the equations of that theory. The shape is made using the complex definitions of the sine and cosine functions: Sin(x)= (eix - e-ix)2i, and Cos(x)=(eix + e-ix)/2. These equations can both be multiplied by a phase factor that shifts them, depending on the value of a certain variable.

form baryonic matter. String theory emerged when General Relativity was applied to four spatial dimensions, which yielded, Maxwell's equations. Hypothetically, the Electromagnetic spectrum and the photon itself represent a key to the oscillations in extra dimensions.[441]

Regarding the wave-particle duality, this is analogous to the pulse modulation (rhythm) of the notes. The particle phenomenon that we observe is due perhaps to the amplitude (mass) present in the Fourier coefficient and the wave is due to the phase(charge) and this may also account for the orbits of electrons and the spin of particles. Musical gravity which holds everything together is achieved by movement toward the tonal center by melodic implication, harmonic resolution or repetition. Gravitation at the string level could also be the result of movement in time and in space toward a fundamental oscillation. Hence, subatomic particles may be the result of harmonic movement within the Higgs-like field.

As in music, the closer you are harmonically, the greater the musical gravity. The gravitational pull of a note decreases with harmonic distance and the gravitational pull of two masses also decreases with distance. Hence, musical gravity is a manifestation and an indication of the type of movement (resonance) within the *Cycle of Fifths* and possibly within the Higgs-like field. Hypothetically, any movement within the Higgs-like field could cause a corresponding movement (resonance) proportional to it in the field and a natural attraction which we observe as gravity but in reality, it is nothing more than resonance as it is in the case of music.

Attraction and Resonance Within Baryonic Matter and Hadrons[442]

Although string theory remains within the confines of mathematics and theoretical physics, another wave phenomenon at the subatomic level was verified using Schrödinger equations for standing waves. In this regard, additional analogies could be made between standing waves and the subatomic level and the vibration of musical strings. For example, if we were to assign a note (frequency) to an 'Up' quark in the case the note C and a 'Down' quark to the note G and we assume that they are played on the same instrument, then certain phenomena occur. First, they have the same waveform only shifted by frequency with different Fourier coefficients. Second, each is a harmonic of the other, thus "mutual" harmonization occurs. Since they both have the same waveform and furthermore harmonics of each other (G the third harmonic

441 I am referring here to the theory of Kaluza and Klein.

442 Hadrons are any of the subatomic particles (such as protons and neutrons) that are made up of quarks and are subject to the strong force.

of C and C the fourth harmonic of G) the "probability" that their harmonics will be aligned is great as demonstrated by the fact that the C scale and the G scale only differ by one note, F# instead of F (a semitone) therefore their mutual attraction is great. As a result, it is easy to modulate from one to the other.[443]

Furthermore, it is important to note that the difference involves the fourth of the previous scale, e.g., F for the key of C. This is because the F in the key of C pulls in C in such a manner as to imply the key of F. By raising the F to F# in the key of G, G has eliminated the competition for a mutual resonance of its fourth which is C. In a similar manner within the key of G, when C is played it implies the Key of C. This also accounts for the steps within the major scale as Bb has been replaced by B in the Key of C because of the mutual resonance between Bb and F, the previous key on the cycle of fifths.

Rotation around the *Cycle of Fifths* in either direction reduces the "probability" of harmonic alignment reaching its maximum directly across from the original scale (6 o'clock) and thus reduces its attraction to the fundamental. Using this analogy, this means that there would be a strong natural attraction between 'Up' and 'Down' quarks similar to the musical attraction and resolution of the notes C and G which resolve each other. This attraction is solely due to the resonance frequencies and harmonic relationships between the two. Although each of the notes in this example was played on the same instrument, they are also harmonious with other instruments that play within the associated scale. Therefore, the waveform content of these other instruments (hadrons and other subatomic particles) may vary, but the "probability spectrum" is dictated by the scale itself. Of course, the presence and amplitudes of the actual harmonics of a given instrument (waveform) are dictated by its Fourier series and it is what makes one instrument sound different from the other and perhaps one subatomic group different from the other. Regarding the relationship between particles and antiparticles, this could be akin to the relationship between major keys and their relative minor keys.

The baryonic matter of the universe is comprised of Up and Down, quarks and their lepton companions, the electron and the neutrino. [444]Quarks occur in clusters, either

[443] I find it most ironic that the note that the G scale includes is tied to the fundamental of the key that is the most inharmonic to C and thus does not align itself to C harmonically. This is true of all scales contained within the Cycle of Fifths, which illustrates the further away you are the less harmonic and the less "gravity" or mutual attraction between the two. It is also ironic that the F# scale is the most inharmonic scale to C being at six o'clock on the cycle of fifths. In addition to the g scale, the F scale also only differs from the C scale by one note.

in pairs or a triplet as if some type of "mutual" resonance is required to bind quarks. No physicist has yet seen a naked quark which may mean that its existence is dependent upon resonance between quarks. Protons and neutrons each contain three quarks. A proton is composed of two 'Up' quarks and one 'Down' quark while neutrons are composed of one 'Up' quark and two 'Down' quarks.[445] It takes two 'Down' quarks to equal one 'Up' quark and therefore the proton has a positive charge, and the neutron is neutral. It should also be noted that the neutron is the complement or negative reflection of the proton as far as the constitution of quarks is concerned. Hence, extending this analogy, a proton could be viewed as CCG and a neutron as CGG. Therefore, by extension, all that is needed to change a neutron into a proton is to change the G into a C by slightly altering its frequency (or one of its harmonics) and in the process creating an electron and a neutrino as shown by the following equation: $CGG \rightarrow CCG + e^- + \nu_e$. As mentioned previously, the harmonic content of a C and a G is almost the same, differing by one harmonic. It is important to note that if the $F^{\#}$ is changed to F and the harmonics rearranged, the former G and the new C would share the same harmonic spectrum. In this case, the $e^- + \nu_e$ corresponds to the difference in frequency between F and F#.

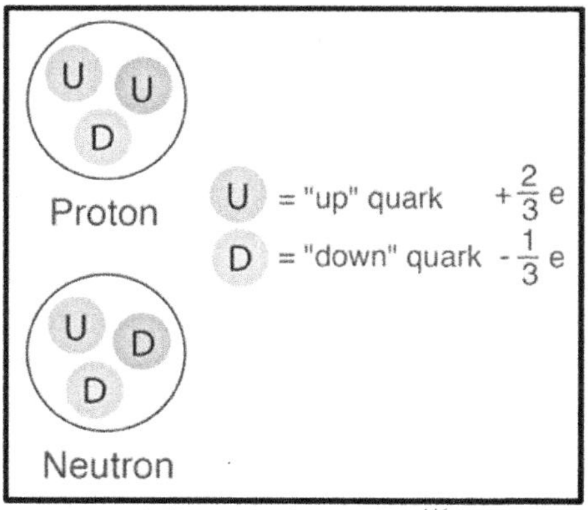

A Proton and a Neutron[446]

444 [Tweed: Essential Elements, p.44]

445 This is what makes a proton positive and a neutron neutral. One may even surmise that the difference in frequency accounts for the presents of the electron and the neutrino as energy must be conserved.

446 Quarks (gsu.edu)

Quark	Symbol	Spin	Charge	Baryon Number	S	C	B	T	Mass*
Up	U	1/2	+2/3	1/3	0	0	0	0	1.7-3.3 MeV
Down	D	1/2	-1/3	1/3	0	0	0	0	4.1-5.8 MeV
Charm	C	1/2	+2/3	1/3	0	+1	0	0	1270 MeV
Strange	S	1/2	-1/3	1/3	-1	0	0	0	101 MeV
Top	T	1/2	+2/3	1/3	0	0	0	+1	172 GeV
Bottom	B	1/2	-1/3	1/3	0	0	-1	0	4.19 GeV(MS) 4.67 GeV(1S)

http://hyperphysics.phy-astr.gsu.edu/hbase/particles/quark.html#c1

This analogy could also account for the presence of electrons and their energy levels. The electron is the complement of the proton from a charge perspective. Let's again, for the sake of discussion, say that whenever the cadence CCG is present an electron *must* appear as a result of resonance which in turn, keeps the atom electrically balanced. In this regard, radioactive decay is a result of modulation of the G in the CGG cadence to a C to form a new CCG cadence or proton/electron pair. The half-life in this case is an indication of the duration of the modulation. Therefore, a CGG cadence cannot be present without a CCG cadence and will eventually migrate toward a CCG cadence thus producing an electron and a neutrino because of its resonance[447]. Free neutrons (CGG cadence) could be viewed in the same manner as a G in the key of C: it wants to resolve itself to a C or in this case a proton/electron pair which, in essence, forms the identity of the atom and its spectrum. In this case, the atom is balanced and is solely based on its CCG cadence. Moreover, the C (which is an up quark) acts like a tonal center.

Similarly, two CCG cadences *cannot* be present without a CGG cadence which is required to counteract the repulsion of two protons. This is why all atoms except for hydrogen (excluding its isotopes of deuterium and tritium) have neutrons. The strong nuclear force may be viewed in a similar way, pions, gluons or gravitons may be the result of mutual resonance between protons and neutrons which have quarks as their fundamental elements of resonance. However, as in the case of gluons, it is a quark antiquark pair. Thus, the CCG cadence is a fundamental element of the

447 Of course, what I am describing is the process of beta decay.

subatomic world, and all baryonic matter is dependent upon a fundamental frequency and its relationships to its harmonic content. With this in mind, an analogy could be made between baryonic matter and the musical scale. However, other notes may be required to account for such things as the weak electrical force, the arrangement of electron orbits, and so forth.

In addition to C and G-type strings, other strings such as E-type strings may be required to complete the atomic structure. For example, the Lambda is a baryon that is made up of three quarks: an "Up', a 'Down" and a 'Strange' or a 'Charm' or a 'Bottom' quark. In this regard, all the known quarks such as the 'Charm,' 'Strange', 'Top' and 'Bottom' may be accounted for by the type of strings they emulate. It is as if these six types of quarks along with their associated colors form a type of "diatonic" scale. I find it most ironic that the relationship between the 'Up' and the 'Down', the 'Charm' and the 'Strange', the 'Top' and the 'Bottom' quarks are basically the same as if a modulation in frequency has occurred.[448] These quarks may provide the correct quantum numbers of electrons and other subatomic phenomena based on resonance with each other. Furthermore, the frequency of the electron orbits may be analogous to the pulse modulation of musical notes, and the locations of the orbits analogous to the notes within a musical key.

Maybe, the up quark acts as a fundamental frequency at the subatomic level and that the down quark is a harmonic of the up quark as illustrated by my analogy of the C and G notes with the G being the third harmonic of the C. It is also my belief that the energy levels of the electron orbits are also dependent upon the frequency of the up quark. An analogy between the energy level and the seats of an auditorium has been made in the past. Perhaps, an analogy between the energy levels and the musical scale is more appropriate because they are both based on standing waves. If the energy levels are harmonics of the fundamental vibration of the up quark, then the "notes" in the scale (which are all harmonics of the fundamental) will provide the probability of where an electron will or could be.

The emission or absorption of energy occurs only when an electron jumps from one orbit to another. The difference between the total energies of electrons in the two stationary orbits is absorbed when the electron jumps from the inner to the outer orbit and emitted when electrons jump from the outer orbit to the inner one. [449] When a photon of the right wavelength strikes an atom, it disappears and imparts all its

448 See the appendix for a table of the know quarks.

449 Energy Level - Bohr's Atomic Model and Postulates of Bohr Theory (vedantu.com)

energy to kicking the electron into a new energy level. Therefore, the quantization of the energy levels could be the result of the harmonization of the photon with the electron. Furthermore, the entire atom is harmonized or tuned to the fundamental vibration of the up quarks. Although some isotopes are stable, the most stable state consists of an equal number of protons and neutrons with the number of C vibrations equal to the number of G vibrations. As I have stated previously, in music it is the harmonics that are close to the fundamental that has the greatest attraction, and this would correspond to the electrons that are at the lowest energy levels. The higher energy levels, still part of the overall scale, will have less attraction to the fundamental and therefore are free to bond with other elements.

All particles have spin and according to the *Pauli Exclusion Principle,* electrons can occupy the same space provided each has opposite spin, i.e., + ½ or – ½ given by the formula $s = +/- \frac{1}{2}\ h/2\pi$. [450] Where *s* is the spin *h* is planks constant (6.62606957 × 10^{-34} joule second) with each spin state counting as a different quantum state. When a photon at a specific wavelength strikes an atom, the electron gains energy and is allowed to move up to the next energy level. This is analogous to bending a guitar string to make a higher sound by changing the overall tension of the string. We must keep in mind that an electron is a standing wave, and the photon in effect bends the tension of the string thus moving it up to a higher energy level.

When the electron returns to its normal position, it emits a photon with energy equal to the difference between the two levels according to $E = (n+ \frac{1}{2})\ h/v$ where v is equal and is related to the resonant frequency of the atom. The increase in the number of up quarks, generates, more protons, which in turn generate more electrons, which in turn occupy more energy levels in accordance with the musical scale thus changing the overall resonance of the atom. This, perhaps, is why energy levels are quantized and why different atoms respond and emit photons at a specific wavelength. This also accounts for the spectral lines of each atom and why no two atoms have the same spectrum or signature.

---

450 Particles obeying the exclusion principle have a characteristic value of spin, or intrinsic angular momentum; their spin is always some odd whole-number multiple of one-half. In the modern view of atoms, the space surrounding the dense nucleus may be thought of as consisting of orbitals, or regions, each of which comprises only two distinct states. The Pauli exclusion principle indicates that, if one of these states is occupied by an electron of spin one-half, the other may be occupied only by an electron of opposite spin or spin negative one-half. An orbital occupied by a pair of electrons of opposite spin is filled: no more electrons may enter it until one of the pair vacates the orbital. An alternative version of the exclusion principle as applied to atomic electrons states that no two electrons can have the same values of all four quantum numbers. http://www.britannica.com/EBchecked/topic/447124/Pauli-exclusion-principle

Furthermore, it is only those atoms that resonate together that are capable of bonding because the electrons at their highest energy level are free to share (overlap) their wave function, regardless of the energy level since all the energy levels are "harmonized" like all the notes in a particular key of music. As part of this scenario, the entire structure of atoms and their ability to resonate (combine) with other atoms to form molecules is a result of the common resonance between them. Thus, a subatomic particle may be analogous to a note, the atom analogous to a chord and the molecule, analogous to a chord progression or, for all intents and purposes, the song with each level revealing a higher state of *Consciousness.*

The internal relationship of chords can also be described as a mutual resonance between the fundamental frequency and its harmonics, where in most cases, notes are stacked atop each other to form triads and extended chords. There are four different types of musical triads: Major Triads, Minor Triads, Diminished Triads, and Augmented Triads each with a distinct sound. These relationships are also dependent upon the cycle of fifths with concentric circles of cycles. Such relationships form the internal and external bonding through the process of resonance.

Likewise, the molecule can then become a new fundamental, the gene the chord and the cell, the song, each in harmony with each other and with nature. The building blocks of life, Hydrogen, Carbon, Oxygen, Nitrogen, etc., may be analogous to chords built around notes that give spacetime its harmony. Just as chords are bound together by resonance, molecules may be bound by a similar resonance not between notes but between groups of atoms.

In effect, the entire periodic table could be seen as a series of "notes" with given resonance based on the number of electrons (voids) in the outer shell. Therefore, the columns and grouping of the Periodic Table could be a result of the common resonance of elements. Since inert gases have no voids in their outer shell, their resonance may be low or inharmonic with respect to the others. Absorption spectral lines may be indicative of the atom's resonant frequency or tonal center. For example, it takes two Hydrogen Atoms to "resonate" with one Oxygen atom to form the H_2O molecule. It is the pattern of atoms and molecules that defined what something is, and that pattern is dependent upon their wave content and their ability to resonate with others like strings on a cosmic guitar. In this regard, the bonding of molecules may be viewed as common chords resonating in harmony with each other. Hence, stars are constantly singing the song of the universe each time they create a new element. Thus, all chemical reactions of elements are in harmony with each other; and suns via their nuclear fusion process function as tonal centers of the cosmic music.

Chemical Formulas of Common Substances

Substance	Chemical Formula	Atoms that make up the molecule
Water	H_2O	2 hydrogen atoms, 1 oxygen atom
Hydrogen Peroxide	H_2O_2	
Carbon Dioxide	CO_2	
Oxygen	O_2	
Ozone	O_3	
Sodium Chloride (Salt)	NaCl	
Acetic Acid (vinegar)	$C_2H_4O_2$	
Nitrous Oxide	N_2O	
Isopropanol (rubbing alcohol)	C_3H_8O	
Sucrose (table sugar)	$C_{12}H_{22}O_{11}$	
Glucose	$C_6H_{12}O_2$	
Alanine (an amino acid)	$C_3H_7O_2N$	
Oleic Acid (a fat)	$C_{12}H_{24}O_2$	
Sodium Bicarbonate (baking soda)	$NaHCO_3$	

Conclusion

Perhaps a different paradigm at the Quantum Mechanics level is in order. One that views various subatomic interactions entirely from the frequency domain using Laplace or Fourier Transforms. The relationship between particles at the subatomic level and the natural phenomenon of music lies at the core. It was perhaps the spirit of the French Revolution imbued in the mathematicians of its day such as Laplace, Fourier and Galois that gave rise to the mathematical genius that was bequeathed to future generations of science.[451]

451 The mathematics of Laplace, Fourier and Galois has proven to be invaluable to the modern era. Galois theory provides a connection between field theory and group theory the later a field resulting from his mathematical genius. It has proven to be an indispensable tool in the study of Quantum Mechanics.

THE NATURE OF OUR EXISTENCE

Just as Pythagoras lacked the understanding of the relationships of harmonics to the unison, we may also be naïve to the ancillary effects of string oscillations in orthogonal dimensions resulting from fundamental vibrations within our own. Science has always been the search to discover the unknown which often results in the unraveling of the Gordian knot concealed in a scientific paradox. However, science is not immune to the movement of waves. Instead, it rides the crests to a greater understanding of the physical world. Newton resolved the paradox of Galileo, and Einstein the paradox of Newton, yet we still ask the question if Schrödinger cat is dead or alive. From our scientific perch, we can see only so far. Schrödinger improved Bohr's model, by extending it to three spatial dimensions while at the same time basing his new paradigm around the wave function. Perhaps Schrödinger like Pythagoras, did not go far enough?

The harmonic-like effects mentioned hitherto, working in concert, form a subatomic symphony that is pleasing to the ear of the cosmos and is the magnum opus of our baryonic and three-dimensional universe. As Leibniz so eloquently stated: "music is the processing of mathematics in real-time." Music is math and math lies at the bedrock of Physics which forms the foundation of Chemistry which forms the foundation of Biology. Therefore, life is a symphony orchestrated by the forces of Nature. Music has the beauty of resonation; and listening to it makes us want to resonate with it as if its vibrations are felt in the abyss of our souls. We have discerned, to some degree, the message hidden within light. Perhaps we should learn to listen to the message in the music that tells us more about how the universe actually works. Music is so aesthetically pleasing and so mathematical precise that it must tell us more about how God rolled His dice.

Maybe, when the French Mathematician Joseph Fourier developed the Fourier series, he discovered one of the basic rules that permeate not only our universe but how Nature uses its needle to thread the fabric of the cosmos.[452] Fourier and Laplace did not foresee how their theories would apply to electronics, yet without their genius the electronic field would not have had a genesis. The idea that the cosmos sings music and moves to sacrosanct rhythms has been around since the early days of philosophy. Music has been the metaphor used to describe what is happening on the larger scale.

452 Fourier analysis has proved to be a most useful tool in discerning how Nature works across a wide range of disciples from heat transfer to electronics and has been a master key that unlock the door of things hidden in Nature's closet. Fourier did not envision how has discovery would be use across such a wide range; he was only interested in solving problems associated with heat.

With the advent of string theory, the musical analogy has moved from the celestial sphere into the smaller sphere of Quantum Mechanics, and I often wonder if we could listen to the music of the denizens of the Quantum World, if the music would be as pleasing to our ears as the universe is to our eyes? Pythagoras thought the cosmos was like an orchestra playing its hymnic song throughout the universe. He saw the planets as celestial instruments playing in perfect harmony. He believed that the music was out there and only the gods could hear it. Although such music, like ambrosia may be fitting for the gods to feast; it is left to us mortals to use Pythagoras' muse as food for thought to help us understand the "music of the string."

TOWARD A HIGHER CONSCIOUSNESS

The key to growth is the introduction of higher dimensions of consciousness into our awareness.

Lao Tzu

It is generally accepted that within our physical reality, energy, time and space are consistent with Einstein's *Theory of Relativity*. However, many inside and outside the intelligentsia of physics believe our reality is not the only or the ultimate reality. The belief that places us as the cynosure of God's eye is primarily rooted in a geocentric paradigm that evolved from our limited knowledge of the cosmos. It was further enhanced by religious affirmations and dogma that had little to do with scientific facts. In times past, present and future, God's eye could wander to different celestial bodies, dimensions and universes. For all we know, billions of years ago, it could have been focused on Mars as the planet of life in our solar system. Then, our planet was a literal hell on Earth, void of water and life and not the blue celestial giant that astronauts gaze upon today. Only in the eye of such a God would our current planet have existed.

I have never been one to ignore the possible role of the supernatural. Perhaps, within science, there is indeed room for the natural and the supernatural. What we observe at the subatomic and cosmic levels may be classified as supernatural since it seems to violate what we have established as natural laws. Phenomena such as event horizons and black holes, photons being in two places at the same time could indeed be considered supernatural.[453] Einstein's view of relativity may only apply to spacetime and the three spatial dimensions we call reality with the speed of light being the maximum speed possible within our dimension. I went further to suggest that superluminal speeds may be achieved outside our dimension and what we see within our dimension is not the phenomena itself but rather the effects of external phenomena or secondary causation. In that case, we are, in some respects, the shadows of a higher dimension. Central to this paradigm is the belief that higher

453 Richard Feynman's sum over histories yields some insight to this phenomenon.

dimensions themselves are void of mass and are not restricted to the limits ascribed by the Lorentz-Fitzgerald equations, i.e., as the velocity of a mass approaches the speed of light its rest mass approaches infinity. For all we know, a photon could be a luminous arrow launched from a bow of conscious energy that pierces through the barriers of many dimensions. "So let there be light" may be a scientific as well as a spiritual statement regarding our existence within the realm of spacetime.

We often use infinity in vague terms, when in essence it is nothing more than the set of possibilities that lie outside of the finite world in which we are all residents. In this regard, it is much like that previous distinction between the natural and the supernatural one lying outside the realm of the other. However, in a higher dimension, such as the fourth spatial dimension, where spacetime has been reduced to a singularity and time as we know it does not exist, the realm of possibilities is boundless and infinite. It is very difficult for someone that has finitude as a constant frame of reference to imagine the realm of the infinite. Although we can readily identify with infinity as a mathematical construct, the practicality of infinity and all that it entails is unfathomable.

In such a world, there is no beginning and there is no end. No time or entropy to battle against, as it is in our world. From the viewpoint of such a dimension any finite dimension, or for that matter an infinite number of finite dimensions, will appear all at once because there is no past or future relative to it. Thus, our future and past would be viewed not as a single frame of the present moving through time but as a collection of finite past, present and future frames from one end to the other. Unlike our world where the past only exists as a reference for the present and the future, in such a realm, our past and future are always in their continual present.

Our world and our dimension are all about energy conversion on one scale or another. For example, on the macro level our Sun and all the stars are constantly converting mass into energy. We in turn absorb this energy in one form or another and use it to survive. In this regard, mass is nothing more than a storage element for energy. We know that the cosmos, like time, is unidirectional expanding and moving toward a higher state of entropy or disorder.[454] And it appears that the conversion of mass into energy is a natural way of deferring the inevitable demise because everything in the cosmos eventually reaches its highest state of entropy. One needs only to

454 Mathematics allows time and entropy to move in the other direction as well.

examine the hierarchal system of nature to verify this phenomenon. Thus, a very simplistic definition of life may be its ability to transform mass into energy and energy into mass. Maybe this lies at the heart of *Consciousness;* thus beings, planets, and stars are all dancing to the rhythm of this conscious cosmic beat. Although its tune is quite familiar, *Consciousness* sings its lyrics in a sacrosanct language that we do not quite understand. Such *Consciousness*, in its pristine form, may best be exemplified by the unification of all forces before and immediately after the Big Bang when energy reigned supreme and mass, as we know it, did not exist.

As time and space progressed, energy was transformed into mass and our cosmos was born. Time, mass and space are all relative constructs and are thus derivatives of a higher and pervasive *Consciousness* of energy. Without such innate *Consciousness*, how would the forces know how and when to divide and how would the cosmos itself come into being? The simple answer is that something in the form of energy has always been and will always be here; neither created nor destroyed. Our concept of time has become the lens though which we view all things, and that lens diffuses our ability to comprehend any existence without it.

Succinctly put, there is no *creatio ex nihilo* (creation out of nothing) because time did not exist in the realm from which our cosmos emerged. Mass itself is a secondary causation of conscious energy, however, we have chosen to make it and all of its derivatives crucial to the criteria of existence. Perhaps Descartes put it best when he placed *Consciousness* primary, and in some ways ultimate, in his aphorism: "I think therefore I am". Hence, *Consciousness* is the only absolute reality, and all other realities are relative. Conceivably, it even permeates death, and the Earth is alive within the realm of *Consciousness.* One may even argue that *Consciousness* is the sole criterion for the existence of anything, e.g., the notion of Plotinus' *Divine Mind.*

Many believe that the universe in which we live is finite but unbounded and that the stars could indeed be counted. For example, if an ant started walking the Earth in any direction it will never come to an end, i.e., a point where it could no longer walk. In an ant's mind, it would probably think that the Earth itself is infinite. However, we all know by now that the Earth is indeed finite. Infinity sometimes wears a disguise but always lies outside of what could be counted. Thus, infinity could be reserved for a realm far beyond our own and beyond spacetime itself. Eternal terms like forever and everlasting lie outside our dimension because they imply for all time. Unfortunately, entropy will simply not allow us to have an eternal universe.

We live in a universe of complimentary and complementary forces that create temporal states of equilibrium that allow us to exist in an entropic milieu. Entropy is the shadow of death from which even our sun reaches its final hour. The life of a star is spent battling the forces of gravity by producing enough energy to establish equilibrium with gravity. We earthlings are the beneficiaries of this epic battle as we receive energy in the form of light that stimulates life on this planet. Suns that are much larger than ours spread their forged and diverse atoms across the heavenly sky. Thus, death is not only intertwined with life but a necessary condition of life. In this regard, we are nothing more than the remnants of such conflicts, the remains of suns in our distant past that sacrificed their lives so that we, in turn, may live.

It has often been said that "the absence of evidence is not evidence of absence." What is not seen could be much more powerful than what is seen. This could also be applied to *Consciousness.* As we gazed upon the universe in the darkness of the night a silent voice of greatness speaks to our inner soul. The nocturnal bliss is indescribable in words. The rising and the setting of our Sun work in concert with *Consciousness*; because only in the absence of its brilliance are the beauty and wonders of the universe revealed. We know our universe is full of stars that we gaze upon, but we do not understand their origins or their purpose. Some form of *Consciousness* chose a slice in time to create what we know as life on this planet. Therefore, we are only a frame in the movie of our universe. Moreover, the Big Bang itself may only be a show in the theater of spacetime.

Consequently, we cannot unequivocally say that we are the main event, for all we know; we may only be a sequel in a string of events such as the Big Bang. If people are born and eventually die, and if stars are born and eventually die; then surely a universe can be born and die, many times, within the confines of spacetime. The only true constant is the pervasive *Consciousness* from which they emerged. Hence, in our world, we are always conscious of something, and it is impossible to think of nothing, there is always something that is on your mind. There is no bigger paradox than to say that nothing is on your mind. The concept of Nothingness is incomprehensible to us and therefore, like infinity, lies outside of our realm of *Consciousness.* Therefore, a coeval world consisting of Nothingness and Infinity, both of which are supernatural to us, could exist in a higher dimension.[455] The closest that we ever get to it, this side of death is within the confines of our dreams or through intense meditation. Our dreams are not only based on our imagination, but the myriad events and experiences stored in our subconscious as they rise to the level of *Consciousness.*

It is this growth in *Consciousness* that has enabled sapiens to reign supreme among the beasts of this Earth and to take the first steps toward the discovery of their being.[456] It is from such *Consciousness* that our species first gazed at the stars and asked the questions: Who am I and where did I come from? Henceforth, humans were released from the realm solely dominated by survival and procreation into a greater quest for knowledge and wisdom. At that moment, we became a complex creature walking alone on a blue planet in search of an inner *Consciousness* that never dies and a sentient weeper gazing up at the majestic heaven and wanting a God to hear his cries. Thus, man's final evolution is not physical but metaphysical toward the path of a higher state of *Consciousness.*

As beautiful as it may seem, spacetime is not a utopia but a reformatory and a repository from which we all would like to escape. Everything within spacetime is temporal, thus, it lives, and it eventually dies. Everything is dependent on space and time as a reference, even light itself is somewhat restricted by both, and its constant speed is relative to distance and time. However, light has a secret message that tells us how far it has traveled, the composition of the star that it emanated from, how big the star is, and so forth. We have reached a point in our *Consciousness* where we can listen to the coded message contained within light and get a glimpse of its inherent *Consciousness.* We are now able to discern different energy levels across the Electromagnetic spectrum contained within it. We have determined that light has a duality sometimes acting like a particle and sometimes like a wave.

---

455 The brilliant Indian mathematician Ramanujan built a theory of reality around Zero and Infinity. Zero to him was Absolute Reality and Infinity was the myriad manifestation of that reality. The Man who knew Infinity: A Life of the Genius Ramanujan p. 66.

456 Scientists have sequenced the genome of the chimpanzee and found that humans are 96 percent similar to the great ape species. The number of genetic differences between humans and chimps is ten times smaller than that between mice and rats.
Yet one cannot deny that there is a much greater difference in the intellect and consciousness of humans. http://news.nationalgeographic.com/news/2005/08/0831_050831_chimp_genes.html

We also know that stars tend to cluster in groups of billions to form galaxies and that our planet is small in the grand scheme of things.[457] Perhaps the greatest revelation lies in the fact that there is nothing special or unique about our particular sun. All of this we have achieved since those nascent days when the first Homo sapiens gazed upon the sky, a long time in the history of humankind but a very short time in the history of the cosmos. So massive is our universe that today we may receive a message from a star that died before our Sun was born over four billion years ago. In fact, we never see our Sun as it truly is but how it was eight minutes ago. Within spacetime, reality is delayed by minutes or by billions of years. We may even ponder if beings once lived on planets around those deceased stars and what level of *Consciousness,* they achieved during their sojourn in spacetime.

I have often wondered if *Consciousness* shares the same duality as light. That in the long term it acts like a wave, however in the short term, we receive packets of *Consciousness*. Take for example the events over history that seems to develop independently of each other yet coalesce at the proper time. [458] A perfect example would be the train, the telegraph and the expansion across the continental United States. Samuel Morse did not invent the telegraph specifically to communicate with trains. He was just following his intuition and if you examine his life closely you will see he was almost born to do it. Yet, his telegraph lines followed the tracks as they went west. The War with Mexico provided the land, the train the means to travel efficiently and the telegraph the way to communicate over long distances. Likewise, when the airplane replaced the train, the radio replaced the telegraph thereby migrating from the land to the sky. Both were developed by independent and different trains of thought. Similar examples could be cited throughout history.

Another example that is more germane to this discussion centers around the early years of the twentieth century. During that time, ideas of time and space were not only explored by Einstein but by prominent artists and even metaphysics.[459] When Einstein stumbled, there were mathematicians, physicists, and astronomers to pick him up[460]. More striking, are the antecedent ideas, discoveries, and formulae that

---

457 There are at least 100 billion stars in our galaxy alone and there are billions of such galaxies.

458 This is analogous to a stream of photons that trace out an interference pattern.

459 Edwin Abbott's Flatland: A romance of many dimensions is a prime example. Another example is the Early 20th Century Modernism of Classical Music with the works of Claude Debussy, Igor Stravinsky, and others.

served as kindle for the fire lit by Einstein. [461] All these events I attribute to a more pervasive *Consciousness* outside and independent of the realm of spacetime. We know that gravity warps spacetime and bends light, but we do not know what bends and warps events, nor do we understand why certain events and discoveries tend to gravitate toward each other.

We do know that just before and right after the turn of the 20th century, the world was full of inventors, scientists, and mystics. The Gilded Age gave birth to a new scientific and creative era. It was a world of mystics such as Madame Blavatsky (1837-1891) and Mary Baker Eddy (1821-1910). It was the world of engineers such as the Wright Brothers (1867-1948), Thomas Edison (1847-1931) and Henry Ford (1863-1947). It was a world of mystic scientists such as Albert Einstein (1879- 1955) and Srinivasa Ramanujan (1887-1920) men who in their own mystical way gave birth to a new insight into science and engineering. It was a world in which the buds of ragtime bloomed into beautiful flowers of Jazz and Cubism with its distortion of time and space emerged. It was as if a new circle of *Consciousness* engulfed the world and circumscribed a new century.

It has often been asked if *Consciousness* is in the mind, of the mind, or beyond the mind. My answer is all three. There can be no doubt that the mind is a major receptor of *Consciousness.* However, it may not be the only one. The ancient Egyptians believed that a person thought with his heart and at the end of his life, the heart was weighed against a feather as a symbol of his judgment. The metaphysical prowess of the ancient Egyptians has been renowned throughout the ages and perhaps there is more to it than what meets the eye.[462] They were the first to chart the heavens and to calculate how long it took our planet to revolve around its closest star, and they were first to conceive the concept of a soul. Perhaps the pyramids stand as silent monuments to the state of *Consciousness* achieved in ancient times. Maybe we do

460 I am alluding primarily to the work by Neils Bohr in Physics and Edwin Hubble in Astronomy. Bohn opened up our minds to the micro giving birth to Quantum Mechanics and Humble to the macro by discovering galaxies. Einstein believed in a static universe which was proven wrong by Hubble's discovery of an expanding universe. Furthermore, Einstein never embraced Quantum Mechanics although it is well accepted by physicist today.

461 Chief among them would be Maxwell Equations, The Michelson-Morley experiment and the mathematics of Lorenz and Fitzgerald.

462 Egyptian art generally depicts the left foot forward. This is because it is the side that the heart is on and the foot is clearing the way for the heart, thus the person is walking with a clear heart or in righteousness.

receive conscious messages through our hearts and the brain is the local interpreter of such thoughts. For example, we seldom say that a person has a loving brain; instead, we say a loving heart. For all we know, the heart could be a repository of conscious energy stored during our lifetime and that is why it, and not the mind, was weighed.

The heart, in many ways, is like our Sun keeping us alive by dispersing energy throughout our body. During our lifetime, it sustains our body and after we die maybe, through *Consciousness*, it tells the tales of our lives recorded in conscious energy that transcends the physical dimension. Our souls maybe the manifestation of such *Consciousness* and like energy cannot be created or destroyed. Therefore, the meaning of life may be about generating conscious energy to be used to determine your fate after physical death. Heaven and Hell are constructs of the mind that have been around for ages.

Maybe, Heaven and Hell are directions (metaphysical vectors) not final destinations. Perhaps when we die, we cross over into a dimension of pure energy. We are not sent to Heaven or Hell but reinserted back into spacetime according to how much conscious energy we accrued previously in spacetime. Therefore, anyone could be inserted anywhere within spacetime and experience myriad realities perhaps even the one previously experienced. This may account for what we call "genius". People, such as Einstein may have returned with conscious energy that far exceeds those that are experiencing the same reality.

Moreover, if a person dies young or unjustly or serving the better good, their conscious energy may be increased to compensate accordingly. In harmony with this paradigm, the wicked will not prosper but the righteous will. Because each good deed or act causes your inner *Consciousness* to vibrate more, and each evil act causes it to vibrate less. In accordance with physics, the higher the vibrations the higher the energy generated. The heart, being the central point of all vibrations, would be a perfect representation of such energy. Nature teaches us that the cycle is always completed, and that energy is always conserved.[463] Therefore, death is more of a transformation from dust back to dust and energy back to energy.

We live in a world where everything is preceded by something else and perhaps our current reality is based on our previous one. In some ways, we are like electrons at the

463 The cosmos has as much energy in it today as it did during the time of the Big Bang.

subatomic level and conscious energy is like a photon that can increase the energy of an electron and thus release it. In such an analogy, spacetime would be represented as the metal surface.[464] Therefore, I believe that religions tend to teach us to be righteous and why I referred previously to spacetime as a reformatory and a repository and not a utopia. We all enter spacetime at one frequency, and we leave at another. Without evil and sin in this world, the mechanism for the transformation of energy would not exist. Perhaps, the truly righteous and sapient among us never return; they have generated enough conscious energy to sustain them in a true afterlife. Therefore, they receive everlasting life in a dimension of Nothingness and Infinity outside of spacetime in a place we call Paradise. Moreover, there may be an evolution beyond that dimension but much too hard to conceive.

Consequently, I believe that the meaning of life is really about the righteous road less traveled along a thin line where physics and metaphysics meet. I also believe, with all my heart, that beyond the stars that we so much adore; lies the path to our utopia evermore!!!

464 This is known as the photoelectric effect of which Albert Einstein was awarded a Nobel Peace Prize for and not his for his Theory of Relativity. It is essential the following: Energy contained within light is absorbed by electrons within metal, giving the electrons sufficient energy to be released from the surface of the metal. We also know that is frequency not intensity that is the predominate factor.

I first became interested in the Dogons in the late 70s. Although their legends lie on the periphery of Astronomy, nonetheless, I found them intriguing. Like the Ancient Egyptians, the Sirius star system rests at the core of their religious beliefs. As part of their folklore, they believe that our planet was visited by extraterrestrials from Sirius B. More shocking is that they seemed to have knowledge of Sirius B's orbit and what it is made of as discussed in the following essay. As the frequency of UFO sightings continue to increase, perhaps we should look deeper into their mythology to see if any of its elements are true.

THE DOGONS AND SIRIUS B

In 1862, it was discovered that Sirius has a binary partner. This blue, dwarf star has been called Sirius B, nicknamed The Pup. One of the great mysteries of the Sirius Star System has to do with the Dogon People of Africa. It was reported in the 20th Century that they knew of the existence of Sirius B, although they would've had no way of knowing about it, since they possessed no distant viewing technology[465]

A CONNECTION BETWEEN WEST AFRICA AND ANCIENT EGYPT IS SELDOM EXPLORED. However, a deep investigation into the customs, practices and rituals will reveal that there is more to the connection than meets the eye. Recent evidence suggests that not only were the Egyptians, Ethiopians and sub-Saharan people linked phenotypically, and culturally but linguistically as well. Nonetheless, the deeper we investigate the past the clearer the connection becomes. For example, according to Professors Catherine Obianuju Acholonu and Erich Fred Legner.[466]

> The grandson of Ra was called Osiris by the Greeks and Asar by the Egyptians. Osiris' was associated with the number 'seven'. No one knows the meaning of his name in Egypt, but in Igbo language Asaa means 'seven'! Osiris represents the cyclical aspect of nature—the physical creation and its cycles of becoming and returning. The universal cyclical number par excellence is seven. The son of Osiris was called Horus. This is a Greek version of a native Egyptian word Heru, which means 'Face', as in 'Face of the Sun'. Its Igbo original is Iru – 'Face'. Horus was known as the Lord of the Horizon. The Horizon being known to the Egyptians as the land of the Rising Sun, a place located in the Southwestern direction from Egypt - the original mythological home of the gods of Egypt. [467]

---

465 https://www.gaia.com/article/star-sirius-and-egypt

466 Professor Catherine Acholonu was a writer, researcher and former lecturer on African Cultural and Gender Studies. She is the author of over 15 books, most of which are used in secondary schools and universities in Nigeria, and in African Studies Departments in USA and Europe. Her works and projects have enjoyed the collaboration and the support of United States Information Service (USIS), the British Council, the Rockefeller Foundation.

Professors Catherine Obianuju Acholonu and Erich Fred Legner contend that language and the precursors of civilization can be traced to West Africa. The Dogons of Mali and the Ancient Egyptians shared a common fascination with the star Sirius located in the Orion constellation. Likewise, they share a common mythology.

The Dogon Sirius B Controversy

The controversy began with a book entitled *Sirius Mysteries* by Robert Temple.[468] Although the nexus between Ancient Egypt and the rest of the African continent has been highlighted by Dr. Diop and Afrocentric historians, except for Budge and a handful of traditional scholars it is virtually ignored by mainstream scholars. Consequently, other civilizations of Africa may have retained ancient knowledge that is unknown to modern science. For example, the cosmic knowledge of the Dogons (Doh-gun) of Mali (Western Africa) about the Sirius star system astonishes the scientific world. Astronomical discoveries confirmed the presence and orbit of Sirius B, a faint star in the Sirius star system, unavailable to the naked eye.

This star was not known to the ancients; in fact, the concept of a binary star system was equally unknown. In 1718, Edmond Halley, (discover of the comet that bears his name), discovered that stars have "proper motion" relative to one another while studying Sirius and other stars. As a result of Halley's work, Alvan Graham Clark discovered Sirius had a faint companion, now known as Sirius B in 1862. [469] The

---

467 https://egypt-tehuti.org/egyptian-cosmic-religion/osiris/
Seven of something frequently makes a complete set—the 7 days of the week, 7 colors of the spectrum, 7 notes of the musical scale, etc. The cells of the human body are totally renewed every 7 years.
http://faculty.ucr.edu/~legneref/igbo/westafricanorigin.htm
The Igbo language is a part of the Niger-Congo language family. It is divided into numerous regional dialects and somewhat mutually intelligible with the larger "Igboid" cluster

468 Born in 1945, Robert K. G. Temple holds an undergraduate degree in Oriental Studies and Sanskrit from the University of Pennsylvania. At the age of 22, he began writing his most famous work, The Sirius Mystery, published in 1976, which claimed that amphibious extraterrestrials from the Sirius star system were responsible for jump starting Sumerian culture.

469 In January 1862, Alvan G. Clark, discovered a speck of light around Sirius A which turned out to be Sirius B Sirius B has a magnitude of 8.6, which is bright enough to see with binoculars if it wasn't completely drowned by the glare of Sirius A. Sirius B has a diameter of three times the earth, yet its mass is 98 percent of our sun. Its luminosity is 1/400th that of our sun. On the other hand, its binary companion, Sirius A has a diameter of twice our sun and is its luminosity is 26 times our sun. It is the brightest star in our sky and resides 8.6 light-years, 50 million miles, away. Together, Sirius A and B function as a binary star system. Astronomically speaking, Sirius is known as Alpha Canis Majoris. In the world of astronomy, the lower the apparent magnitude of a celestial body, the brighter the object. Sirius has an apparent magnitude of -1.46. By comparison, the second brightest star in the sky is Canopus,

discovery of Sirius B opened the eyes of astronomers to new cosmic phenomena such as binary star systems, white dwarfs, neutron stars, and of course; black holes. During the century following Clark's discovery, advancements in technology such as high-powered optical and X-ray telescopes and mathematics helped to confirm the phenomena associated with Sirius B and other stars. Such high technology instruments are specifically designed to detect and measure cosmic phenomena difficult to see with conventional telescopes and impossible to see with the naked eye.[470] Even with such tools, modern-day scientists had to rely on Einsteinium physics to explain this phenomenon.

The Dogons of Mali without such tools somehow had advanced information about the Sirius system. Hundreds of years ago, they were aware of the phenomena of Sirius B hitherto unknown to Western scientists. In fact, they plotted its orbit, up to the year 2,000 CE. According to the Dogons, the Sirius system has three stars, thus it contains a Sirius C. This legend might be of little interest to anybody but the two French anthropologists, Marcel Griaule and Germain Dieterlen, who recorded it from four Dogon priests in the 1930s. The mystery lies in how Africans who lacked any kind of astronomical devices know so much about an invisible star? Sirius B was not photographed until it was done by a large telescope in 1970. According to their oral traditions, a race of people from the Sirius system called the Nommos visited Earth thousands of years ago. The Nommos were ugly, amphibious beings that resembled mermen and mermaids.

---

appearing in the constellation Centaurus. Canopus' apparent magnitude is -.72, roughly half as bright as Sirius. Sirius is 40 times brighter than our sun and is only 8.6 light years away, making it a close neighbor. https://www.gaia.com/article/star-sirius-and-egypt

470 A search for black holes began in earnest in 1970 with the launch of Explorer 42 from the coast of Kenya on the equator. This satellite consisted of two X-ray telescopes which gave us our first comprehensive look at the X-ray sky. Out of gratitude to the Kenyan people, Explorer 42 was christened Uhuru, which means "freedom" in Swahili. [Universe, p. 140]

Sirius A and Sirius B taken by Hubble Space Telescope 2016

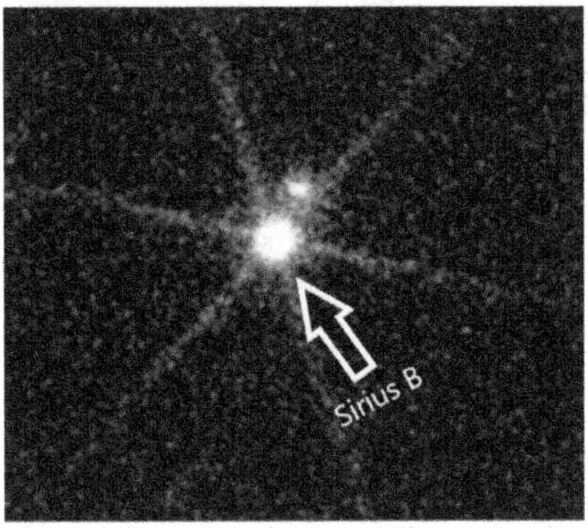

Sirius A and Sirius B taken by A Chandra X-ray Observatory image the bright star
Images of Sirius A and B Sirius A is approximately
10 magnitudes brighter than Sirius B. Only at the shortest far-UV wavelengths
or in the EUV/soft X-ray band does Sirius B become brighter than Sirius A as shown above[471]

Furthermore, the Dogons knew about the rings of Saturn, the spiral structure of the Milky Way Galaxy, and where our solar system lies.[472] They also knew the most important aspect of Sirius B, or any white dwarf, is its heavy mass relative to its visible size. White dwarfs are small, dense stars that burn dimly. Sirius B is, in fact, smaller than the planet Earth. Professor Kenneth Brecher of the Massachusetts Institute of Technology (MIT) stated in his article entitled *Sirius Enigmas*: "Many myths conflict with scientific theory, a few do not. We are dealing here with the myth that is most

471 https://chandra.harvard.edu/photo/2000/0065/more.html
472 Jackson: Ages of Gold and Silver, p. 203

nearly correct by our standards of truth." How the Dogons acquired their knowledge remains a mystery. The late Carl Sagan believed that a group of Jesuit priests gave them the theory. However, there are many flaws in Sagan's hypothesis.

Chief among them is the fact that the Dogons possess knowledge about the cosmos that was unknown at the time of this alleged visit by the Jesuit priests. [473] Sirius A, the Dog Star, is the brightest star in the sky. It has a mass of 2.3 times our sun but is 23 times as luminous. Sirius is not that far from our sun in relative terms, in fact, it is our eighth closes neighbor. It is not an average star, for every Sirius in the Milky Way there are fifteen average stars like our sun based on mass and luminosity. This somewhat unique aspect of Sirius coupled with its proximity and binary properties makes it attractive to modern scientists and some believe that it may be the Rosetta Stone required to understand the Cosmos.[474]

When we look at Sirius B, we are in effect looking at how our sun will die in about 5 billion years from now. At 4.5 billion years, the Sun is about halfway through its life. It will eventually, like Sirius B, become a white dwarf exhibiting all the characteristics that we observe in Sirius B from the perspective of a star, except we are not part of a binary star system. However, for all intents and purposes, it will go down the same death path as Sirius B and at some point, it will become so large that it will consume the radius of the Earth from the Sun before shrinking to less than the size of the Earth. Sirius B although comparable to the Sun in mass is now the size of the Earth. To achieve this, Sirius B must have a density 90,000 times that of the Sun. The force of gravity on Sirius B is 350,000 times stronger than on Earth. A teaspoon of Sirius B material would weigh about 2 tons. The original size of Sirius B is estimated at 5 solar masses. Sirius A is expected to have completely exhausted the store of hydrogen at its core within a billion years. At that point, it will pass through its own red giant stage, then settle down to become a white dwarf. The fact that Sirius B is no longer a main sequence star indicates that it was by far the more massive of the binary stars.

---

473 According to Count Voley, the name Dog Star was given to Sirius by the Ethiopians inventors of the zodiac [Ruins of Empires, pp. 120-123]. John G. Jackson tells us that the Egyptian word for Sirius (Sothis) means barker or monitor [Man, God, and Religion, p. 112]. In ancient Egypt, it was associated with Isis-Sothis (giver of life) and Osiris (Orion) symbols of yearly renewal, of the regeneration of the Nile of cyclic death and rebirth [Egyptian Mysteries, p. 29]. On a Ptolemaic text Isis addresses Osiris: Thy sacred image, Orion in heaven, rises and sets every day; I am Sothis following him, and I will not forsake him. [The Myth of the Goddess, p. 239]

When Temple first published his book in the 1970s, there was no solid evidence of a Sirius C. In 1995, however, two French researchers, Daniel Benest and J.L. Duvent, authored an article in the prestigious journal Astronomy and Astrophysics with the title Is Sirius a Triple Star? and suggested (based on observations of motions in the Sirius system) there is a small third star there. They thought the star was probably of a type known as a "red dwarf" and only had about .05 the mass of Sirius B.[475] If this proves to be true and confirmed by NASA, it will add validity to the Dogon myth and would be viewed by many as the smoking gun. However, when we are dealing with a myth, there could be kernels of truth, but we must remain cautious because some elements may be distorted by time, circumstance, or misinterpretation.

For example, the Greek myth of the Trojan War helped rediscover the Mycenean and Minoan civilizations. However, there were flaws in their myths. Although the Sirius B myth should be taken with a bit of scientific salt, we should not discard it because deep within their metaphors and oral traditions may lay kernels of truth yet to be discovered or understood. It appears that NASA has taken the claims of the Dogons as well as the fascination of Ancient Egypt with the Orion constellation as points of interest as its peers deep into the abyss of the Sirius star system. Moreover, in recent times, several UFOs have been reported diving into our oceans which make up about 71 percent of the Earth's surface and hold about 96.5 percent of all Earth's water. The Dogons may, after all, be completely wrong, but just think of the implications if they are right?

[475] It is also important to remember that although many parts of the Dogon legends seem to ring true, other portions are clearly mistaken. One of the Dogon's beliefs is that Sirius B occupied the place where our Sun is now. Physics clearly prohibits this. Also, if the Dogon believe that Sirius B orbits Sirius A every 50 years, why do they hold their celebrations every 60 years?

APPENDIX

Glossaries

GLOSSARY OF MUSICAL TERMS

Glossary of Music

1. Articulation - Is the way a note begins, sustains and end
2. Blues Scale - Is a seven note scale counting the octave. In relationship to the C Major Scale the Blues Scale is C- Eb- F- Gb- G- Bb- C
3. **Cadence**- Is the harmonic, melodic and rhythmic conclusion to a phrase
4. **Chord Progression** - The playing of a group of chords one after the other
5. **Chord** - Refers to 3 or more notes which are played together (simultaneously).
6. **Chorus** - Choir (group of singers); section of a song/composition which is repeated after every verse.
7. **Chromatic Scale** - A scale that contains all 12 notes of an octave with all pitches space a half step apart.
8. **Cycle of fifths (fourths)** - A chart that organizes key signatures and tells us about the natural laws of music progression and structure, it depicts the relationship of tonal centers to each other.
9. **Diatonic** - Music is created by dividing the octave into 7 notes – with intervals of 5 whole and 2 half notes. Of a scale, interval, etc.) involving only notes proper to the prevailing key without chromatic alteration.
10. Diatonic Scale - A scale with eight notes in an octave; all but two are separated by whole tones
11. **Double Flat** - Lowers the pitch by two half steps while retaining the same letter name
12. **Duration** - The length of time that sound or silence lasts
13. **Enharmonic** - Intervals sound the same but are spelled differently
14. **Envelop** - Of sound is composed of a sound's attack, sustain and release
15. Flat - Denoted by the symbol 'b' shows a decrease in a note by one semitone
16. **Grand Staff** - Is s system of two staves- the top staff uses the treble clef, and the bottom uses the bass clef
17. **Harmonic** - A component of a periodic quantity, such as a musical tone, with a frequency that is an integral multiple of the fundamental frequency. The first harmonic is the fundamental, the second harmonic (twice the fundamental frequency) is the first overtone, the third harmonic (three times the fundamental frequency) is the second overtone, etc..

18. **Harmony** - Refers to a mix or combination of two or three tones which are played together in the background while a melody is being played
19. **Interval** - Is the distance between two notes
20. **Key signatures** - Is written at the beginning of every staff which shows which pitches are to be sharp or flat and determines the key or tonal center
21. **Major Scale** - Refers to a form of scale with specific intervals. Music based on the Major keys have a positive and refreshing sound. In the major or diatonic scale, the pattern or steps followed are – whole, whole, half, whole, whole, whole, half – is e.g., C-D-E-F-G-A-B-C. All major scales share the same pattern.
22. **Middle C** - There are eight Cs on the piano. Middle C is designated C^4
23. **Minor Scale** - Refers to a form of the tonal system and music based on the minor keys, generally having a melancholic and sad sound. In this diatonic scale the pattern or steps followed is – whole-half-whole-whole-half-whole-whole – C-D-Eb-F-G-Ab-Bb-C. All minor scales share the same pattern.
24. **Octave** - The interval between two notes with half or double the frequency. i.ie from one C to another.
25. **Pentatonic Scale** - Such a scale is made of 5 notes. The contain no half steps and take five consecutive pitches from the Circle of Fifths and then rearrange them in order to fit within an octave
26. **Pitch** - Determines how high or low a sound is and is dependent upon the wavelength or frequency of the note
27. **Relative Minor and Major** - The Major or Minor scales having the same key signature.
28. **Rhythm** - Refers to the arrangement or organization of notes based on time or their duration.
29. **Scale** - An ascending or descending lineup of notes of a key. An ordered collection of pitches in whole and half-step patterns.
30. **Seventh Chord** - Contains four notes—the root the third the fifth and the seventh. All seventh cords are dissonant and hence unstable.
31. **Sharp** - Denoted by the symbol '#' shows an increase in a note by one semitone

32. **Timbre** - Is the quality of sound and is dependent on the shape and form of the sound wave and is made up of more than on frequency often involving harmonics or overtone
33. **Time Signature** - Helps to determine the number of beats to a measure and is depicted by a numeric sign in sheet music.
34. **Tonality** - Refers to the tonic pitch or the relation between the notes and their tone.
35. **Tonic** - Is the 1st tone or keynote in a scale. It is also equal to the fundamental frequency
36. **Triad** - A chord made up of three tones, called chord factors, of the diatonic scale: root, third, and fifth. The system of diatonic triads is the basis of tonal

Musical Frequency Table

	C	C#	D	Eb	E	F	F#	G	G#	A	Bb	B
0	16.35	17.32	18.35	19.45	20.60	21.83	23.12	24.50	25.96	27.50	29.14	30.87
1	32.70	34.65	36.71	38.89	41.20	43.65	46.25	49.00	51.91	55.00	58.27	61.74
2	65.41	69.30	73.42	77.78	82.41	87.31	92.50	98.00	103.8	110.0	116.5	123.5
3	130.8	138.6	146.8	155.6	164.8	174.6	185.0	196.0	207.7	220.0	233.1	246.9
4	261.6	277.2	293.7	311.1	329.6	349.2	370.0	392.0	415.3	440.0	466.2	493.9
5	523.3	554.4	587.3	622.3	659.3	698.5	740.0	784.0	830.6	880.0	932.3	987.8
6	1047	1109	1175	1245	1319	1397	1480	1568	1661	1760	1865	1976
7	2093	2217	2349	2489	2637	2794	2960	3136	3322	3520	3729	3951
8	4186	4435	4699	4978	5274	5588	5920	6272	6645	7040	7459	7902

Note	Ratio to Fundamental (Equal Temperament)
1	1.0000
2	1.05946
3	1.12246
4	1.18921
5	1.25992
6	1.33483
7	1.41421
8	1.49831
9	1.58740
10	1.68179
11	1.78180
12	1.88775
13	2.0000

Fourier Series and Coefficients

$$g(t) = a_0 + \sum_{m=1}^{\infty} a_m \cos\left(\frac{2\pi\, mt}{T}\right) + \sum_{n=1}^{\infty} b_n \sin\left(\frac{2\pi\, nt}{T}\right)$$

$$= \sum_{m=0}^{\infty} a_m \cos\left(\frac{2\pi\, mt}{T}\right) + \sum_{n=1}^{\infty} b_n \sin\left(\frac{2\pi\, nt}{T}\right)$$

$$a_0 = \frac{1}{T}\int_0^T f(t)dt$$

$$a_m = \frac{2}{T}\int_0^T f(t)\cos\left(\frac{2\pi mt}{T}\right)dt$$

$$b_n = \frac{2}{T}\int_0^T f(t)\sin\left(\frac{2\pi nt}{T}\right)dt$$

GLOSSARY OF SCIENTIFIC, TECHNOLOGY ENGINEERING AND MATHEMATICAL (STEM) TERMS

A

1. **Aberration of starlight** - Difference between a star's observed position and its true direction. The displacement of the apparent path of light from a star, resulting in a displacement of the apparent position of the star from its true position; discovered by the English astronomer James Bradley and explained by him in 1729.
2. **Absolute value** - A value that tells you how far you are away from zero. The magnitude of a real number without regard to its sign
3. **Absolute zero** - The lowest theoretical temperature (0K = -273.16°C) where all molecular activity ceases measured in degrees Kelvin. The point where no more heat can be removed from a system, according to the absolute or thermodynamic temperature scale. This corresponds to zero Kelvin, or minus 273.15 C. This is zero on the Rankine scale and minus 459.67 F.
4. **Absorption** - When light is used to push an electron from a ground state to an excited state
5. **Absorption line** - When heated every element gives off light. When this light is decomposed using a prism it is found to be made up of a series of ``lines", that is, the output from the prism is not a smooth spectrum of colors, but only a few of them show up. This set of colors is unique to each element and provides a unique fingerprint: if you know the color lines which make up a beam of light (and you find this out using a prism), you can determine which elements were heated up in order to produce this light
6. **Absorption spectrum** - A spectrum obtained when white light is passed through a substance that selectively absorbs light of certain wavelengths.
7. **Acceleration** - Rate of change of velocity (Delta-V), divided by the change in time, (Delta-t). So, $a = \Delta v/\Delta t$. Acceleration is also related to force. $a = F/m$, where F is the net force applied to a mass, m
8. **Acid** - A compound that yields hydrogen ions (H+) when in aqueous solution. Acids have a sour taste and turn blue litmus red
9. **Acid rain** - Any form of precipitation that contains high levels of nitric and sulfuric acids. It can also occur in the form of snow, fog, and tiny bits of dry material that settle to Earth
10. **Acoustics** - The branch of physics that deals with sound and sound waves. The science concerned with the production, control, transmission, reception, and effects of sound

11. **ActiveX Programs** - ActiveX programs, often written using Visual Basic, are used for active contents that are called and used by other programs and the operating system
12. **Ada** - A Pascal-based programming language, is a structured, statically typed programming language. It is positioned to address much the same tasks as C or C++, but with the type-safety of a language like Java. Ada was designed to enable computers to control automatic equipment but is now also used for other applications. Ada was named after Ada, Lady Lovelace
13. **Age of the Dinosaur** - 248- 65 million years ago(mya). Dinosaurs went extinct about 65 million years ago (at the end of the Cretaceous Period), after living on Earth for about 165 million years. If all of Earth time from the very beginning of the dinosaurs to today were compressed into 365 days (one calendar year), the dinosaurs appeared January 1 and became extinct the third week of September. (Using this same time scale, the Earth would have formed approximately 18.5 years earlier.) Using the same scale, people (Homo sapiens) have been on Earth only since December 31 (New Year's Eve).
14. **Alleles** - Different forms of a gene, some of which may be dominant and some recessive. Any of the alternative forms of a gene that may occur at a given locus
15. **Alloy** – A metal in which several elements are present. An alloy is a mixture of chemical elements of which at least one is a metal.
16. **Alpha Centauri** - Is the closest star system to the Sun, at a distance of 4.37 light-years, is the Alpha Centauri system. Of the three stars in the system, the dimmest, called Proxima Centauri, is actually the nearest star. The bright stars Alpha Centauri A and B form a close binary as they are separated by only 23 times the Earth- Sun distance – slightly greater than the distance between Uranus and the Sun.
17. **Alpha Decay** - Is one process that unstable atoms can use to become more stable. During alpha decay, an atom's nucleus sheds two protons and two neutrons in a packet that scientists call an alpha particle. Since an atom loses two protons during alpha decay, it changes from one element to another. For example, after undergoing alpha decay, an atom of uranium (with 92 protons) becomes an atom of thorium (with 90 protons)
18. **Alpha Test** - Is the first formal test of a newly developed hardware or software product by internal people. The key objectives of the alpha test are functionality confirmation and bug identification. When the first round of bugs has been fixed, the product goes into beta test with actual users and customers
19. **Alternating Current (AC)** - A voltage switching polarity or as a current switching direction back and forth

20. Alternating Series - Series whose consecutive terms alternate between positive and negative
21. Alternative histories – A formulation of quantum theory in which the probability of any observation is constructed from all the possible histories that could have led to that observation.
22. Amino Acid - Small molecule comprising Nitrogen containing units used to synthesize proteins. Amino acids are organic compounds containing the basic amino groups (-NH2) and carboxyl groups (-COOH). The ingredients present in proteins are amino acids. Both peptides and proteins are long chains of amino acids. Altogether, there are twenty amino acids, which are involved in the construction of proteins.
23. Amplitude - A measurement that indicates the movement or vibration of something (such as a sound wave or a radio wave)
24. Amplitude Modulation - One of the principal methods of transmitting audio, visual, or other types of information using radio waves, the relevant signal being superimposed onto a radio-frequency carrier wave
25. Analog to Digital Conversion (ADC) - An electronic process in which a continuously variable (analog) signal is changed, without altering its essential content, into a multi-level (digital) signal
26. Analytical chemistry - Spans nearly all areas of chemistry but involves the development of tools and methods to measure physical properties of substances and apply those techniques to the identification of their presence (qualitative analysis) and quantify the amount present (quantitative analysis) of species in a wide variety of settings. Computational chemistry uses computer simulation to assist in solving chemical problems
27. Anatomy - Is the science that studies the structure of the body. On this page, you'll find links to descriptions and pictures of the human body's parts and organ systems from head to toe
28. Andromeda Galaxy - Also known as M31, is the nearest major galaxy to our own Milky Way at 2.5 million light-years away. At 2.5 million light-years, it's the most distant thing most of us humans can see with the unaided eye. Although it does appear that this galaxy is approaching our Milky Way, it's nothing to lose sleep over. According to NASA, if they collide, it'll be four billion years from now and our sun will be on its deathbed.
29. Antenna – A specialized transducer that converts radiofrequency (RF) fields into alternating current (AC) or vice-versa.: A metallic device (such as a rod or wire) for radiating or receiving radio waves

30. Anthropic Principle - It essentially states that we will only find ourselves in a universe that's capable of giving rise to us. Put another way, observations of the universe must be compatible with the conscious life that observes it.
31. Anthropoid - Resembling humans or a higher primate, especially an ape or ape-man Resembling a human, especially in shape or outward appearance. Resembling or characteristic of a nonhuman ape; apelike:
32. Anthropology - The study of humankind in all its aspects, especially human culture or human development. Physical anthropology is the branch that concentrates on the biology and evolution of humanity
33. Anthropomorphic - Having the form of a person or a human being·
34. Antiparticle - Subatomic particle having the same mass as one of the particles of ordinary matter but opposite electric charge and magnetic moment
35. Aphelion - The point in the orbit of a planet or comet where it is farthest from the Sun. The opposite of perihelion. The dates of perihelion and aphelion change slightly every year, but they often fall in January and July respectively. The US Navy has noted that in 2022 perihelion occurred on 4 January and aphelion will take place on 4 July
36. Applet - Is a little application that can be a utility or another simple program. Applets are often written in Java language which are attached to HTML documents. An applet runs in the context of another program, for example a web browser
37. Application - In computer software refers to a complete and self-contained program that helps the user accomplish a specific task. Application programs should be distinguished from system programs such as operating systems, which control the computer and run those application programs
38. Application-Specific Integrated Circuit (ASIC) - Is a microchip designed for a special application, such as a particular kind of transmission protocol or a hand-held computer
39. Archeology - The scientific study of human remains and artifacts
40. Archimedes' Principle- States that the apparent upward force (buoyancy) of a body immersed in a fluid is equal to the weight of the displaced fluid
41. ARM processor - Is one of a family of CPUs based on the RISC (reduced instruction set computer) architecture developed by Advanced RISC Machines (ARM). ARM processors are extensively used in consumer electronic devices such as smartphones, tablets, multimedia players and other mobile devices, such as wearables
42. Arnthropod - A group of invertebrate animals such as the insects, crustaceans, arachnids, centipedes, etc. that are characterized by an exoskeleton and a segmented body with jointed appendages

43. **Artificial Intelligence (AI)** - is a branch of computer science on how to endow computers with capabilities of human intelligence. The scientific study of AI aims to develop computer hardware and software that emulates activities in the human brain, including the processing of photographic images and human languages. It also enables computers and machines to think and learn from experience and carry out human-like duties. AI can be seen as an attempt to model aspects of human thought on computers. It is also sometimes defined as trying to solve by computer any problem that a human can solve faster. Examples of AI include speech and pattern recognition.
44. **Assembly Language** - Is a type of computer CPU programming language, which consists entirely of numbers and are almost impossible for humans to read and write. Each type of CPU has its own machine language and assembly language, so an assembly language program written for one type of CPU won't run on another. In the early days of programming, all programs were written in assembly language. Now, most programs are written in a high-level language such as C and C++. Programmers still use assembly language when speed is essential or when they need to perform an operation that isn't possible in a high-level language
45. **Asteroid** - A loosely defined term generally meaning a small solar system body that is orbiting the Sun, does not show a comet-like appearance, and is not massive enough to be a dwarf planet. The term is usually restricted to bodies with orbits interior or similar to Jupiter's
46. **Asteroid belt** - Is a region of the solar system falling roughly between the planets Mars and Jupiter where the greatest concentration of asteroid orbits can be found. The asteroid belt may also be divided into the inner and outer belts, with the inner belt formed by asteroids orbiting nearer to Mars than the 3:1 Kirkwood gap (2.5 AU), and the outer belt formed by those asteroids closer to Jupiter's orbit
47. **Astrobiology** - Is the study of life in the universe. The search for life beyond the Earth requires an understanding of life, and the nature of the environments that support it, as well as planetary, planetary system and stellar processes. To provide this understanding, astrobiology combines the knowledge and techniques from many fields, including astronomy, biology, chemistry, geology, atmospheric science, oceanography and aeronautical engineering. Astrobiologists can work alone on particular scientific questions, but often astrobiologists from different scientific disciplines work together to examine complex questions that no one field can answer alone

48. **Astrodynamics** – The branch of Physics that studies the dynamics of natural and human-made bodies in outer space. Celestial mechanics applied to space vehicles
49. **Astronomical unit** - A unit of length, equal to the mean distance of the Earth from the sun: approximately 93 million miles (150 million km)
50. **Astronomy/Astrophysics** - Astronomy is the study of the sun, moon, stars, planets, comets, gas, galaxies, gas, dust and other non-Earthly bodies and phenomena Astrophysics is a field of intersection between astrophysics and biophysics concerned with the influence of the astrophysical phenomena upon life on planet Earth or some other planet in general. Astrophysics involves the study of the physics of astronomy and concentrates on the behavior, properties, and motion of objects out there
51. **Atom** - The smallest particle of an element which can take part in chemical reactions and display the chemical properties of that element. It also is the smallest unit of matter that has the characteristic properties of a chemical element. As such, the atom is the basic building block of chemistry. Composite particles of protons, neutrons and electrons; the smallest part of a substance that can take part in a chemical reaction
52. **Atomic mass** - The atomic mass of an element (more correctly the relative atomic mass, Ar) is the mass of one atom of that element, divided by 1/12th of the mass of one atom of the Carbon isotope 12C. In older textbooks, the relative atomic mass was called the atomic weight.
53. **Atomic mass unit** - The atomic mass unit is 1/12th of the mass of one atom of the Carbon isotope 12C. It is sometimes called the Dalton
54. **Atomic number** - The atomic number of an element is the number of protons in its nucleus
55. **Atomic symbol** - The letters representing each of the elements. The mass number is the sum of the number of neutrons and protons. Atomic number is the number of protons. The electrical charge represents the net gain (for anions) or loss (for cations) of electrons. X represents the alphabetical symbol.
56. **Atomic weight** - The Atomic weight is defined as the average atomic mass of isotopes of an element in a given sample. It is the weighted average atomic mass calculated using the relative abundance of isotopes of an element. The atomic weight is also known as the relative atomic mass. Unlike the atomic mass, the atomic weight does not have any unit.
57. **Atoms Attenuation** - A reduction in signal strength commonly occurring while transmitting analog or digital signals over long distances
58. **Audio Video Interleaved (AVI) file** - Is a sound and motion picture file that conforms to the Microsoft Windows Resource Interchange File Format (RIFF)

specification. AVI files (which end with an .avi extension) require a special player that may be included with your Web browser or may require downloading

59. **Aufbau Principle** - Is a method of explaining the arrangements of electrons within atoms of different chemical elements. According to this principle the first electron in an atom occupies the quantum state with lowest possible energy and then the second electron goes to the next quantum state having higher energy. The model of the many electron atom which Niels Bohr constructed in order to explain the structure of the periodic table was based on detailed spectroscopic and chemical evidence
60. **Australopithecines** - (Latin: "southern ape") group of extinct creatures closely related to, if not actually ancestors of, modern human beings and known from a series of fossils found at numerous sites in eastern, central, and southern Africa. The various species of Australopithecus lived during the Pliocene (5.3 to 2.6 million years ago) and Pleistocene (2.6 million to 11,700 years ago) epochs
61. **AutoCAD** - Is a popular computer-aided drafting (CAD) software package for 2D and 3D design and drafting. Initially a general-purpose 2D drafting program, AutoCAD has been extended into a family of products, used by land developers, architects, mechanical engineers, and other design professionals. The AutoCAD family of products, taken as a whole, is by far the most widely used CAD software in the world
62. **Automatic Gain Control** - A circuit that modulates an amplifier's gain, in response to the relative strength of the input signal, in order to maintain the output power
63. **Azimuth and elevation** - Are angles used to define the apparent position of an object in the sky, relative to a specific observation point. The observer is usually (but not necessarily) located on the Earth's surface

B

64. **Backplane** - Is like a personal computer's motherboard and is sometimes a term improperly describing the computer's motherboard. The backplane is a printed circuit board containing connections (slots) for expansion boards and allows for communication between all connected boards

65. **Bacteria** - A domain that contains prokaryotic cells that are not multicellular. Bacteria are unicellular organisms belonging to the prokaryotic group where the organisms lack a few organelles and a true nucleus"
66. **Bandwidth** - A measurement of the ability of an electronic communications device or system (such as a computer network) to send and receive information. Am measurement of the frequency range
67. **Baryon**- A subatomic particle belonging to a group that undergoes strong interactions, has a mass greater than or equal to that of the proton, and consists of three quarks. Baryons, the protons and neutrons, are the basic building blocks of ordinary matter. They interact via the strong and electromagnetic forces and constitute the material out of which all luminous material is composed.
68. **BASIC** - Is a simple computer language designed by John G. Kemeny and Thomas E. Kurtz at Dartmouth College in 1963. It first ran on an IBM 704 on 1964-05-01. It was designed for quick and easy programming by students and beginners. BASIC exists in many dialects, and is popular on microcomputers with sound and graphics support
69. **Batch Job** - Is a computer program or set of programs processed in batch mode. This means that a sequence of commands to be executed by the operating system is listed in a file (often called a batch file, command file, or shell script) and submitted for execution as a single unit. The opposite of a batch job is interactive processing, in which a user enters individual commands to be processed immediately
70. **Baud Rate** - Refers to the number of signal or symbol changes that occur per second. A symbol is one of several voltage, frequency, or phase changes. The baud rate is the rate at which information is transferred in a communication channel. Baud rate is commonly used when discussing electronics that use serial communication. In the serial port context, "9600 baud" means that the serial port is capable of transferring a maximum of 9600 bits per second.
71. **Bays Shuffle** - A shuffling algorithm used in a class of random number generators.
72. **Bees algorithm** - Is a method of problem solving that mimics the behavior of honeybees to find the optimum solution. Based on the behaviors bees employ to search and prioritize, the algorithm is a classic example of swarm intelligence, in which many individuals work together to solve problems or optimize scenarios
73. **Beer's Law** - States that the quantity of light absorbed by a substance dissolved in a nonadsorbing solvent is directly proportional to the

concentration of the substance and the path length of the light through the solution

74. **Bell Curve** - The bell shape created when a line is plotted using data points for an item that meets the criteria of normal distribution. The center of a bell curve contains the highest value points.
75. **Bernoulli Effect** - Also known as Venturi effect, means that decrease in the pressure of fluid increases the velocity of the fluid. The Bernoulli's principle and the equation of continuity form the basis of Bernoulli's effect. Bernoulli's principle states that as air moves around an object, it creates different pressures on that object. Faster air means less pressure. Slower air means more pressure. The key to flight is creating pressure upwards on the bird's wing to keep the bird in the air
76. **Beta Decay** - Any of three processes of radioactive disintegration by which some unstable atomic nuclei spontaneously dissipate excess energy and undergo a change of one unit of positive charge without any change in mass number. The three processes are electron emission, positron (positive electron) emission, and electron capture
77. **Beta Particle** - An electron or positron emitted from an atomic nucleus in a certain type of radioactive decay. a positively charged particle consisting of two protons and two neutrons, emitted in radioactive decay or nuclear fission
78. **Beta test** - Is the computer system test prior to commercial release. Beta testing is the last stage of testing, and normally can involve sending the product to beta test sites outside the company for real-world exposure or offering the product for a free trial download over the Internet. Beta testing is often preceded by a round of testing called alpha testing
79. **Big Bang** - A theory that deduces a cataclysmic birth of the universe (big bang) from the observed expansion of the universe, cosmic background radiation. In 1965 Arno A. Penzias and Robert W. Wilson of the Bell Telephone Laboratories discovered that the space of the universe is bathed with a diffuse radiation that is assumed to persist following the "big bang" expansion of the universe from an original fireball.
80. **Big Crunch** - The theory that the universe will stop expanding and start shrinking
81. **Binary coded decimal (BCD)** - Is a system of writing numerals that assigns a four-digit binary code to each digit 0 through 9 in a decimal (base-10) numeral. The four-bit BCD code for any particular single base-10 digit is its representation in binary notation
82. **Binary Number System** – A number system based on two numbers 0 and 1 from which all of the other numbers can be represented. It is the system used

by computers for computations. For example, the decimal number 5 binary equivalent is 101 and 8 is 1000

83. **Binary Star System** - A binary star is a star system consisting of two stars orbiting around their center of mass. For each member of a pair, the other is called its companion star. The closest star system, Alpha Centauri, is a binary star system, consisting of two closely orbiting Sun-sized stars orbited in turn by a red dwarf
84. **Biochemistry** - Is the branch of science that explores the chemical processes within and related to living organisms. It is a laboratory-based science that brings together biology and chemistry. By using chemical knowledge and techniques, biochemists can understand and solve biological problems. Biochemistry focuses on processes happening at a molecular level. It focuses on what's happening inside our cells, studying components like proteins, lipids and organelles. It also looks at how cells communicate with each other, for example during growth or fighting illness. Biochemists need to understand how the structure of a molecule relates to its function, allowing them to predict how molecules will interact
85. **Biological Anthropology** - The branch of anthropology that deals with human evolutionary biology, physical variation, and classification. Also called somatology. It is the branch of anthropology concerned with the origin, evolution, and diversity of people. Biological anthropologists work broadly on three major sets of problems: human and nonhuman primate evolution, human variation and its significance and the biological bases of human behavior. The course that human evolution has taken and the processes that have brought it about are of equal concern. In order to explain the diversity within and between human populations, biological anthropologists must study past populations of fossil hominins as well as the nonhuman primates. Much light has been thrown upon the relation to other primates
86. **Biomarker** - A biological molecule used as a marker for a substance or process of interest. Biological markers (biomarkers) are characteristics that can be objectively measured and used as an indicator of normal biological processes, disease processes, or pharmacologic responses to a therapy. Biomarkers are important for the development of drugs, biologics, and certain medical devices.
87. **Biomimetic** - Refers to human-made processes, substances, devices, or systems that imitate nature. The art and science of designing and building biomimetic apparatus is also known as biomimicry because they mimic biological systems. The field is of special interest to researchers in nanotechnology, robotics, artificial intelligence (AI), the medical industry, and the military

88. **Biophysics** - Is the application of physics to biological systems. This interdisciplinary field is quickly growing in the modern scientific community, searching for ways that the laws of physics can be applied to living systems at all levels, from individual cells and microbes to animals, plants, and even entire ecosystems. Biophysicists study life at every level, from atoms and molecules to cells, organisms, and environments. As innovations come out of physics and biology labs, biophysicists find new areas to explore where they can apply their expertise, create new tools, and learn new things
89. **BIOS** - Stands for Basic Input/Output System. It gives the computer a platform to run the software using a floppy disk or a hard disk. BIOS is responsible for booting a PC
90. **Biosphere** - The part of the Earth's crust, waters, and atmosphere that supports life. The ecosystem comprising the entire Earth and the living organisms that inhabit it
91. **Biosynthesis** - The production of cellular material. The production of complex molecules within living organisms or cells. Biosynthesis - Biosynthesis is a multi-step, enzyme-catalyzed process where substrates are converted into more complex products in living organisms.
92. **Bit Rate** - The speed of the data is expressed in bits per second (bits/s or bps). Bitrate can also describe the quality of an audio or video file. For example, an MP3 audio file that is compressed at 192 Kbps will have a greater dynamic range and may sound slightly clearer than the same audio file compressed at 128 Kbps
93. **Black Hole** - A region of spacetime from which gravity prevents anything, including light, from escaping. A black hole's "surface," called its event horizon, defines the boundary where the velocity needed to escape exceeds the speed of light, which is the speed limit of the cosmos. Matter and radiation fall in, but they can't get out. Black holes are one of the most interesting and terrifying objects in the universe. Scientifically speaking, rather than being an object, a black hole is a spherical region in space characterized by a tremendous gravitational field. Black holes can be of different types based on their mass. They can also be classified based on whether they are charged or exhibit rotation. The simplest black hole is a static, non-rotating, and electrically neutral black hole, called the Schwarzschild black hole
94. **Blackbody** - An ideal body or surface that completely absorbs all radiant energy falling upon it with no reflection and that radiates at all frequencies with a spectral energy distribution dependent on its absolute temperature
95. **Blackbody radiation** - Or "cavity radiation" refers to an object or system which absorbs all radiation incident upon it and re-radiates energy which is

characteristic of this radiating system only, not dependent upon the type of radiation which is incident upon it

96. **Bluetooth** - A technology that allows voice and data connections between a wide range of mobile and stationary devices through short-range digital two-way radio. For instance, it specifies how mobile phones, Wireless Information Devices (WIDs), computers and PDAs interconnect with each other, with computers, and with office or home phones
97. **Bode's Law** - Titius-Bode law, empirical rule giving the approximate distances of planets from the Sun. It was first announced in 1766 by the German astronomer Johann Daniel Titius but was popularized only from 1772 by his countryman Johann Elert Bode.
98. **Bohr Radius** - The radius of the electron orbit having the lowest energy. Also called atomic unit of length. The Bohr radius is based on the so-called Bohr model of the atom, named after the Danish physicist and philosopher Niels Bohr (1884-1962). Bohr envisioned atoms as consisting of small, dense nuclei with positive electric charge, around which negatively charged electrons orbit in circular paths
99. **Boltzmann's Constant** - Having dimensions of energy per degree of temperature, the Boltzmann constant has a value of $1.38064852 \times 10-23$ joule per kelvin (K), or $1.38064852 \times 10-16$ erg per kelvin. The total kinetic energy (E) in joules is related to temperature (T) in kelvins according to the equation $E = kBT$. For scale, 1 joule is the amount of energy expended by a 100-watt light bulb in 0.01 second, or a 1-watt bulb in 1 second.
100. **Boolean Algebra** - A set of rules formulated by the English mathematician George Boole describe certain propositions whose outcome would be either true or false. Regarding digital logic, these rules are used to describe circuits whose state can be either, 1 (true) or 0 (false). In order to fully understand this, the relation between the AND gate, OR gate and NOT gate operations should be appreciated
101. **Bootstrap** - Bootstrapping is a term used in Computer Programming to describe the techniques involved in writing a compiler (or assembler) in the target programming language which it is intended to compile. The boot sequence defines which devices a computer should check for the operating system's boot files. It also specifies the order devices are checked. The list can be changed and re-ordered in the computer's BIOS, as shown in the example below.
102. **Borromean rings** - Are composed of three circles interlocking each other, forming a Brunnian Link. This basically means that if one were to cut or take away one ring, the other two would fall apart. As such, Borromean Rings were

used, since the early times, as symbols for strength in unity. Although Borromean Rings are usually composed of circles, other shapes have also been used, to represent this symbol, such as triangles

103. **Boson** - An elementary particle that carries force. In particle physics, a boson is a type of particle that obeys the rules of Bose-Einstein statistics. These bosons also have a quantum spin with contains an integer value, such as 0, 1, -1, -2, 2, etc. (By comparison, there are other types of particles, called fermions, that have a half-integer spin, such as 1/2, -1/2, -3/2, and so on.)
104. **Botany** - Is the branch of biology dealing with the scientific study of plant life. It is also sometimes referred to as plant science(s) or plant biology
105. **British Thermal Unit (BTU)** - Is the amount of heat energy needed to raise the temperature of one pound of water by one-degree F. This is the standard measurement used to state the amount of energy that a fuel has as well as the amount of output of any heat generating device
106. **Broadband** - A transmission medium with enough bandwidth to carry multiple voice, video, or data channels simultaneously. This technique is used, for example, to provide fifty CATV channels on one coaxial cable; or to provide Internet access over cable TV; or to add DSL to a voice-grade telephone line
107. **Brownian Motion** - Also called Brownian movement, any of various physical phenomena in which some quantity is constantly undergoing small, random fluctuations. Random motion such as particles in fluid
108. **Buffer** - Is a location(s), especially in RAM, for storage of temporary data to speed up an operation such as disk access or printing
109. **Bus** - Is a set of wires that enables flow of data from one location of the computer to another. A Bus could be serial (one bit after the other) or parallel (all bits flow at one time)
110. **Byte** - a unit of data that is eight binary digits long. A unit of storage capable of holding a single character. A byte is a unit of memory data equal to either seven or eight bits, depending on whether it needs error correction (parity). You can think of a byte as one letter. For example, the letter 'h' is one byte or eight bits, and the word 'hope' is four bytes or 32 bits (4*8).

C

111. **C Programming Language** - The C programming language (often, just "C") is a general-purpose, procedural, imperative computer programming language

developed in the early 1970s by Dennis Ritchie for use on the Unix operating system. The first major program written in C was the UNIX operating system. It has since spread to many other operating systems. Although originally designed as a systems programming language, C has proved to be a powerful and flexible language that can be used for a variety of applications, from business programs to engineering. C is a particularly popular language for personal computer programmers because it is relatively small, it requires less memory than other languages

112. C# - Is an object-oriented programming language developed by Microsoft as part of their .NET initiative, and later approved as a standard by ECMA and ISO. C# has a procedural, object-oriented syntax based on C++ that includes aspects of several other programming languages (most notably Delphi, Visual Basic, and Java) with a particular emphasis on simplification (fewer symbolic requirements than C++, fewer decorative requirements than Java)
113. C++ - originally named "C with Classes, is a high-level programming language developed by Bjarne Stroustrup at Bell Labs. C++ adds object-oriented features to its predecessor, C. C++ is a statically typed free-form multi-paradigm language supporting procedural programming, data abstraction, object-oriented programming, and generic programming. C++ is one of the most popular programming languages. The C++ programming language standard was ratified in 1998 as ISO/IEC 14882:1998, and the current version of which is the 2003 version, ISO/IEC 14882:2003. A new version of the standard (known informally as C++0x) is being developed
114. **CAAD: Computer Aided Architectural Design** - Computer Aided Architectural Design (CAAD) software programs are the repository of accurate and comprehensive records of buildings and are used by architects and architectural companies. In a more general sense, CAAD also refers to the use of any computational technique in the field of architectural design other than by means of architecture-specific software. For example, software which is specifically developed for the computer animation industry (e.g. Maya and 3DStudio Max), is also used in architectural design
115. **Cache (pronounced CASH)** - Is a place to store something temporarily in a computing environment. In computing, active data is often cached to shorten data access times, reduce latency and improve input/output (I/O). Because almost all application workload is dependent upon I/O operations, caching is used to improve application performance
116. **CAD: Computer Aided Design** - Computer-aided design (CAD), also known as Computer Assisted Design, is the use of a wide range of computer-based tools that assist engineers, architects and other design professionals in

their design activities. It is the main geometry authoring tool within the Product Lifecycle Management process and involves both software and sometimes special-purpose hardware. Current packages range from 2D vector based drafting systems to 3D parametric surface and solid design modellers

117. **CADD: Computer-Aided Design and Drafting** - Computer-Aided Design and Drafting (CADD), one type of computer aided design (CAD), is a form of automation that helps designers prepare drawings, specifications, parts lists, and other design-related elements using special graphics- and calculations-intensive computer programs. The technology is used for a wide variety of products in such fields as architecture, electronics, and aerospace, naval, and automotive engineering
118. **CAE: Computer-Aided Engineering** - Computer-aided engineering (CAE) analysis is the application of computer software in engineering to analyze the robustness and performance of components and assemblies. It encompasses simulation, validation and optimization of products and manufacturing tools
119. **CAID: Computer-Aided Industrial Design** - Computer-aided industrial design (CAID) is a subset of computer-aided design (CAD) that includes software that directly helps in product development
120. **Calendar** - A system of reckoning time in units of solar days. The days are enumerated according to their position in cyclic patterns usually involving the motions of the Sun and/or the Moon
121. **Calendar, Gregorian** - The calendar introduced by Pope Gregory XIII in 1582 to replace the Julian calendar. This calendar is now used as the civil calendar in most countries. In the Gregorian calendar, every year that is exactly divisible by four is a leap year, except for centurial years, which must be exactly divisible by 400 to be leap years. Thus 2000 was a leap year, but 1900 and 2100 are not leap years
122. **Calendar, Julian** - The calendar introduced by Julius Caesar in 46 B.C. to replace the Roman calendar. In the Julian calendar a common year is defined to comprise 365 days, and every fourth year is a leap year comprising 366 days. The Julian calendar was superseded by the Gregorian calendar
123. **Calorie** - The amount of heat required at a pressure of one atmosphere to raise the temperature of one gram of water one degree Celsius that is equal to about 4.19 joules calorie. A unit of heat, equal to 4.1868 joules (International Table calorie): formerly defined as the quantity of heat required to raise the temperature of 1 gram of water by 1°C under standard conditions. It has now largely been replaced by the joule for scientific purposes
124. **CAM: Computer-Aided Manufacturing** - Computer-aided manufacturing (CAM) refers to using software to generate the instruction codes for a

Computer Numerical Control (CNC) machine in order for it to cut out a shape designed in a computer-aided design (CAD) system. Sometimes the CAM software is integrated with the CAD system, but not always. In electronic design automation, CAM tools prepare printed circuit board (PCB) and integrated circuit design data for manufacturing

125. **Cambrian Explosion** - Relates to an abrupt appearance of a wide range of organisms, mainly invertebrates, with hard (fossilizable) parts in Cambrian strata which mainstream scientists date from about 540 million years ago. Around 530 million years ago, a wide variety of animals burst onto the evolutionary scene in an event known as the Cambrian explosion. In perhaps as few as 10 million years, marine animals evolved most of the basic body forms that we observe in modern groups. Among the organisms preserved in fossils from this time are relatives of crustaceans and starfish, sponges, mollusks, worms, chordates, and algae, exemplified by these taxa from the Burgess Shale.
126. **Capacitor** - A device used to store an electric charge, consisting of one or more pairs of conductors separated by an insulator They are commonly used in computers or electronic systems.
127. **Carbohydrates** - Energy rich molecules composed of Carbon, Hydrogen, Oxygen. There are three main types of carbohydrates: Sugar. Sugar is the simplest form of carbohydrate. It occurs naturally in some foods, including fruits, vegetables, milk and milk products. Types of sugar include fruit sugar (fructose), table sugar (sucrose) and milk sugar (lactose). Added sugars can be found in many foods, such as cookies, sugary drinks and candy. Starch. Starch is a complex carbohydrate. This means it is made of many sugar units bonded together. Starch occurs naturally in vegetables, grains, and cooked dry beans and peas. Fiber. Fiber also is a complex carbohydrate. It occurs naturally in fruits, vegetables, whole grains, and cooked dry beans and peas
128. **Carbon Cycle** - The cycle where Carbon-dioxide is taken in and converted to organic compounds by photosynthesis or chemosynthesis, after which it is partially incorporated into sediments, and then returned to the atmosphere by respiration or combustion
129. **Carbon-14 Dating** - Also called radiocarbon dating, method of age determination that depends upon the decay to Nitrogen of radiocarbon (Carbon-14). Carbon-14 is continually formed in nature by the interaction of neutrons with Nitrogen-14 in the Earth's atmosphere; the neutrons required for this reaction are produced by cosmic rays interacting with the atmosphere
130. **Carnot engine** - The Carnot Efficiency is the theoretical maximum efficiency one can get when the heat engine is operating between two temperatures: The efficiency is always less than 100%.

131. **Carrier Signal**- Is a transmitted electromagnetic pulse or wave at a steady base frequency of alternation on which information can be imposed by increasing signal strength, varying the base frequency, varying the wave phase, or other means. This variation is called modulation. With the advent of laser transmission over optical fiber media, a carrier can also be a laser-generated light beam on which information is imposed
132. **Cartesian** - Of or relating to the methods or philosophies of Rene' Descartes
133. **Cascading Style Sheets (CSS)** - Was first developed in 1997, as a way for Web developers to define the look and feel of their Web pages. It was intended to allow developers to separate content from design so that HTML could perform more of the function that it was originally based on - the markup of content, without worry about the design and layout .CSS is also used to define how Web pages should look when viewed in other media than a Web browser. For example, you can create a print style sheet that will define how the Web page should print out and another style sheet to display the Web page on a projector for a slide show.
134. **CCD Charged-coupled devices** - Sensors used in digital cameras and video cameras to record still and moving images. A type of semiconductor that's sensitive to light
135. **cDNA** - Stands for complementary DNA, a synthetic type of DNA generated from messenger RNA, or mRNA, the molecule in the cell that takes information from protein-coding DNA - the genes - to the protein-making machinery and instructs it to make a specific protein. By using mRNA as a template, scientists use enzymatic reactions to convert its information back into cDNA and then clone it, creating a collection of cDNAs, or a cDNA library. These libraries are important to scientists because they consist of clones of all protein-encoding DNA, or all of the genes, in the human genome
136. **Celestial sphere** - An imaginary sphere of arbitrary radius upon which celestial bodies may be considered to be located. As circumstances require, the celestial sphere may be centered at the observer, at the Earth's center, or at any other location
137. **Cell** - The smallest independent part of an organism. A single cell is often a complete organism in itself, such as a bacterium or yeast. Other cells acquire specialized functions as they mature. These cells cooperate with other specialized cells and become the building blocks of large multicellular organisms, such as animals and humans. Although cells are much larger than atoms, they are still very small. The smallest known cells are a group of tiny bacteria called mycoplasmas; some of these single-celled organisms are spheres about 0.3 micrometer in diameter, with a total mass of 10−14 gram—equal to

that of 8,000,000,000 hydrogen atoms. Cells of humans typically have a mass 400,000 times larger than the mass of a single mycoplasma bacterium, but even human cells are only about 20 micrometers across. It would require a sheet of about 10,000 human cells to cover the head of a pin, and each human organism is composed of more than 75,000,000,000,000 cells

138. **Cell biology** – Is the study of cells. basic membrane-bound unit contains the fundamental molecules of life and of which all living things are composed
139. **Celsius temperature** - A temperature scale used in most countries, based on 100-degree difference between the melting point of ice, 0 °C, and the boiling point of water, 100 °C, at standard pressure
140. **Cenozoic Era** - Cenozoic (from the Greek *kainos* meaning "new" and *zoe* meaning "life") era is an interval of about 65 million years defined on the geologic timescale as spanning roughly from 65 million years ago (mya) to the present and ongoing into the foreseeable future. The Cenozoic era began after the Cretaceous-Tertiary extinction event at the end of the Cretaceous period that marked the demise of the last dinosaurs and the end of the Mesozoic era
141. **Center of gravity** - A point from which the weight of a body or system may be considered to act. In uniform gravity, it is the same as the center of mass
142. **Centrifugal force** - An outward force on a body rotating about an axis, assumed equal and opposite to the centripetal force and postulated to account for the phenomena seen by an observer in the rotating body
143. **Centripetal force** - The component of force acting on a body in curvilinear motion that is directed toward the center of curvature or axis of rotation
144. **Cercopithecoid** - Of, relating to, or belonging to the primate superfamily Cercopithecoidea (Old World monkeys)
145. **Cerenkov radiation** - Light produced by charged particles when they pass through an optically transparent medium at speeds greater than the speed of light in that medium
146. **Ceres** - Ceres is a dwarf planet, the only one located in the inner reaches of the solar system; the rest lie at the outer edges, in the Kuiper Belt. While it is the smallest of the known dwarf planets, it is the largest object in the asteroid belt. It was the first member of the asteroid belt to be discovered when Giuseppe Piazzi spotted it in 1801. When Dawn arrived in 2015, Ceres became the first dwarf planet to receive a visit from a spacecraft. Called an asteroid for many years, Ceres is so much bigger and so different from its rocky neighbors that scientists classified it as a dwarf planet in 2006. Even though Ceres comprises 25 percent of the asteroid belt's total mass, tiny Pluto is still 14 times more massive. Unlike other rocky bodies in the asteroid belt, Ceres is an oblate

spheroid, rounded with a rotational bulge around its equator. Scientists think Ceres may have an ocean and possibly an atmosphere

147. **Chain reaction** - A series of events in which each event causes the next one: a series of events caused by one single event. A chemical or nuclear change that causes other changes of the same kind to happen. Polymerization initiated by the bonding of a free radical with a monomer.
148. **Chandrasekhar limit** - If a dying star has a mass lower than the Chandrasekhar Limit, then it will become a white dwarf. Above this limit, it will explode as a supernova. Chandrasekhar limit is established at a point when the mass at which the pressure from the degeneration of electrons is not able to balance the self-attraction of the gravitational field. The limit that has been established these days is 1.39 $M_\odot$.
149. **Chaos Theory** - Is the study of nonlinear dynamics, in which seemingly random events are actually predictable from simple deterministic equations
150. **Character** - In computer and machine-based telecommunications terminology, a character is a unit of information. An example of a character is a letter, numeral, or punctuation mark. This includes all the ASCII and extended ASCII characters, including the space and control characters
151. **Charles' law** - The principle that all gases expand equally for the same rise of temperature if they are held at constant pressure: also, that the pressures of all gases increase equally for the same rise of temperature if they are held at constant volume. The law is now known to be only true for ideal gases
152. **CHNOPS** - Although more than 25 types of elements can be found in biomolecules, six elements are most common. These are called the CHNOPS elements; the letters stand for the chemical abbreviations of Carbon, Hydrogen, Nitrogen, Oxygen, Phosphorus, and Sulfur
153. **Circumsellar shell** - Shell of gas surrounding star, generally red giant or supergiant, composed of dust grains; grains possibly formed in star's atmosphere
154. **Classical mechanics** - Classical mechanics is the study of the motion of bodies (including the special case in which bodies remain at rest) in accordance with the general principles first enunciated by Sir Isaac Newton. Moreover, classical mechanics has many important applications in other areas of science, such as Astronomy (e.g., celestial mechanics), Chemistry (e.g., the dynamics of molecular collisions), Geology (e.g., the propagation of seismic waves, generated by earthquakes, through the Earth's crust), and Engineering (e.g., the equilibrium and stability of structures). Classical mechanics is also of great significance outside the realm of science

155. **Clean Room** - Is a room designed to have minimal levels of static electricity, dirt, dust particles, and/or airborne chemicals. It provides a sterile environment for the research and development of sensitive electronic devices like computer processors. The environment must be controlled to ensure the product is not damaged or contaminated.
156. **Client (Software)** - Is a program that asks for information from other processes or programs. Outlook Express is a great example of a client
157. **Clouds of Magellan** - Two dwarf galaxies located in the far southern sky and visible to the unaided eye; they are classified as irregular because they show no definite symmetry or nucleus
158. **CMOS** - Is an abbreviation for Complementary Metal-Oxide-Semiconductor. It is the battery powered chip that is situated on the Motherboard that retains system information such as date and time. It is also a technology used in the fabrication of hardware
159. **Code** - In Computer Programming refers to source code or machine code. Source code is any series of statements written in some human-readable computer programming language, while machine code refers to instructions for a computer processor in some machine language. The word "code" is often used to distinguish instructions from data
160. **Codec** - A device or computer program capable of encoding or decoding a digital data stream or signal. They can convert an analog signal to a digital signal and vice versa as well as compress and decompress data
161. **Coefficient** - A number multiplied by a variable, e.g., 5 is the coefficient of y in the term 5y.
162. **Coevolution** - Is the mutual evolutionary influence between two or more interdependent species, reflected in structural, physiological, or behavioral changes in the species related to their interaction,
163. **Color Force** - A property of quarks labeled color is an essential part of the quark model. The force between quarks is called the color force. Since quarks make up the baryons, and the strong interaction takes place between baryons, you could say that the color force is the source of the strong interaction, or that the strong interaction is like a residual color force which extends beyond the proton or neutron to bind them together in a nucleus
164. **Combustion** - When organic molecules combine with Oxygen to form Carbon Dioxide water vapor and a large quantity of heat
165. **Comet** - A celestial body, observed only in that part of its orbit that is relatively close to the Sun, having a head consisting of a solid nucleus surrounded by a nebulous coma up to 2.4 million kilometers (1.5 million miles) in diameter and an elongated curved vapor tail arising from the coma when

sufficiently close to the Sun. Comets are thought to consist chiefly of ammonia, methane, Carbon Dioxide, and water

166. **Common factor** - Factors that are common to two or more numbers. A common factor is a factor that is shared between two different numbers. It can also be referred to as a common divisor. As an example. The factors of 16 include: 1, 2, 4, 8, and 16. The factors of 12 include: 1, 2, 3, 4, 6, and 12.
167. **Compiler** - Is a type of computer program that translates source code into object code. A compiler is likely to perform many or all of the following operations: lexing, preprocessing, parsing, semantic analysis, code optimizations, and code generation
168. **Complex Number** - A combination of a Real Number and an Imaginary Number. the field $\mathbb{C}$of numbers of the form $x+iy$, where $x$and $y$are real numbers and i is the imaginary unit equal to the square root of -1, $\sqrt{-1}$
169. **Composite number** - Whole number larger than one that isn't a prime. A positive integer that can be generated by multiplying two smaller positive integers is a composite number. All composite numbers can be written as the product of two or more primes. For example, an integer 21 is a composite number as it is the product of the two smaller integers 3 and 7.
170. **Compton effect**- Is the result of a high-energy photon colliding with a target, which releases loosely bound electrons from the outer shell of the atom or molecule
171. **Computer Bus** - Is a set of physical connections (cables, printed circuits, etc.) which can be shared by multiple hardware components in order to communicate with one another
172. **Computer Science** - Is an academic field that studies the theoretical foundations of information and computation and their implementation and application in computer systems. Computer science has many sub-fields; some emphasize the computation of specific results (such as computer graphics), while others (such as computational complexity theory) relate to properties of computational problems. Still others focus on the challenges in implementing computations. For example, programming language theory studies approaches to describing computations, while computer programming applies specific programming languages to solve specific computational problems
173. **Computer Software Configuration Item (CSCI)** - Computer Software Configuration Item (CSCI) is a group of software treated as a single entity by a configuration management (CM) system. CSCI is for configuration control at the system architectural design, by contractual specification, or as development products. CSCI entities include software products to be developed or used

under contract, and certain elements required from the software development environment

174. **Condensation** - The process in which a gas becomes a liquid. Condensation is the change of the state of matter from the gas phase into the liquid phase, and is the reverse of vaporization
175. **Condensed matter physics** - Explores the macroscopic and microscopic properties of matter. Condensed Matter physicists study how matter arises from many interacting atoms and electrons, and what physical properties it has as a result of these interactions
176. **Conduction** - The movement of heat by atom-to-atom contact. The transfer of heat (internal energy) by microscopic collisions of particles and movement of electrons within a body
177. **Confinement (Quark)** - The moment after the Big Bang when Quarks bound together to form protons and neutrons
178. **Conjunction** - The phenomenon in which two bodies have the same apparent ecliptic longitude or right ascension as viewed from a third body. Conjunctions are usually tabulated as geocentric phenomena. For Mercury and Venus, geocentric inferior conjunctions occur when the planet is between the Earth and Sun, and superior conjunctions occur when the Sun is between the planet and Earth
179. **Conservation of Angular Momentum** - If no net external torque acts on a system, the total angular momentum of the system remains constant.
180. **Conservation of Energy** - Stated that the total energy of a system remains constant
181. **Conservation of Mass** - The weights in a chemical reaction are the same as the weights of their products. "The mass in an isolated system can neither be created nor be destroyed but can be transformed from one form to another"
182. **Conservation of Momentum**- Conservation of momentum is a fundamental law of physics which states that the momentum of a system is constant if there are no external forces acting on the system. It is embodied in Newton's first law (the law of inertia)
183. **Constellation** - Is a group of stars that, when seen from Earth, form a pattern. The stars in the sky are divided into 88 constellations. The brightest constellation is Crux (the Southern Cross). The constellation with the greatest number of visible stars in it is Centaurus (the Centaur - with 101 stars). The largest constellation is Hydra (The Water Snake) which extends over 3.158% of the sky. There are also asterisms, smaller apparent star patterns within a constellation, like the Big Dipper (in Ursa Major), the Little Dipper (in Ursa Minor), Keystone (in Hercules), and the Pleiades (in Taurus)

184. **Convection** - The transfer of heat by the circulation or movement of the heated parts of a liquid or gas. You may have noticed that if you leave a cup of hot coffee on a table, it gets cold in some time. This is because heat is lost to the surroundings. Thus, heat transfer is the transfer of heat or thermal energy between physical systems. So when there is a temperature difference between two bodies, heat is transferred from the hot body to the colder body.
185. **Convergence series** - An infinite series for which the sequence of partial sums converges
186. **Coordinate**- Part of an ordered pair that designates a point or position
187. **Copenhagen Interpretation** - On the atomic level a physical phenomenon expresses itself differently depending on the experimental setup used to observe it. Thus, light appears sometimes as waves and sometimes as particles.
188. **Coprocessor** - A special purpose processor that helps the CPU perform special operations such as math operations, encryption, and computer graphics. A math coprocessor helps with floating point operations and allows the computer to perform more complicated mathematical instructions.
189. **Core** - A core, or CPU core, is the "brain" of a CPU. It receives instructions, and performs calculations, or operations, to satisfy those instructions. A CPU can have multiple cores. Each CPU core can perform operations separately from the others. Multiple cores may also work together to perform parallel operations on a shared set of data in the CPU's memory cache
190. **Cosmic background radiation** - Is radiation left over from early development of the universe, and is a landmark proof of the Big Bang theory
191. **Cosmology** - The branch of philosophy dealing with the origin and general structure of the universe
192. **Cosmos** - The world or universe regarded as an orderly, harmonious system. 2. a complete, orderly, harmonious system
193. **Coulomb's Law** - The fundamental law of electrostatics stating that the force between two charged particles is directly proportional to the product of their charges and inversely proportional to the square of the distance between them. The force between charged particles is directly related to the amount of charge carried by each particle. Coulomb's law describes a force of infinite range which obeys the inverse square law, and is of the same form as the gravity force
194. **Covalent Bond** - Is a chemical bond that involves the sharing of electron pairs between atoms. These electron pairs are known as shared pairs or bonding pairs, and the stable balance of attractive and repulsive forces between atoms, when they share electrons, is known as covalent bonding

195. Cyber security - is the process of protecting data from unauthorized users or hackers. It also represents a huge sector of the tech industry as more and more companies race to stay ahead of cyber criminals and security threats

D

196. Dalton's Law - The total pressure of a mixture of gases is equal to the sum of the partial pressures of the individual gases
197. Dark Energy - Is a hypothetical form of energy that permeates all of space and tends to accelerate the expansion of the universe. repulsive force that is the dominant component (69.4 percent) of the universe. The remaining portion of the universe consists of ordinary matter and dark matter. Dark energy, in contrast to both forms of matter, is relatively uniform in time and space and is gravitationally repulsive, not attractive, within the volume it occupies. The nature of dark energy is still not well understood.
198. Dark Matter- Is a type of matter hypothesized to account for a large part of the total mass in the universe. Almost all of the universe — 96 percent — is invisible stuff called dark matter and dark energy
199. Data Converter - A/D or D/A converter: An electronic circuit that converts analog signals to digital, or vice-versa. An analog signal is a continuously varying voltage or current. Its digital counterpart is a stream of digital numbers, each representing the amplitude of the analog signal at a moment in time
200. Data mining (DM) - Also known as Knowledge-Discovery in Databases (KDD) or Knowledge-Discovery and Data Mining (KDD), is the process of automatically searching large volumes of data for patterns. Data mining uses automated data analysis techniques to uncover previously undetected relationships among data items
201. Database (DB) - Is a collection of organized information and data. In computer industry, database is an electronic filing system. Various techniques are used to model data structure. Most database systems are built around one particular data model, although it is increasingly common for products to offer support for more than one model. Traditional databases are organized by fields, records, and files. A field is a single piece of information; a record is one complete set of fields; and a file is a collection of records
202. De Morgan's Theorem - State the same equivalence in "backward" form: that inverting the output of any gate results in the same function as the opposite type of gate (AND vs. OR) with inverted inputs

203. **Debugger** - Is a computer program that is used to debug other programs. Sometimes, a debugger may also include testing and optimizing functions. For example, when the program crashes, the debugger shows the position in the original code if it is a source-level debugger or symbolic debugger, commonly seen in integrated development environments. If it is a low-level debugger or a machine-language debugger, it shows the line in the disassembly
204. **Decibel** - Is the unit used to measure the intensity of a sound. The smallest audible sound (near total silence) is 0 db. A sound 10 times more powerful is 10 dB. A sound 100 times more powerful than near total silence is 20 dB A sound 1,000 times more powerful than near total silence is 30 dB
205. **Decoder** - Is a device which does the reverse operation of an encoder, undoing the encoding so that the original information can be retrieved. The same method used to encode is usually just reversed in order to decode.
206. **Decompiler** - Is a computer program that translates executable programs (the output from a compiler) into an equivalent (relatively) high level language (source code). By comparison, a disassembler translates an executable program into assembly language
207. **Density** - The mass per unit volume in a substance
208. **Deoxyribonucleic Acid (DNA)** - The nucleic acid molecule that contains all the genetic information regarding the development and functioning of all living organisms is known as deoxyribonucleic acid._DNA is also known as the blueprint of life. It was discovered by James Watson and Francis Crick in 1953, as the double stranded molecule containing even the minutest of information regarding the building of an organisms
209. **Dependent Variable** -Two related variable that are dependent on each other are known as dependent variables. The variables that are free to roam are known as independent variables. The independent variable and dependent variables are plotted against each other in a two-dimensional graph when carrying out a scientific experiment. The vertical axis of the graph is used to plot the dependent variable
210. **Deposition**- The process by which a solvent is added to a solution to make the solution less concentrated
211. **Deuterium**- An isotope of hydrogen with one proton and one neutron in the nucleus having an atomic weight of 2.014. This stable isotopic variation of hydrogen gas has only one proton and one neutron. Protium, the basic hydrogen atom, which does not have a neutron, makes up 99.9% of naturally occurring marine hydrogen, while Deuterium makes up only 0.02 percent.
212. **Device driver** - Or a software driver, is a specific type of computer software, developed to allow interaction between the computer and hardware devices.

Each device driver is typically a piece of operating system-specific as well as hardware-dependent software that makes it possible for application software to attach to, read and write data from, and change the behavior of the peripheral device

213. **Dew Point** - The temperature to which air must be cooled to bring about the condensation of water vapor
214. **Digital Signal Processor (DSP)** - Is a special-purpose digital circuit that acts on digitized signals, such as audio. DSP circuits can replace traditional analog functions, such as filtering and more complex functions that are difficult to accomplish in the analog domain
215. **Diode** - A two-terminal device that rectifies signals (passes current in only one direction). Most commonly, a semiconductor consisting of a P-N junction, but diodes can also be realized using vacuum tube, point-contact, metal-semiconductor junction (Shockley) and other technologies
216. **Direct Current (DC)** - Electricity flowing in a constant direction, and/or possessing a voltage with constant polarity. DC is the kind of electricity made by a battery (with definite positive and negative terminals), or the kind of charge generated by rubbing certain types of materials against each other
217. **DirectX** - Provided by Microsoft, is a suite of multimedia application programming interfaces (APIs) built into Microsoft Windows OS. DirectX provides a standard development platform for Windows-based PCs by enabling software developers to access specialized hardware features without having to write hardware-specific code
218. **DNA Sequence** - 95% of gene-containing part of human sequence finished to 99.99% accuracy
219. **Doppler Effect** - The difference between the frequency of a wave (as of sound or light) as measured at its source and as measured by an observer in relative motion. When the source and observer are moving relative to each other, the frequency observed by the observer (fa) is different from the actual frequency produced by the source (f0). This is basically the definition of the Doppler effect in Physics. Here, when the source of waves is moving towards the observer, they will have an upward shift in frequency. As for the observers from whom the source is receding, there will be a downward shift in frequency.
220. **Double bond**- A covalent bond between two atoms resulting from the sharing of two pairs of electrons. Normally written as C=C, C=O, C=N, N=O.
221. **Double Slit Experiment** - The classic experiment first conducted by Thomas Young to determine the characteristics of waves. Until we observe the soon to be particle, it's a wave

222. **Dynamic Memory Access (DMA)** - Is a feature of computer systems that allows certain hardware subsystems to access main system memory independently of the central processing unit (CPU)
223. **Dynamic RAM (DRAM)** - Random-Access Memory that uses a continuous clock. Unlike SRAM, when DRAM is no longer clocked, its data is lost. DRAM, or Dynamic Random Access Memory, implies, this form of memory technology is a type of random access memory. It stores each bit of data on a small capacitor within the memory cell. The capacitor can be either charged or discharged and this provides the two states, "1" or "0" for the cell. Since the charge within the capacitor leaks, it is necessary to refresh each memory cell periodically. This refresh requirement gives rise to the term dynamic - static memories do not have a need to be refreshed
224. **Dynamic-Link Library** - Is Microsoft's implementation of the shared library concept in the Microsoft Windows operating systems. These libraries usually have the file extension DLL, OCX (for libraries containing ActiveX controls), or DRV (for legacy system drivers)
225. **Dynamics** - Is the branch of physics concerned with the motion of bodies under the action of forces. Compare with statics. The study of the motion of objects subjected to forces Developmental biology – Is the study of the process by which organisms grow and develop

E

226. **Eclipse** - The obscuration of a celestial body caused by its passage through the shadow cast by another body
227. **Eclipse, annular** - A solar eclipse in which the solar disk is not completely covered but is seen as an annulus or ring at maximum eclipse. An annular eclipse occurs when the apparent disk of the Moon is smaller than that of the Sun
228. **Eclipse, lunar** - An eclipse in which the Moon passes through the shadow cast by the Earth. The eclipse may be total (the Moon passing completely through the Earth's umbra), partial (the Moon passing partially through the Earth's umbra at maximum eclipse), or penumbral (the Moon passing only through the Earth's penumbra)
229. **Eclipse, solar** - Actually an occultation of the Sun by the Moon in which the Earth passes through the shadow cast by the Moon. It may be total (observer

in the Moon's umbra), partial (observer in the Moon's penumbra), annular, or annular total

230. **Ecliptic** - The mean plane of the orbit of the Earth-Moon barycenter around the solar system barycenter. 2. The apparent path of the Sun around the celestial sphere
231. **Einstein Cosmological Constant** - The supposed repulsion force value that may contribute to the expansion of universe and first predicted by Einstein in 1917 is known as Einstein Cosmological Constant. During this period, the universe was thought to be constant, and it was predicted that it may one day collapse under its own gravity. It was later proven to be expanding constantly.
232. **Einstein Ring** - According the Theory of General Relativity gravity causes spacetime to bend resulting in some cases in rings. An Einstein Ring forms when the observer, source, and lens all happen to align perfectly with each other. For example, if there is a galaxy located behind another galaxy, and both objects happen to be aligned with the Earth, then the gravitational lensing observed will come in the form of an Einstein Ring
233. **Electrical Numerical Integrator And Calculator (ENIAC**)- Was the first programmable general-purpose electronic digital computer, built during World War II by the United States. In the United States, government funding during the war went to a project led by John Mauchly, J. Presper Eckert, Jr., and their colleagues at the Moore School of Electrical Engineering at the University of Pennsylvania; their objective was an all-electronic computer. The ENIAC contained 17,468 vacuum tubes, along with 70,000 resistors, 10,000 capacitors, 1,500 relays, 6,000 manual switches and 5 million soldered joints. It covered 1800 square feet (167 square meters) of floor spacc, weighed 30 tons, consumed 160 kilowatts of electrical power
234. **Electrochemistry** - Is the science dealing with the relationship between electricity and chemical changes. Of principal interest are the reactions that take place between electrode terminal through which electric current passes between metallic and nonmetallic parts of an electric circuit. In most familiar circuits current is carried by metallic conductors, but in some circuits the current passes for some distance through a nonmetallic conductor., as well as the reactions that take place in an electrolyte as electricity passes through it. The principles of electrochemistry are applied in a variety of ways, e.g., in electroplating and in the generation of electricity by magnetohydrodynamics
235. **Electromagnetic Interference, EMI** - is a disturbance that occurs when two or more magnetic fields interact with one another. These fields can cause problems in any electric device with a circuit board. A good example of EMI

would be to place an electric fan close to your monitor and the monitor flickers or jumps.

236. **Electromagnetic radiation** - Radiation whereby energy is transmitted in a wave form through space. Examples of electromagnetic radiation would include X-rays, ultraviolet rays, visible light, infrared radiation, and radio waves
237. **Electromagnetic spectrum** – Is the range of wavelengths of electromagnetic radiation. Each variety of radiation has its own frequency, thus giving it its own "signature." These waves include radio, micro, infrared, visible, ultraviolet, x-rays, and gamma-rays. For example, a radio wave is used to send signals to devices such as an AM Radio to listen to a broadcast or music.
238. **Electromagnetic spectrum** - The arrangement of electromagnetic radiations in order of increasing frequencies. The electromagnetic spectrum is comprised of all frequencies of electromagnetic radiation that propagate energy and travel through space in the form of waves. Longer wavelengths with lower frequencies make up the radio spectrum. Shorter wavelengths with higher frequencies make up the optical spectrum. The portion of the spectrum that we can see is called the visible spectrum.
239. **Electromagnetic waves** - Waves with both an electric and magnetic component. They are radio, micro, infra-red, visible light, ultraviolet, X and gamma rays
240. **Electromotive force** - Electromotive Force (EMF) in electromagnetism is defined as the amount of electricity passing through an electric source like a generator that gets converted into work done. This work done in the energy transformation (or conversion) is characteristic of any energy source that is capable of moving electric charges around the circuit. Electromotive force is measured in Volts or equivalent to 1 Coulomb of electric charge, and it is denoted by the symbol ε (or E).
241. **Electron** - A particle of mass 9.11 x 10-31 kg, carrying the fundamental negative electric charge, -1.60 x 10-19 C.; a subatomic particle with a negative elementary electric charge. An electron has no known components or substructure
242. **Electron-positron Annihilation** - The collision of an electron and its anti-particle, a positron, leads to emission of gamma-ray photons of each with an energy of 511 keV that leads to their annihilation. This is known as electron-positron annihilation.
243. **Electrostatic Charge** - The charge at rest on the surface of an insulated body
244. **Elongation** - The angular distance between a planet and the Sun
245. **Emulator** - is a software or hardware device with emulation program to imitate another program or device. A software emulator allows computer

programs to run on a platform (computer architecture and/or operating system) other than the one for which they were originally written. A hardware emulator is an emulator which takes the form of a hardware device. Examples include printer emulators inside the ROM of the printer, and FPGA-based emulators

246. **Encoder**- A device, circuit, transducer, software program, algorithm or person that converts information from one format or code to another, for the purposes of standardization, speed, secrecy, security, or compressions.
247. **Endothermic reaction** - A reaction in which heat is absorbed ie: melting or boiling
248. **Endothermy**- An organism that generates heat to maintain its body temperature, typically above the temperature of its surroundings.
249. **Energy** - The capacity to do work. Work is done by transferring energy from one form to another. For example, the chemical energy in a fuel is converted to thermal energy as it burns
250. **Enthalpy** - (Symbol H) A thermodynamic function of a system, equivalent to the sum of the internal energy of the system plus the product of its volume multiplied by the pressure exerted on it by its surroundings.
251. **Entropy** - A measure of the disorder in a system. The state of disorder in a thermodynamic system: the more energy the higher the entropy. 1. Symbol S. For a closed thermodynamic system, a quantitative measure of the amount of thermal energy not available to do work $S = k \log W$. k is Boltzmann's constant and W is the number of possible molecules. 2. A measure of the disorder or randomness in a closed system. 3. A measure of the loss of information in a transmitted message 4. The tendency for all matter and energy in the universe to evolve toward a state of inert uniformity. 5. Inevitable and steady deterioration of a system or society
252. **Enzymes** - Consist of various types of proteins that work to drive the chemical reaction required for a specific action or nutrient. Enzymes can either launch a reaction or speed it up. The chemicals that are transformed with the help of enzymes are called substrates. In the absence of enzymes, these chemicals are called reactants. There are an estimated 75,000 different enzymes in the human body
253. **Epidemiology** - Is the science concerned with the study of the factors determining and influencing the frequency and distribution of disease, injury, and other health-related events and their causes in a defined human population for the purpose of establishing programs to prevent and control their development and spread

254. **Epistasis** - The interaction between genes at different loci, as a result of which one hereditary character is unexpressed, or is masked by the superimposition of another upon it
255. **Equation** - The mathematical definition of an equation states that it is an equality of two expressions that are written in a linear manner. The expressions are represented by symbols and are separated into left and right sides that is joined by an equal to (=) sign. Equation is defined as a representation of chemical reaction written in a linear array with symbols and quantities of the reactants. The reactants and products are separated with an equal (=) sign, an arrow (← /→) or a set of opposing arrows (< — >)
256. **Equinox** - Equal day and night: either of the two annual crossings of the equator by the Sun, once in each direction, when the length of day and night are approximately equal everywhere on Earth. The equinoxes occur around March 21 and September 23
257. **Escape Velocity** - The speed that an object needs to be traveling to break free of a planet or Moon's gravity well and leave it without further propulsion. On Earth, the escape velocity is around 40,270 kmph, which is around 11,186 m/s. For example, when a spacecraft is launched into outer space, the velocity attained by this should be greater than the escape velocity so that the rocket doesn't fall back onto Earth. And guess what, the escape velocity is different at the poles of the Earth compared to that from the equator because the radius is slightly more at the equator.
258. **Ethernet** - The original 10-megabit/second local area network, later increased to 100 gigabits/second, connecting the host with a coaxial cable developed by Xerox is called Ethernet. The transmission conflicts are minimized by backing off and re-sending later.
259. **Eukaryotic cells** – Are cells with a distinct nucleus containing a number of chromosomes. Any organism having as its fundamental structural unit a cell type that contains specialized organelles in the cytoplasm, a membrane-bound nucleus enclosing genetic material organized into chromosomes, and an elaborate system of division by mitosis or meiosis, characteristic of all life forms except bacteria, blue-green algae, and other primitive microorganisms
260. **Event Horizon** - The event horizon of a black hole is the boundary ('horizon') between its 'outside' and its 'inside'; those outside cannot know anything about things. The event horizon of a black hole is the boundary ('horizon') between its 'outside' and its 'inside'; those outside cannot know anything about things inside. It is the limit of influence of a black hole.. The event horizon of a black hole can be thought of as the boundary separating the black hole from the rest of the universe. The escape velocity within the event

horizon is exactly equal to the velocity of light. No event occurring within the event horizon of a black hole is visible outside; it is the horizon beyond which no occurring events can reach the world outside.

261. **Evolution** - Natural selection, the survival of the fittest, is the driving force behind evolution and is measured by a species viability and fecundity. Governed by Darwin's theory of evolution by natural selection. Evolution is change in the heritable characteristics of biological populations over successive generations.

262. **Excited state** - An atom is said to be in an excited state if an electron has absorbed energy sufficient to promote that electron to an energy level which is higher than that in which it finds itself in the ground state. For example, the single electron of hydrogen in the ground state is in the level $n = 1$. If the electron is in the level $n = 2$ or higher, the atom is in an excited state

263. **Exoplanet** - Is a planet located outside the solar system. The first confirmed detection of exoplanets was announced in 1992, with two planets found orbiting a pulsar. They come in a wide variety of sizes, from gas giants larger than Jupiter to small, rocky planets about as big around as Earth or Mars. They can be hot enough to boil metal or locked in deep freeze. They can orbit their stars so tightly that a "year" lasts only a few days; they can orbit two suns at once.

264. **Exothermic reaction** - A reaction from which heat is lost e.g.: combustion

265. **Exponent** - The exponent of a number says how many times to use the number in a multiplication. In 8^2 the "2" says to use 8 twice in a multiplication

266. **Extended ASCII** - Also known as high ASCII, is a set of codes using 8 bits for each character, instead of 7 bits as the basic ASCII set. The basic ASCII set provides it a total of 128 unique symbols. The extended ASCII character set gives it an additional 128 characters. The extra characters represent characters from foreign languages and special symbols for drawing pictures

267. **Extensible Hypertext Markup Language (XHTML)** - is HTML 4.0 Written in XML Format. HTML 4 is a markup language used for displaying text and documents across different platforms and machines. XML is an extensible markup language that was developed to retain the flexibility and power of HTML while reducing most of the complexity

268. **Extreme Ultraviolet (EUV)** - The spectrum between the ultraviolet and X-ray regions is known as extreme ultraviolet (EUV). It is observed in the region of spectrum between 10-100 nm

F

269. **Factorial** - Operation that multiples a whole number by every counting number smaller that it. For example, the factorial of 4 is 4 × 3 × 2 × 1, which is equal to 24. It is represented using the symbol '!' So, 24 is the value of 4!.
270. **Fahrenheit temperature** - A temperature scale still used in some countries, based on 180-degree difference between the melting point of ice, 32 °F, and the boiling point of water, 212 °F
271. **Faraday Cage**- A Faraday Cage, sometimes known as a Faraday Shield, is an enclosure that is used to shield things from electromagnetic fields (both static and non-static).
272. **Faraday's Law** - Any change in the magnetic environment of a coil of wire will cause a voltage (emf) to be "induced" in the coil
273. **FAT** - Is an acronym for File Allocation Table. It resembles a table of contents so that files can be located on a computer
274. **Fermat's principle** - States that light travels between two points along the path that requires the least time, as compared to other nearby paths
275. **Ferromagnetic** - Of or characteristic of substances such as iron, nickel, or cobalt and various alloys that exhibit extremely high magnetic permeability, a characteristic saturation point and magnetic hysteresis
276. **Fiber Optic (Optical fiber)** - Refers to the medium and the technology associated with the transmission of information as light pulses along a glass or plastic strand or fiber. Optical fiber carries much more information than conventional copper wire and is in general not subject to electromagnetic interference and the need to retransmit signals. Most telephone company long-distance lines are now made of optical fiber. Transmission over an optical fiber cable requires repeaters at distance intervals. The glass fiber requires more protection within an outer cable than copper
277. **Fibre Channel** - A highly reliable, gigabit interconnect technology that allows concurrent communications among workstations, mainframes, servers, data storage systems, and other peripherals using SCSI and IP protocols. It provides interconnect systems for multiple topologies that can scale to a total system bandwidth on the order of a terabit per second. (The standardized spelling is "fibre channel" but often misspelled as "fiber channel"
278. **Fibonacci sequence** - Is a series of numbers where a number is found by adding up the two numbers before it. Starting with 0 and 1, the sequence goes 0, 1, 1, 2, 3, 5, 8, 13, 21, 34, and so forth. Written as a rule, the expression is xn = xn-1 + xn-2. A sequence of numbers defined recursively by the Fibonacci

rule: after two starting values, each number is the sum of the two numbers before it. For example, the first Fibonacci numbers for n = 0, 1, 2, 3, are - 0, 1, 1, 2, 3, 5, 8, 13, 21, 34, 55, 89, 144, 233, 377, 610, 987, 1597, 2584, 4181, 6765, 10946. The Gold mean is related to the Fibonacci series as the ratio between two consecutive Fibonacci terms converge on the Golden mean https://math.temple.edu/~reich/Fib/fibo.html

279. **Field** - A region in space that is defined by a vector function. Common fields are: gravitational, electric and magnetic
280. **Field Programmable Gate Array (FPGA)** - A family of general-purpose logic devices that can be configured by the end user to perform many, different, complex logic functions. It can be programmed after it is manufactured rather than being limited to a predetermined, unchangeable hardware function. The term "field-programmable" refers to the ability to change the operation of the device "in the field," while "gate array" is a somewhat dated reference to the basic internal architecture that makes this after-the-fact reprogramming possible. An FPGAs can implement any logical function that an ASIC (application-specific Integrated Circuit) can do
281. **FIFO** - First-In First Out: A type of memory that stores data serially, where the first bit read is the first bit that was stored
282. **Firewall** - Is used to protect a networked server from damage by those who log in to it. This can either be a computer equipped with security features, software protection, or both. A firewall allows only certain messages from the Internet to flow in and out of the internal network.
283. **FireWire** - Apple Computer trademarked name for the IEEE 1394 serial interface standard: A high-speed interface between computers and peripherals such as external disk drives, cameras, and camcorders. Also referred to by Sony trademarked name, "I-Link."
284. **Firmware** - Is software that is embedded in a hardware device. It is often provided on flash ROMs or as a binary image file that can be uploaded onto existing hardware by a user. The firmware typically implements a portion of the network protocols, security mechanisms and administrative capabilities of the hardware device
285. **First Law of Motion** - States that a body at rest will remain at rest unless an outside force acts on it, and a body in motion at a constant velocity will remain in motion in a straight line unless acted upon by an outside force
286. **First Law of Thermodynamics** - Neither matter nor energy can be created or destroyed. An observable law of nature that begs the question of origins.
287. **Fission** - Splitting the nucleus of an atom into smaller units

288. **Flash storage** - Is storing data using electricity in surface-mounted chips on a printed circuit board (PCB). There are no moving mechanical parts involved, which reduces power consumption. Operating more like RAM than ROM, flash drives have no mechanical limitation for file access, which enables nanosecond seek times rather than the milliseconds required by mechanical hard drives, several orders of magnitude less latency. The size and complexity of flash-based systems varies for storage in wearable computing devices, embedded systems, smartphones, portable USB drives and more, all the way up to enterprise-class all-flash arrays
289. **Flow Chart** - Are easy-to-understand diagrams that show how the steps of a process fit together. They are extremely useful in outing steps in computer programming
290. **Fluid** - A liquid or gas; a substance that has no fixed shape and yields easily to external pressure; a gas or (especially) a liquid
291. **Fluid Mechanics (Dynamics)** - The study of fluids - liquids and gases. Involves various properties of the fluid, such as velocity, pressure, density and temperature, as functions of space and time
292. **Fluorescence** - The emission of light or other radiation from atoms or molecules that are bombarded by particles, such as electrons, or by radiation from a separate source. The bombarding radiation produces excited atoms, molecules, or ions and these emit photons as they fall back to the ground state
293. **Force** - An action (transfer of energy) that will accelerate a body in the direction of the applied force F= ma
294. **FORTRAN** - Is the oldest high-level programming language, designed by John Backus for IBM in the late 1950s. FORTRAN is a general-purpose, procedural, imperative programming language that is especially suited to numeric computation and scientific computing
295. **Fourier series** - Named after Joseph Fourier, is an expansion of a periodic function f(x) in terms of an infinite sum of sines and cosines One of the main uses of Fourier series is in solving some of the differential equations that arise in mathematical physics, such as the wave equation and the heat equation. Fourier series play a role in the analysis and synthesis of musical sounds[476]

$$F_n(x) = a_0 + \sum_{k=1}^{k=n} \left(a_k \cos(kx) + b_k \sin(kx)\right).$$

476 http://www.sosmath.com/fourier/fourier1/fourier1.html
http://www.stewartcalculus.com/data/CALCULUS%20Early%20Transcendentals/upfiles/FourierSeries5ET.pdf

296. **Fourier Transform** - Named after Joseph Fourier, is an integral transform that re-expresses a function in terms of sinusoidal basis functions, i.e. as a sum or integral of sinusoidal functions multiplied by some coefficients ("amplitudes"). Fourier Transform technique is used to evaluate the importance of various frequency cycles in a time series pattern
297. **Frequency** - The rate as which periodic motion repeats itself
298. **Frequency Modulation (FM)** - Is the encoding of information in a carrier wave by varying the instantaneous frequency of the wave. (Compare with amplitude modulation, in which the amplitude of the carrier wave varies, while the frequency remains constant.)
299. **Friction** - The interaction between surfaces: a measure of the resistance felt when sliding one body over another. The atoms inside a solid material can experience friction as well. For instance, if a solid block of metal gets compressed, all the atoms inside the material move, creating internal friction
300. **Fundamental Particles** - The particles like leptons, quarks and gauge bosons that do not contain a smaller component are known as fundamental particles.
301. **Fuzzy Logic** - Is an approach to computing based on "degrees of truth" rather than the usual "true or false" (1 or 0) Boolean logic on which the modern computer is based

G

302. **Galactic cannibalism** - A term used to describe when smaller cluster members because of their dominating gravitational fields and have absorbed the other galaxies into their own structures. NASA's Webb Space Telescope captured two galaxies merging together nearly 500 million light-years from Earth
303. **Galaxy** - A massive, gravitationally bound system consisting of stars, stellar remnants, and interstellar medium of gas and dust; there are over 100,000 stars in the Milky Way galaxy
304. **Game Theory** - The branch of applied mathematics that provides tools for analyzing situations in which parties, called players, make decisions that are interdependent. It is the study of mathematical models of negotiation, conflict and cooperation between individuals, organizations and governments. The study has direct applications in contract theory, economics, sociology and psychology Game theory is applied in various areas of study to understand why

an individual makes a particular decision and how the decisions made by one individual affect others

305. **Gamma Decay** – Is when a nucleus changes from a higher energy state to a lower energy state through the emission of electromagnetic radiation (photons). The number of protons (and neutrons) in the nucleus does not change in this process, so the parent and daughter atoms are the same chemical element
306. **Gamma Function** - As the improper integral of another function. Thus, the gamma function provides a way of giving a meaning to the "factorial" of any positive real number
307. **Gamma Rays** - Have the smallest wavelengths and the most energy of any wave in the electromagnetic spectrum. They are produced by the hottest and most energetic objects in the universe, such as neutron stars and pulsars, supernova explosions, and regions around black holes
308. **Gene** - A unit of inheritance. A section of DNA comprising a sequence of four bases: Guanine, Cytosine, Adenine and Thymine. (GCAT). Every person has two copies of each gene, one inherited from each parent. Most genes are the same in all people, but a small number of genes (less than 1 percent of the total) are slightly different between people. Alleles are forms of the same gene with small differences in their sequence of DNA bases. These small differences contribute to each person's unique physical features
309. **Genetic Code** - The information on the DNA, which is required for the synthesis of proteins. The genetic code is the sequence of three nucleotides in RNA that determines the sequence of amino acids for the synthesis of proteins. In other words, the genetic code is defined as the set of rules by which information encoded in the DNA or RNA sequences is translated into proteins by the living cell
310. **Genetic Drift** - Alteration in the genetic makeup of a particular population, which mostly takes place by chance alone. Genetic drift refers to random changes across generations in the genetic makeup (genome) of a population. A genome refers to the diversity of forms of genes (alleles) that occur in a population and how these alleles are organized within individual animals to provide allele-combinations that determine animal characteristics.
311. **Genetic Engineering** - Is any process by which genetic material (the building blocks of heredity) is changed in such a way as to make possible the production of new substances or new functions. As an example, biologists have now learned how to transplant the gene that produces light in a firefly into tobacco plants. The function of that gene—the production of light—has been added to the normal list of functions of the tobacco plants

312. Genetics - Is the study of heredity. Heredity is a biological process where a parent passes certain genes onto their children or offspring. Every child inherits genes from both of their biological parents and these genes in turn express specific traits. Some of these traits may be physical for example hair and eye color and skin color etc. On the other hand, some genes may also carry the risk of certain diseases and disorders that may pass on from parents to their offspring.
313. Genetics, Nanotechnology, Robotics (GNR) – Is a set of emerging technologies that will significantly alter life in the 21st century and blur the line between human and machine. Some experts predict that we won't experience 100 years of technological advance in the 21st; we will witness on the order of 20,000 years of progress (again, when measured by today's progress rate), or progress on a level of about 1,000 times greater than what was achieved in the 20th century
314. Genome - The collective noun for a set of genes. Our bodies are made up of millions of cells? (100,000,000,000,000), each with their own complete set of instructions for making us, like a recipe book for the body. This set of instructions is known as our genome and is made up of DNA?. Each cell in the body, for example, a skin cell or a liver cell, contains this same set of instructions:
315. Geology - The study of the Earth, the materials of which it is made, the structure of those materials, and the processes acting upon them
316. Geothermal Energy - Heat from the Earth's interior that can be transformed into various renewable forms of energy
317. Gibbs Free Energy - The energy associated with a chemical reaction that can be used to do work. The free energy of a system is the sum of its enthalpy (H) plus the product of the temperature (Kelvin) and the entropy (S) of the system: $G= H -TS$
318. Gigahertz, abbreviated GHz - Is a unit of alternating current (AC) or electromagnetic (EM) wave frequency equal to one thousand million hertz (1,000,000,000 Hz). The gigahertz is used as an indicator of the frequency of ultra-high-frequency (UHF) and microwave EM signals and also, in some computers, to express microprocessor clock speed
319. Global Positioning System (GPS) - A satellite-based navigation system in which two or more signals, received from satellites, are used to determine the receiver's position on the globe. PS is a multi-use, space-based radionavigation system owned by the US Government and operated by the United States Air Force to meet national defense, homeland security, civil, commercial, and scientific needs.

320. **Glue logic** - Is a special form of digital circuitry that allows different types of logic chips or circuits to work together by acting as an interface between them
321. **Gluon**- The so-called messenger particle of the strong nuclear force, which binds subatomic particles known as quarks within the protons and neutrons of stable matter
322. **Gödel's Incompleteness Theorems** - States that in any consistent formal system containing a minimum of arithmetic (symbols and the usual rules for handling them) a formally undecidable proposition can be found. First Incompleteness Theorem**-** Any consistent formal system F within which a certain amount of elementary arithmetic can be carried out is incomplete, i.e., there are statements of the language of F which can neither be proved nor disproved in F. Second Incompleteness Theorem - For any consistent system F within which a certain amount of elementary arithmetic can be carried out, the consistency of F cannot be proved in F itself.
323. **Golden mean** - Is the division of a given unit of length into two parts such that the ratio of the shorter to the longer equals the ratio of the longer part to the whole. 1: 1.618 or .618 to 1. The Gold mean is related to the Fibonacci series as the ratio between two consecutive Fibonacci terms converge on the Golden mean
324. **Goldilocks Zone** - The Goldilocks zone is a place where planetary temperatures allow liquid water to exist. The distance Earth orbits the Sun is just right for water to remain a liquid. This distance from the Sun is called the habitable zone, or the Goldilocks zone. Rocky exoplanets found in the habitable zones of their stars, are more likely targets for detecting liquid water on their surfaces. Why is liquid water so important? Life on Earth started in water, and water is a necessary ingredient for life (as we know it).
325. **Google Chromebook** - is any laptop or netbook computer running the Google Chrome OS. Chromebooks are different from other laptops because they require the Internet (although their offline capabilities have improved) and run almost everything from the cloud. Their functionalities are mainly limited to the use of the Google Chrome browser and related extensions, supported Android applications, and a file browsing system.
326. **Grand Unified Theory (GUT)** - A theory of elementary forces that unites the weak, strong, electromagnetic, and gravitational interactions into one field theory and views the known interactions as low-energy manifestations of a single unified interaction
327. **Graphic Processing Unit (GPU)** - Also occasionally called visual processing unit (VPU), is a specialized electronic circuit designed to rapidly manipulate and alter memory to accelerate the creation of images in a frame buffer intended for

output to a display. graphics processing unit, GPU is an electronic circuit used to speed up the creation of both 2D and 3D images. GPUs can either be integrated, meaning they are built into the computer's CPU or motherboard, or dedicated, meaning they are a separate piece of hardware known as a video card. By having a separate processor, the GPU allows the computer's CPU resources to be used for other important tasks.

328. **Graphical User Interface (GUI)** - Is a method of human interacting with a computer through direct manipulation of graphical images and widgets in addition to text. A program with a GUI runs under some windowing system (e.g. The X Window System, Apple Macintosh, Microsoft Windows, Acorn RISC OS, NEXTSTEP). The program displays certain icons, buttons, dialogue boxes etc. in its windows on the screen and the user controls it mainly by moving a pointer on the screen (typically controlled by a mouse) and selecting certain objects by pressing buttons on the mouse while the pointer is pointing at them
329. **Gravitational wave** - a wave-like motion in a gravitational field, produced when a mass is accelerated or otherwise disturbed; they travel through spacetime at the speed of light and the strongest sources are those with the strongest gravitational fields, although the waves are very weak; predicted by Einstein in 1916 but not first observed directly until 2015.
330. **Graviton**- A theoretical particle having no mass and no charge that mediates (carries) the gravitational force. The graviton is a boson. The existence of a graviton has not yet been confirmed string heory predicts the existence of gravitons as closed strings with the minimum possible energy. It is also theorized that gravitons interact with leptons and quarks.
331. **Greatest Common Factor** - The largest number by which all the terms of an expression can be divided. For example, the factors of 18 and 27 and then we find out the common factors. Factors of 18: 1, 2, 3, 6, 9, 18Factors of 27: 1, 3, 9, 27. The common factors of 18 and 27 are 1, 3, and 9. Among these numbers, 9 is the greatest (largest) number. Thus, the GCF of 18 and 27 is 9. This is written as: GCF (18, 27) = 9.
332. **Greenhouse Effect**- The warming of the Earth's surface by the atmospheric gases, notably Carbon Dioxide, trapped by infrared radiation

H

333. **Hadrons** - Quark composites: mesons and baryons. Protons and neutrons are the most common hadrons. In general, particles that participate in strong interactions are called hadrons: protons and neutrons are hadrons. The hadrons are further sub-divided into baryons and mesons, according to the number of quarks they contain. Protons and neutrons each contain three quarks; they belong to the family of particles called the baryons. Other baryons are the lambda, sigma, xi, and omega particles. On the other hand mesons, and bosons are composed of two quarks: a quark and an antiquark.
334. **Half Life** - The time required for a quantity to reduce to half its initial value. For example, the radioactive isotope cobalt-60, which is used for radiotherapy, has, for example, a half-life of 5.26 years. Thus, after that interval, a sample originally containing 8 g of cobalt-60 would contain only 4 g of cobalt-60 and would emit only half as much radiation. After another interval of 5.26 years, the sample would contain only 2 g of cobalt-60. Neither the volume nor the mass of the original sample visibly decreases, however, because the unstable cobalt-60 nuclei decay into stable nickel-60 nuclei, which remain with the still-undecayed cobalt.
335. **Half-Duplex** - Data transmission over a circuit capable of transmitting in either direction, but not simultaneously
336. **Heat** - The internal energy of a body (substance). Heat is the term we use to refer to specifically the transfer of thermal energy from one object or a system to another, transfer being the key. The thermal energy is the energy possessed within the object or within the system due to movement of particles
337. **Heat capacity** – The heat capacity of an object is the energy required to raise the temperature of that object by 1-degree K
338. **Heat of Fusion**- The energy that must be put into a solid to melt it
339. **Heat of Vaporization**- The energy put into a liquid to turn it into a gas
340. **Heisenberg's uncertainty principle** - At the atomic level, the more certain one is about the position of a particle, the greater the uncertainty in its momentum
341. **Heliocentric Universe** - Having or representing the Sun as the center, as in the accepted astronomical model of the solar system. Compare with geocentric
342. **Herbivore -** Animals that feed directly on plants
343. **Hertz** - The unit of frequency. 1 Hz = 1 s^{-1}

344. **Hess's Law** - States that regardless of the multiple stages or steps of a reaction, the total enthalpy change for the reaction is the sum of all changes. This law is a manifestation that enthalpy is a state function.
345. **Hexapod** – A six-legged arthropod of the class Insect a (formerly Hexapoda); an insect having six feet.
346. **Higgs boson** – A particle that gets its mass like all other particles: by interacting with ("swimming in") the Higgs field. The Higgs boson is the fundamental force-carrying particle of the Higgs field, which is responsible for granting other particles their mass. This field was first proposed in the mid-sixties by Peter Higgs for whom the particle is named The particle was finally discovered on July 4, 2012, by researchers at the Large Hadron Collider (LHC)
347. **High Energy Physics (also known as particle physics)** - Determines the most fundamental building blocks of matter and to understand the interactions between these particles. The underlying theoretical construct in particle physics is called the Standard Model and it contains 6 quarks, 6 leptons, 4 gauge bosons, and one scalar boson (the Higgs boson), which interact through three interactions (strong force, weak force, and electromagnetism)
348. **Hominin** Any member of the zoological "tribe" Hominini (family Hominidae, order Primates), of which only one species exists today—Homo sapiens, or human beings. The term is used most often to refer to extinct members of the human lineage, some of which are now quite well known from fossil remains: Homo neanderthalensis (the Neanderthals), Homo erectus, Homo habilis, and various species of Australopithecus. In addition, many authorities place the genera Ardipithecus, Orrorin, and Kenyanthropus
349. **Hominoid** - Of or belonging to the superfamily Hominoidea, which includes apes and humans and several their extinct ancestors.
350. ***Homo erectus*** - Homo erectus ("upright man") is an extinct species of the genus Homo. It lived from about 1.8 million years ago (mya) to 50-70,000 years ago
351. ***Homo habilis*** - Latin: "able man" or "handy man") extinct species of human, the most ancient representative of the human genus, Homo. H. habilis inhabited parts of sub-Saharan Africa from perhaps 2 to 1.5 million years ago (mya).
352. ***Homo sapiens*** - The species that you and all other living human beings on this planet belong to Homo sapiens. During a time of dramatic climate change 200,000 years ago, Homo sapiens (modern humans) evolved in Africa
353. **Homology** - A similarity often attributable to common origin
354. **Hox genes** - Found in all animals, Hox genes act to "regionalize" the body along the embryo's anterior-to-posterior (head-to-tail) axis. In a fruit fly, for

example, Hox genes lay out the various main body segments—the head, thorax, and abdomen

355. **H-R diagram** - The Hertzsprung-Russell Diagram is a graphical tool that astronomers use to classify stars according to their luminosity, spectral type, color, temperature and evolutionary stage
356. **HTML Linking Tags** - Allow you to jump from one page to another by clicking on the link text. You can also jump to places on the same page (called fragments), to different sections of your site, or to another web site altogether
357. **Hubble's Law** - The phenomenon that galaxies tend to move away from each other and the farther away they are, the faster they are separated is known as Hubble Law. The velocities of these galaxies increase proportionally to their distance.
358. **Human Genome Project (HGP)** - an international collaboration that successfully determined, stored, and rendered publicly available the sequences of almost all the genetic content of the chromosomes of the human organism, otherwise known as the human genome
359. **Human Genome Project**- A project that provided a detail map of the distribution of genes on the 23 human chromosomes and the sequence of bases
360. **Hund's rule** - States that every orbital in a subshell is singly occupied with one electron before any one orbital is doubly occupied, and all electrons in singly occupied orbitals have the same spin
361. **Huygen's Principle** - Of wave analysis basically states that: Every point of a wave front may be considered the source of secondary wavelets that spread out in all directions with a speed equal to the speed of propagation of the waves
362. **Hydraulics** - The use of a liquid that when moved from one place to another accomplishes by it motion some type of function
363. **Hydrogen bond**- Is the electrostatic attraction between two polar groups that occurs when a Hydrogen (H) atom covalently bound to a highly electronegative atom such as Nitrogen (N) or Oxygen (O) experiences the electrostatic field of another highly electronegative atom nearby
364. **Hydrostatic equilibrium** - Is equilibrium that occurs when compression, due to gravity, is balanced by a pressure gradient, which creates a force in the opposite direction. In stars, the pressure gradient appears as a result of the huge quantity of thermal energy (which acts outward) created by nuclear fusion reactions. It is gravity and this thermal energy that are in equilibrium. When we consider stars, this means that the larger the mass of the star, the higher the temperature must be to achieve this balance. Larger stars will use up their supply of hydrogen more quickly and live a shorter life

365. **HyperText Markup Language (HTML)** - First developed by Tim Berners-Lee in 1990, HTML is short for. HTML is used to create electronic documents (called pages) that are displayed on the World Wide Web. Each page contains a series of connections to other pages called hyperlinks. Every web page is actually a HTML file. Each HTML file is just a plain text file, but with a .html file extension instead of .txt and is made up of many HTML tags as well as the content for a web page. A web site will often contain many html files that link to each other. You can edit HTML files with your favorite editor

I

366. **Immunology** - The branch of biomedical science encompassing the study of the structure and function of the immune system (basic immunology); immunization, organ transplantation, blood banking, and immunopathology (clinical immunology); laboratory testing of cellular and humoral immune function (laboratory immunology); and the use of antigen– antibody reactions in other laboratory tests (serology and immunochemistry). (deals with the components of the immune system)
367. **In-Circuit Emulator (ICE)** - Is a hardware interface that allows a programmer to change or debug the software in an embedded system. The ICE is temporarily installed between the embedded system and an external terminal or personal computer so that the programmer can observe and alter what takes place in the embedded system, which has no display or keyboard of its own
368. **Inductor (Inductance)**- The property of an electric conductor or circuit that causes an electromotive force to be generated by a change in the current flowing inductance. The property of an electric circuit as a result of which an electromotive force is created by a change of current in the same circuit (self-inductance) or in a neighboring circuit (mutual inductance). It is usually measured in Henries
369. **Inflationary universe** – A version of the Big Bang theory in which the universe underwent very rapid growth during the first fraction of a second before it settled down to its current rate of expansion
370. **Information Technology (IT)** - Is a broad subject concerned with technology and other aspects of managing and processing information, especially in large organizations. In particular, IT deals with the use of electronic computers and computer software to convert, store, protect, process, transmit, and retrieve information. For that reason, computer professionals are

often called IT specialists or Business Process Consultants, and the division of a company or university that deals with software technology is often called the IT department. Other names for the latter are information services (IS) or management information services (MIS), managed service providers (MSP)

371. **Infrared radiation** - Electromagnetic radiation with a frequency range 1011 - 1014 Hz. The portion of the electromagnetic spectrum whose wavelength ranges from 0.75 microns to 1 millimeter
372. **Inorganic chemistry** - Inorganic chemistry is concerned with the properties and behavior of inorganic compounds, which include metals, minerals, and organometallic compounds. While organic chemistry is defined as the study of Carbon-containing compounds and inorganic chemistry is the study of the remaining subset of compounds other than organic compounds, there is overlap between the two fields (such as organometallic compounds, which usually contain a metal or metalloid bonded directly to Carbon)
373. **Instruction** - In computer technologies, instruction, also known as computer instruction, refers to a basic command at the level of computer's machine language. It is typically a single operation of a processor within a computer architecture
374. **Intellectual property (IP) core** - Is a block of logic or data that is used in making a field programmable gate array (FPGA) or application-specific integrated circuit (ASIC) for a product. IP cores fall into one of three categories: hard cores, firm cores, or soft cores. Hard cores are physical manifestations of the IP design. These are best for plug-and-play applications and are less portable and flexible than the other two types of cores. Like the hard cores, firm (sometimes called semi-hard) cores also carry placement data but are configurable to various applications. The most flexible of the three, soft cores exist either as a netlist (a list of the logic gate s and associated interconnections making up an integrated circuit) or hardware description language (HDL) code
375. **Interface** - In computer programming, is a defined means for a system to communicate with other systems. It is a boundary between a system and its environment providing ways of providing the system inputs and receiving outputs. In Object Oriented programming, class definitions and method signatures provide interfaces. Application program interfaces (APIs) form the interface of a system to applications and often consist of collections of functions or commands in a scripting language. Interfaces may be hidden (available only to the system developer) or exposed (available to others)
376. **Interference** - The process in which two or more light, sound, or electromagnetic waves of the same frequency combine to reinforce or cancel

each other, the amplitude of the resulting wave being equal to the sum of the amplitudes of the combining waves

377. **Internal Combustion Engine** - Is an engine in which the burning of a fuel occurs in a confined space called a combustion chamber. This exothermic reaction of a fuel with an oxidizer creates gases of high temperature and pressure, which are permitted to expand
378. **Internet** - Is a global computer network providing a variety of information and communication facilities, consisting of interconnected networks using standardized communication protocols. The Internet includes commercial, educational, governmental, and other networks, all of which use the same set of communications protocols
379. **Inverse Square Law** - The principle in physics that the effect of certain forces, such as light, sound, and gravity, on an object varies by the inverse square of the distance between the object and the source of the force. Any point source which spreads its influence equally in all directions without a limit to its range will obey the inverse square law
380. **Ionic Bond**- Is a chemical link between two atoms caused by the electrostatic force between oppositely charged ions in an ionic compound. Examples: There is an ionic bond between the sodium and chloride ions in table salt, NaCl
381. **Ionization Energy** - The energy required to remove an electron from an atom
382. **Irrational Number** - Number with no fractional equivalent whose decimal never repeats or terminates. (pi) is an irrational number. π=3·14159265
383. **Isomer** - Chemical compounds with the same composition but different shapes. Two compounds are said to be isomers if they have the same molecular formula, but different structural formulae
384. **Isomeric structure**. - The shape of a molecule. The isomeric structure is determined by the order in which the atoms are bonded together.
385. **Isotope** - Two or more different forms of an element differing only in their mass numbers. Any of two or more forms of a chemical element, having the same number of protons in the nucleus, or the same atomic number, but having different numbers of neutrons in the nucleus, or different atomic weights. There are 275 isotopes of the 81 stable elements, in addition to over 800 radioactive isotopes, and every element has known isotopic forms. Isotopes of a single element possess almost identical properties

J

386. **Java** - Is an object-oriented programming language developed by Sun Microsystems. It resembles C++ but was designed to avoid some of C++'s most notorious flaws. The Java language is used extensively on the World Wide Web, particularly because of its cross-platform nature, and its sandbox security concept
387. **Joint Test Action Group, JTAG** - Is the name given for the IEEE 1149.1 Standard Test Access Port and Boundary-Scan Architecture. This architecture is used in most computer processors since Intel released the first processor with JTAG, the 80486.
388. **Joule -** A joule is the amount of work done when a force of one Newton is applied through a distance of one meter.1 Joule = 1 Newton-meter = 1 kg-m2/s2
389. **Jovian Planet** - One of the four gas giant planets in the solar system, Jupiter, Saturn, Uranus, and Neptune, which have very large masses and are farther from the Sun than the terrestrial planets
390. **JPEG** - Joint Photography Experts Group; more commonly, files that are compressed using the JPEG standard.

K

391. **Kelvin temperature scale** - A temperature scale starting at 0 K (-273.2 °C), with degree size the same as the degree of the Celsius scale measured on the Kelvin scale. The fundamental unit of temperature that is not calibrated in terms of boiling point and freezing point of water but energy. 0 K is equal to the lowest temperature known as absolute zero. The freezing point of water on the Kelvin scale is -273 K and boiling point of water is 373 K.
392. **Kepler's law** - Three laws governing planetary motion: each planet revolves in an ellipse, with the Sun at one focus; the line connecting a planet to the Sun sweeps out equal areas in equal periods of time (law of areas) the square of the period of revolution of each planet is proportional to the cube of the semi major axis of the planet's orbit (harmonic law)

393. **Kernel** - Is a program called when a computer system is started. Kernel is responsible for setting up system calls in order to manage hardware and system services, and allocate resources to applications
394. **Kinetic energy** - The energy possessed by a body in motion. The energy of a body acquired by virtue of its motion
395. **Klein Bottle** - Is a closed non-orientable surface of Euler characteristic 0 (Dodson and Parker 1997, p. 125) that has no inside or outside
396. **Krebs Cycle** - Is a step wise cyclic process which is used to oxidize the pyruvate formed during the glycolytic breakdown of glucose into Carbon Dioxide and Water. The Krebs cycle occurs in the mitochondrial matrix and generates a pool of chemical energy (ATP, NADH, and FADH2) from the oxidation of pyruvate, the end product of glycolysis

L

397. **LAN** - Is an acronym for Local Area Network that spans small area. A LAN can be connected to another LAN to accommodate more computers
398. **Latency** - Is the measure of time between entering an input and the returned output. The higher the latency, the longer it takes. This is an important factor for web applications and interfaces where a noticeable delay may impact user satisfaction.
399. **Law of inertia** - The law of inertia states that: A body will preserve its velocity and direction so long as no force in its motion's direction acts on it
400. **Law of reflection** - The principle that when a ray of light, radar pulse, or the like, is reflected from a smooth surface the angle of reflection is equal to the angle of incidence, and the incident ray, the reflected ray, and the normal to the surface at the point of incidence all lie in the same plane
401. **LCD** - Abbreviation for Liquid Crystal Display. A technology popular in watches and calculators for displaying information, which also may be used for computer display screens—especially in portable computers.
402. **Leptons** - A class of six particles, including electrons and neutrinos, that do not occur in the atom's nucleus. There are six leptons in the present structure, the electron, muon, and tau particles and their associated neutrinos
403. **Lewis Dot Structures** - Are diagrams that show the bonding between atoms of a molecule and the lone pairs of electrons that may exist in the molecule. A Lewis structure can be drawn for any covalently bonded molecule, as well as coordination compounds

404. **Light** - The visible part of the electromagnetic spectrum. that can be detected by the human eye: red, orange, yellow, green, blue, indigo and violet. White light is a combination of all the above colors. Electromagnetic radiation occurs over an extremely wide range of wavelengths, from gamma rays, with wavelengths less than about 1 × 10−11 meter, to radio waves measured in meters. Within that broad spectrum the wavelengths visible to humans occupy a very narrow band, from about 700 nanometers (nm; billionths of a meter) for red light down to about 400 nm for violet light
405. **Light year** - The distance traversed by light in one mean solar year, about 5.88 trillion mi. (9.46 trillion km): used as a unit in measuring stellar distances
406. **Light-emitting Diode, LED** - is a semiconductor that illuminates when an electrical charge passes through it. LEDs are commonly green, amber, or red, but they can be an assortment of other colors as they've become popular with case lighting. Below are examples of how an LED could be used with a computer.
407. **Linear Algebra** - The study of linear sets of equations and their transformation properties
408. **Linear expansivity** - A measure of the temperature-dependent expansion or contraction of a solid. It is the fractional increase/decrease in length for a rise/fall of 1-degree Kelvin
409. **Linnaean system** - The system of nomenclature that assigns a specific name to each kind of organism. Living organisms are classified into groups depending on their structure and characteristics. This system was developed in the eighteenth century by Carl Linnaeus. The classification of species allows the subdivision of living organisms into smaller and more specialized groups.
410. **Linux** - Also known as GNU/Linux, is a free and open source Unix-like computer operating system. Unlike proprietary operating systems such as Windows or Mac OS, all of Linux underlying source code is available to the general public for anyone to use, modify, and redistribute freely. Linux has gained the support of major corporations such as IBM, Sun Microsystems, Hewlett-Packard, and Novell for use in servers and is gaining popularity in the desktop market. It is used in systems ranging from supercomputers to mobile phones
411. **Lipids** - Biomolecules including fats, oils and waxes the building blocks of cell membranes
412. **Logarithm** - The power to which a base must be raised to produce a given number. If nx = a, the logarithm of a, with n as the base, is x. Logarithm is the opposite of exponentiation.

413. **Logic gate** - Is an elementary building block of a digital circuit. Most logic gates have two inputs and one output. At any given moment, every terminal is in one of the two binary conditions low (0) or high (1), represented by different voltage levels. The logic state of a terminal can, and generally does, change often, as the circuit processes data. In most logic gates, the low state is approximately zero volts (0 V), while the high state is approximately five volts positive (+5 V). There are seven basic logic gates: AND, OR, XOR, NOT, NAND, NOR, and XNOR
414. **Lorentz-Fitzgerald contraction** - The shortening of an object along its direction of motion as its speed approaches the speed of light, as measured by an observer at rest with respect to the object
415. **Lucas sequence** - Describes a pattern of integers that follows a specific recurrence relation. The number series that falls into the category of a Lucas sequence conforms to the formula $X(n) = PX(n-1) - QX(n-2)$, where P and Q are fixed integers. The Fibonacci number series is a famous example of a Lucas sequence
416. **Lucy** - While hunting for fossils in Ethiopia's Afar Triangle on November 24, 1974, paleoanthropologist Donald Johanson and graduate student Tom Gray stumbled upon the partial remains of a previously unknown species of ape-like hominid. Nicknamed "Lucy," after the Beatles' song "Lucy in the Sky With Diamonds", the mysterious skeleton was eventually classified as a 3.2 million-year-old one of the pioneers of upright walking
417. **Lunar phases** - Cyclically recurring apparent forms of the Moon. New moon, first quarter, full moon and last quarter are defined as the times at which the excess of the apparent ecliptic longitude of the Moon over that of the Sun is 0°, 90°, 180° and 270°, respectively

M

418. **Machine Code** - Also known as machine language, is a system of instructions and data directly understandable by computer central processing unit. Every CPU model has its own machine code, or instruction set, although there is considerable overlap between some. If CPU A understands the full language of CPU B it is said that CPU A is compatible with B. CPU B may not be compatible with CPU A, as A may know a few codes that B does not
419. **Magnitude, stellar** - A measure on a logarithmic scale of the brightness of a celestial object. Since brightness varies with wavelength, often a wavelength

band is specified. A factor of 100 in brightness is equivalent to a change of 5 in stellar magnitude, and brighter sources have lower magnitudes. For example, the bright star Sirius has a visual-band magnitude of -1.46 whereas the faintest stars detectable with an unaided eye under ideal conditions have visual-band magnitudes of about 6.0

420. **Main Sequence** - A narrow band in the Hertzsprung-Russell diagram. A star will live the majority of its live in the main sequence phase. This is where nuclear fusion of hydrogen into helium is happening in its core, and the light pressure of this energy balances out the gravitational collapse of the star. About 90 percent of the stars in the universe, including the Sun, are main sequence stars
421. **Marine biology** - Is the scientific study of organisms in the ocean or other marine bodies of water. Given that in biology many phyla, families and genera have some species that live in the sea and others that live on land
422. **Materials physics** - The study of the characteristics and uses of the various materials, such as metals, ceramics, and plastics, that are employed in science and technology
423. **Materials science** - Materials science is a relatively new and very broad field. It involves applications from a number of scientific disciplines that contribute to the creation of new materials. Chemists play a predominant role in materials science because chemistry provides information about the structure and composition of materials, as well as the processes to synthesize and use them
424. **Matter** - Anything that occupies space and has mass. Material substance that constitutes the observable universe and, together with energy, forms the basis of all objective phenomena.
425. **Mechanics** – The branch of Physics that deals with physical energy and forces and their effect on objects
426. **Median** - The median is the "middle value" in a series of numbers ordered from least to greatest. When the total number of values in a list is odd, the median is the middle entry. When the total number of values in a list is even, the median is equal to the sum of the two middle numbers divided by two
427. **Meiosis** - Chromosomes in a diploid cell resegregate, producing four haploid daughter cells. It is this step-in meiosis that generates genetic diversity
428. **Melting point** - The temperature in which a solid change to a liquid at standard pressure
429. **Memory management** - Is the process of controlling and coordinating computer memory, assigning portions called blocks to various running programs to optimize overall system performance. Memory management resides in hardware, in the OS (operating system), and in programs and

applications. In hardware, memory management involves components that physically store data, such as RAM (random access memory) chips, memory caches, and flash-based SSDs (solid-state drives). In the OS, memory management involves the allocation (and constant reallocation) of specific memory blocks to individual programs as user demands change

430. **Meridian** - A great circle passing through the celestial poles and through the zenith of any location on Earth. For planetary observations, a meridian is half the great circle passing through the planet's poles and through any location on the planet
431. **Mesozoic Era** - Is divided into three time periods: The Triassic (251-199.6 million years ago), the Jurassic (199.6-145.5 million years ago), and the Cretaceous (145.5-65.5 million years ago). One of the most striking events in the Mesozoic Era was the rise to dominance of dinosaurs
432. **Messenger RNA** - Abbr. mRNA is the form of RNA that mediates the transfer of genetic information from the cell nucleus to ribosomes in the cytoplasm, where it serves as a template for protein synthesis. It is synthesized from a DNA template during the process of transcription
433. **Metallic bond** – A chemical bond that forms when all atoms release electronics creating a sea of negatively charged elections with positively-charged atoms interspersed
434. **Michelson–Morley experiment** - An experiment to try to detect a difference in the speed of light in two different directions: parallel to, and perpendicular to, the motion of the Earth around the Sun. Much to their dismay, they found no difference. It proved that the speed of light is constant which was fundamental to Einstein's theory of relativity
435. **Microarray** - Microarrays are devices used in many types of large-scale genetic analysis. They can be used to study how large numbers of genes are expressed as messenger RNA in a particular tissue, and how a cell's regulatory networks control vast batteries of genes simultaneously. In microarray studies, a robot is used to precisely apply tiny droplets containing functional DNA to glass slides. Researchers then attach fluorescent labels to complementary DNA (cDNA) from the tissue they are studying. The labeled cDNA binds to its matched DNA sequence at a specific location on the slide. The slides are put into a scanning microscope that can measure the brightness of each fluorescent dot. The brightness reveals how much of a specific cDNA fragment is present, an indicator of how active a gene is.
436. **Microbiology** - Is the study of microorganisms, including algae, bacteria, fungi, viruses, and protozoa. The science concerned with microorganisms, including fungi, protozoa, bacteria, and viruses

437. **Microcode** - Also known as microprogram, refers to the instruction set of a CPU as a sequence of microcode instructions (microinstructions), each of which typically consists of several bit fields and the address of the next microinstruction to execute. Microcode is the translation layer between machine instructions and the elementary operations of a computer
438. **Microcontroller** - Is a compact microcomputer designed to govern the operation of embedded systems in motor vehicles, robots, office machines, complex medical devices, mobile radio transceivers, vending machines, home appliances, and various other devices. A typical microcontroller includes a processor, memory, and peripherals
439. **Microprocessor** - A computer processor that incorporates the functions of a computer's central processing unit on a single integrated circuit, or at most a few integrated circuits
440. **Millikan's Experiment** - An experiment performed by Robert Millikan in 1909 determined the size of the charge on an electron
441. **Minute of Arc** - A unit of angular measurement equal to 1 / 60 of one degree
442. **Mitochondrial DNA** - Mitochondrial DNA is found in every living thing and only the female passes it
443. **Mitosis** - The process that facilitates the equal partitioning of replicated chromosomes into two identical groups. The process by which a eukaryotic cell separates the chromosomes in its cell nucleus into two identical sets, in two separate nuclei
444. **Modem** - A network hardware device that modulates one or more carrier wave signals to encode digital information for transmission and demodulates signals to decode the transmitted information
445. **Molecular biology** - Is the study of biology and biological functions at the molecular level, some cross over with biochemistry
446. **Molecule** - A group of atoms bonded together. It is the smallest part of a substance that retains the chemical properties of the whole
447. **Moment of inertia** - Measures the extent to which an object resists rotational acceleration about an axis, and is the rotational analogue to mass
448. **Momentum** - The product of mass times velocity. Momentum is conserved in any system of particles
449. **Moore's Law** - An axiom of microprocessor and digital development usually holding that processing power doubles about every 18 months especially relative to cost or size
450. **Multitasking** - In computing, multitasking is technique used in an operating system for sharing a single processor between several independent jobs. In the

case of a computer with a single CPU, only one task is said to be running at any point in time, meaning that the CPU is actively executing instructions for that task. Multitasking solves the problem by scheduling which task may be the one running at any given time, and when another waiting task gets a turn

451. **Multiverse** - A set of parallel universes of which our universe is one part
452. **Mutation** - A sudden departure from the parent type in one or more heritable characteristics, caused by a change in a gene or a chromosome

N

453. **Nanobiology** – Is the study of how nanotechnology can be used in biology, and the study of living organisms and parts on the nanoscale level of organization
454. **Nanorobotics** - Is the technology of creating machines or robots at or close to the scale of a nanometer (10^{-9} meters).
455. **Nanotechnology** - Or, as it is sometimes called, molecular manufacturing, is a branch of engineering that deals with the design and manufacture of extremely small electronic circuits and mechanical devices built at the molecular level of matter. Nanotechnology is often discussed together with micro-electromechanical systems (MEMS), a subject that usually includes nanotechnology but may also include technologies higher than the molecular level
456. **Natural selection** - Is the process by which changes in the natural environment favor, for survival and reproductive success, individual organisms
457. **Nebula** - A diffuse mass of interstellar dust or gas or both, visible as luminous patches or areas. Nebulae are made up of gas — primarily hydrogen and helium — and fine cosmic dust. These clouds are part of the interstellar medium of extremely low-density gas and dust that exists between stars in the void of space
458. **Neuroscience** - Is the study of how the nervous system develops, its structure and what it does. Neuroscientists focus on the brain and its impact on behavior and cognitive functions
459. **Neuroscience** - Is the study of how the nervous system develops, its structure and what it does. Neuroscientists focus on the brain and its impact on behavior and cognitive functions
460. **Neutralization** - A reaction in which the characteristics of an acid or base disappear. The reaction of an acid with a base leading to the formation of a salt

461. Neutrino - Any of the massless or nearly massless electrically neutral leptons, they exist across the universe. A neutral subatomic particle with a mass close to zero and half-integral spin, rarely reacting with normal matter. Three kinds of neutrinos are known, associated with the electron, muon, and tau particle
462. Neutron Star - A neutron star is a type of stellar remnant that can result from the gravitational collapse of a massive star during a Type II, Type Ib or Type Ic supernova event. Neutron stars are formed when large stars run out of fuel and collapse. To get a neutron star, you need to have star that's larger than about 1.5 solar masses
463. Neutrons - Particles with zero charge forming part of an atomic nuclei. 3 quarkhadrons. An atomic particle having a relative atomic mass of 1, and zero charge. an uncharged elementary particle that has a mass nearly equal to that of the proton and is present in all known atomic nuclei except the hydrogen nucleus
464. New General Catalog (NGC) - A Catalogue of Nebulae and Clusters of Stars, being the Catalogue of the Late Sir John F. W. Herschel, Bart., revised, corrected, and enlarged, basic reference list of star clusters, nebulas, and galaxies
465. Newcomb's Paradox - A paradox in decision theory. Given two boxes, B1 which contains $1000 and B2 which contains either nothing or a million dollars, you may pick either B2 or both. However, at some time before the choice is made, an omniscient Being has predicted what your decision will be and filled B2 with a million dollars if he expects you to take it, or with nothing if he expects you to take both
466. Newton's Laws of Motion - Classical laws which enable the prediction of the path of any object from a grain of sand to entire galaxies: a. A body will remain at rest or move with a constant velocity unless acted upon by an outside force. b. The acceleration of a body is proportional to the applied force. This is expressed by the universal formula: Force = mass × acceleration ($F=ma$) c. For every action, there is an equal and opposite reaction
467. Newtonian – Of or relating to the methods and theories of Sir Isaac Newton
468. Nitrogen Cycle - The cycle where Nitrogen is used by a living organism, then after the organism dies is restored to soil, followed by its final conversion to its original state of oxidation
469. No hair theorem - States that a black hole can be characterized by a small number of parameters that are visible from distance - mass, angular momentum and electric charge. According to the no-hair theorem, static black holes are described by a Schwarzschild spacetime provided there are no other sources of the gravitational field

470. **Noble gases** - The six noble gases are found in group 18 of the periodic table. These elements were considered to be inert gases until the 1960's, because their oxidation number of 0 prevents the noble gases from forming compounds readily
471. **Nuclear Fission** - A process in nuclear physics in which the nucleus of an atom splits into two or more smaller nuclei as fission products, and usually some by-product particles
472. **Nuclear Fusion** - Process by which nuclear reactions between light elements form heavier elements (up to iron. Nuclear Fusion reactions power the Sun and other stars. In a fusion reaction, two light nuclei merge to form a single heavier nucleus. The process releases energy because the total mass of the resulting single nucleus is less than the mass of the two original nuclei. The leftover mass becomes energy
473. **Nuclear physics** - Nuclear physicists probe nuclei with complex devices called accelerators, gather information from the probes with sophisticated detector systems, and record and analyze that information through advanced computing
474. **Nuclear Reactor** - Release nuclear energy to generate heat, which most frequently is then used in steam turbines to produce electricity in a nuclear power plant
475. **Nucleus** - The nucleus is a pivotal organelle responsible for regulating almost all forms of cellular activities. Mostly, every type of cell that exists is categorized on the basis of the absence or presence of the nucleus within its cell (categorized either as a prokaryotic or eukaryotic cell.). The positively charged central portion of an atom that comprises nearly all of the atomic mass and that consists of protons and usually neutrons. The central portion in the head of a comet, consisting of small solid particles of ice and frozen gases, which vaporize on approaching the Sun to form the coma and tail
476. **Nyquist Rate** - The lowest sampling rate that will permit accurate reconstruction of a sampled analog signal often defined as twice the highest frequency. If the sampling rate is less than 2fmax, some of the highest frequency components in the analog input signal will not be correctly represented in the digitized output. When such a digital signal is converted back to analog form by a digital-to-analog converter, false frequency components appear that were not in the original analog signal

O

477. **Object code** - Also known as object file, is an intermediate representation of code generated by a compiler after it processes a source code file. Object files contain compact, pre-parsed code, often called binaries, that can be linked with other object files to generate a final executable or code library. An object file is mostly machine code that can be directly executed by a computer's CPU
478. **Objective** C - is a general-purpose, object-oriented programming language used by the Apple operating system. It powers Apple's OS X and iOS, as well as its APIs, and can be used to create iPhone apps, which has generated a huge demand for this once-outmoded programming language.
479. **Octet rule** - Is a chemical rule of thumb that reflects observation that atoms of main-group elements tend to combine in such a way that each atom has eight electrons in its valence shell, giving it the same electronic configuration as a noble gas
480. **Olber's Paradox** - The paradox that if the universe consisted of an infinite number of stars equally distributed through space, then every line of sight would come from a star and the night sky would glow uniformly, which is observationally not true
481. **Operating System** - Provides the software platform required for various applications to run on. Its responsibility is to manage memory storage and security of Data. Operating System lies in the category of system software. It basically manages all the resources of the computer. An operating system acts as an interface between the software and different parts of the computer or the computer hardware. The operating system is designed in such a way that it can manage the overall resources and operations of the computer
482. **Optics** - The scientific study of sight and the behavior of light, or the properties of transmission and deflection of other forms of radiation
483. **Orbital** - A region in space around the nucleus of an atom where an electron with a given energy may move. Orbitals are designated with symbols, s, p, d and f
484. **Organic chemistry** -The chemistry of the compounds of Carbon and life. Carbon can form a chemical bond with a wide variety of chemical elements and other Carbon atoms. This allows a nearly unlimited number of combinations; because all known organisms, or living things, are made up of water and Carbon compounds. Organic chemistry is also the study of the structure, properties, composition, reactions, and preparation of Carbon-containing compounds, which include not only hydrocarbons but also compounds with

any number of other elements, including Hydrogen (most compounds contain at least one Carbon–hydrogen bond), Nitrogen, Oxygen, Halogens, Phosphorus, Silicon, and Sulfur. This branch of chemistry was originally limited to compounds produced by living organisms but has been broadened to include human-made substances such as plastics. The range of application of organic compounds is enormous and also includes, but is not limited to, pharmaceuticals, petrochemicals, food, explosives, paints, and cosmetics

485. **Organic compounds** - Substances that contain Carbon. Organic compounds are a type of chemical compounds where one or more than one Carbon covalently bonded with each other and with other atom like Nitrogen, Oxygen, Halogen etc. Such as, methane (CH_4), ethane (C_2H_6), benzene (C_6H_6) etc.
486. **Ozone hole** - A region over the Antarctic with seasonal reduced ozone
487. **Ozone layer** - A region of the atmosphere containing small amounts of ozone which absorbs much of the Sun's harmful ultraviolet radiation

P

488. **Packet** - Sections in which message or data are divided to transfer it over a network
489. **Paleozoic Era** - The Paleozoic Era is the beginning of an explosion of life forms. During the Paleozoic Era multi-celled living things acquired hard body parts, bones, vertebral columns, mandibles, and teeth. Major interval of geologic time, c. 542–251 million years ago. It is the first era of the Phanerozoic Eon and is followed by the Mesozoic Era. It is divided into six periods: (from oldest to youngest) the Cambrian, Ordovician, Silurian, Devonian, Carboniferous, and Permian.
490. **Pangaea** - Was a supercontinent that existed during the late Paleozoic and early Mesozoic eras. Pangaea, in early geologic time, a supercontinent that incorporated almost all the landmasses on Earth. Pangea was surrounded by a global ocean called Panthalassa, and it was fully assembled by the Early Permian Epoch (some 299 million to about 273 million years ago). The supercontinent began to break apart about 200 million years ago, during the Early Jurassic Epoch (201 million to 174 million years ago), eventually forming the modern continents and the Atlantic and Indian oceans.
491. **Parallel lines** - Lines that never intersect and are always the same distance apart. parallel lines continue forever in two directions. They don't have to be straight lines, as long as they are always the same distance apart. A single

existing line can have an infinite number of parallel lines. To be parallel lines, both lines must exist in the same plane; they must be coplanar

492. **Particle physics** - Is a branch of physics that studies the elementary constituents of matter and radiation, and the interactions between them. It aims to determine the fundamental laws that control the make-up of matter and the physical universe
493. **Pascal programming language** - Is a high-level programming language developed by Niklaus Wirth in the late 1960s based on ALGOL, and named after Blaise Pascal, a seventeenth-century French mathematician who constructed one of the first mechanical adding machines. Pascal is best known for its affinity to structured programming techniques
494. **Patent** - A document granting an inventor sole rights to an invention
495. **Pauli Exclusion Principle** - Two electrons can only occupy an orbital if they have opposite spins. The Pauli exclusion principle is the principle, formulated by Wolfgang Pauli in 1925, which states that: No two electrons can occupy the same quantum-mechanical state within a given quantum system simultaneously. No two electrons in an atom can have the same values of all four quantum numbers
496. **Peking Man** - Is an extinct hominin of the species Homo erectus, known from fossils found at Zhoukoudian near Beijing. Peking man was identified as a member of the human lineage by Davidson Black in 1927 on the basis of a single tooth
497. **Perfect number** - A number that equals the sum of its factors. Example 6 =1+2+3
498. **Perigee** - The point in the orbit of the moon or a satellite at which it is nearest to the Earth. The opposite of apogee
499. **Perihelion** - The point in the orbit of a planet or comet at which it is nearest to the Sun. For some bodies, like all the planets of the solar system, the difference between the perihelion and aphelion is quite negligible. For others, many comets and asteroids, for example, the difference can be huge.
500. **Period of oscillation** - The time taken for one complete wave to pass through a given point. The period of oscillation is the reciprocal of the frequency
501. **Periodic table** - A table in which elements are arranged in increasing values of their atomic numbers. The rows in the table are called periods, while the columns are called groups
502. **Perpendicular lines** - Lines that form a 90-degree angle at their intersection

503. **Phenotype** - The physical or functional characteristics of an organism, produced by the interaction of genotype and environment during growth and development.
504. **Photoelectric Effect** - Under the right circumstances light can be used to push electrons, freeing them from the surface of a solid. The rate at which photoelectrons are ejected is directly proportional to the intensity of the incident light, for a given frequency of incident radiation and metal
505. **Photolithography** - Is the standard method of printed circuit board (PCB) and microprocessor fabrication. The process uses light to make the conductive paths of a PCB layer and the paths and electronic components in the silicon wafer of microprocessors
506. **Photons**. - Fundamental quantum particles. A quantum of electromagnetic radiation, usually considered as an elementary particle that is its own antiparticle and that has zero rest mass and charge and a spin of one. It is the interaction of photons with other particles that drives the universe
507. **Photosynthesis** - The conversion of water and carbon-dioxide by plants into glucose and oxygen. Light is used as an energy source. Plants, however, take in and use carbon dioxide gas for photosynthesis. Carbon dioxide enters through tiny holes in a plant's leaves, flowers, branches, stems, and roots. Plants also require water to make their food
508. **PHP** (Hypertext Processor) - Is a free, server-side scripting language designed for dynamic websites and app development. It can be directly embedded into an HTML source document rather than an external file, which has made it a popular programming language for web developers. PHP powers more than 200 million websites, including Wordpress, Digg and Facebook.
509. **Phylum** - The primary subdivision of a taxonomic kingdom, grouping together all classes of organisms that have the same body plan
510. **Physical chemistry** - Is the study of how matter behaves on a molecular and atomic level and how chemical reactions occur. Based on their analyses, physical chemists may develop new theories, such as how complex structures are formed. Physical chemists often work closely with materials scientists to research and develop potential uses for new materials
511. **Physiology** - The branch of biology that deals with the functions and activities of life or of living matter (as organs, tissues, or cells) and of the physical and chemical phenomena involved — compare anatomy. It is also the study of the processes which sustain life in organisms, such as nutrition, respiration and reproduction
512. **Piezoelectrics** - Are a class of materials where a strain results in an electric field inside the material. In most cases, this can lead to a voltage out of the

material. Piezos have two modes. In direct mode, a strain is applied in a voltage is obtained. In converse mode, the opposite is true. A voltage is applied and results in a small movement or deflection of the material

513. **Pixel** - Pixel is formed by combining the two words: *Picture Element.* It represents one point within an image. The smallest element of an image that can be individually processed in a video display system
514. **Planck length** - Is the scale at which classical ideas about gravity and spacetime cease to be valid, and quantum effects dominate
515. **Planck mass** - It is about 10^{-5} g, same order of magnitude of one eyebrow hair or a flea egg
516. **Planck temperature** - The Standard Model, postulates that this is the theoretical highest temperature, and it equals about 100 million million million million million degrees, or 10^{32} Kelvin
517. **Planck time**- Before a time classified as a Planck time, 10^{-43} seconds, all the four fundamental forces are presumed to have been unified into one force
518. **Planck's constant** - Planck's constant, symbolized h, relates the energy in one quantum (photon) of electromagnetic radiation to the frequency of that radiation. In the International System of units (SI), the constant is equal to approximately 6.626176 x 10-34 joule-seconds
519. **Planck's law** - A law that is the basis of quantum theory, which states that the energy of electromagnetic radiation is confined to indivisible packets (quanta), each of which has an energy equal to the product of the Planck constant and the frequency of the radiation
520. **Plasma physics** - Is the study of charged particles and fluids interacting with self-consistent electric and magnetic fields. It is a basic research discipline that has many different areas of application — space and astrophysics, controlled fusion, accelerator physics and beam storage
521. **Pleiotropy** - The single gene controlling or influencing multiple (and possibly unrelated) phenotypic traits
522. **Pollen** - The fertilizing element of flowering plants, consisting of fine, powdery, yellowish grains or spores, sometimes in masses
523. **Polygon** - A plane figure with at least three straight sides and angles, and typically five or more
524. **Polymer chemistry** - Polymer chemists study large, complex molecules. They understand how the smaller building blocks (monomers) combine to form polymers, and they manipulate both their molecular structure and chemical or other processing to develop specific functional characteristics in an end product

525. **Polymer physics** - Is the branch of physics that studies polymers, their fluctuations, mechanical properties, as well as the kinetics of reactions involving degradation and polymerization of polymers and monomers
526. **Polymers** - Long chain molecules such as PVC, nylon or DNA produced by the polymerization of monomers
527. **Port (I/O)** - Is a connecting component mainly a hardware that enables computers and other peripherals to allow data sharing physically. Examples are USB and HDMI
528. **Positron** - An elementary particle having the same mass and magnitude of charge as an electron but exhibiting a positive charge; a positive electron
529. **Potential difference** - The voltage difference between two points. Electricity flows from a high to low level of potential
530. **Potential energy** - Amount of useable energy within a body at rest. The energy that an object possesses as a result of its position or state.
531. **Power** - Amount of work done per second; the rate of energy utilization or production. The SI unit of power is the watt (W), where 1 watt = 1 joule.sec-1
532. **Precession of Equinoxes** - Consists of a cyclic wobbling in the orientation of Earth's axis of rotation with a period of 25,772 years. Precession is caused by the gravitational influence of the Sun and the Moon acting on Earth's equatorial bulge. The sky makes one rotation around the North Celestial Pole each night. The North Celestial Pole rotates around the North Ecliptic Pole approximately once in 26,000 years
533. **Precipitation** - A reaction taking place in solution, whereby soluble reactants form an insoluble product, the precipitate
534. **Pressure** - Force exerted on a unit area. Pressure is measured in pascals, P, where 1 P = 1 N.m2.
535. **Primate** - Member of the mammalian order Primates, which includes humans, apes, monkeys, and prosimians, or lower primates
536. **Prime Number** - A prime number can be divided, without a remainder. The only even prime number is 2. All other even numbers can be divided by 2 only by itself and by 1. For example, 17 can be divided only by 17 and by 1
537. **Primordial Soup** - Nickname given to the theory that life started in the oceans from the synthesis of organic molecules; a mixture of organic molecules in evolutionary theory from which life on Earth originated
538. **Principal quantum number** - An integer, 1, 2, 3..., which characterizes the main electronic levels of an atom
539. **Printed Circuit Board** - Or PC board, or PCB, is a non-conductive material with conductive lines printed or etched. Electronic components are mounted on the board and the traces connect the components together to form a

working circuit or assembly. A PC board can have conductors on one side or two sides and can be multi-layer — a sandwich with many layers of conductors, each separated by insulating layers.

540. **Programming Language** - Is an artificial language that can be used to control the behavior of a machine (often a computer). Programming languages have syntactic and semantic rules used to define meaning. Programming languages are used to facilitate communication about the task of organizing and manipulating information, and to express algorithms precisely. Programming language usually refers to high-level languages, such as BASIC, C, C++, COBOL, FORTRAN, Ada, and Pascal.
541. **Prokaryote** - Are mostly unicellular organisms without a nucleus, in contrast to eukaryotes; the simplest living things: bacteria and archaea
542. **Proteins** - Amino acid polymers with specific biological functions, especially the growth, regeneration and repair of cells
543. **Protocol** - Refers to a set of rules that are followed by two devices while interacting with each other
544. **Proton** - An atomic particle with relative atomic mass of 1, and unit positive charge, found in the nucleus of the atom Positively charged particles forming part of atomic nuclei
545. **Protostar** - Is a baby star. Stars form when vast clouds of cold molecular hydrogen and helium collapse under mutual gravity. This collapse could have been triggered by a galaxy collision, or the shockwave of a nearby supernova. As the cloud collapses, it breaks into fragments, each of which will eventually become a star of some size. As the cloud contracts, it begins to increase in temperature. This comes from the conversion of gravitational energy into kinetic energy. The cloud continues to heat up, and the conservation of momentum of all the different particles causes the protostar to spin
546. **Public-Domain Software** - Refers to any program that is not copyrighted. Public-domain software is free and can be used without restrictions. The term public-domain software is often used incorrectly to include freeware, free software that is nevertheless copyrighted
547. **Pulsar** - A pulsar (portmanteau of pulsating star) is a highly magnetized, rotating neutron star that emits a beam of electromagnetic radiation. . one of several hundred known celestial objects, generally believed to be rapidly rotating neutron stars, that emit pulses radiation
548. **Python** - Is a high-level, server-side scripting language for websites and mobile apps. It's considered a fairly easy language for beginners due to its readability and compact syntax, meaning developers can use fewer lines of code

to express a concept than they would in other languages. It powers the web apps such as Instagram, and is used by Google, Yahoo! and NASA

Q

549. **Quadratic Equation** - Equation in which the highest power is two
550. **Quantum** - A small, discrete amount of energy. (Plural: quanta)
551. **Quantum chemistry** - Is the application of quantum mechanical principles and equations to the study of molecules. In order to understand matter at its most fundamental level, we must use quantum mechanical models and methods
552. **Quantum Electrodynamics (QED)** - Describes the properties of electromagnetic radiation and its interaction with electrically charged matter in the framework of quantum theory
553. **Quantum Mechanics** - Is the body of scientific principles that explains the behavior of matter and its interactions with energy on the scale of atoms and subatomic particles. It also deals with the behavior of matter and light on the atomic and subatomic scale. At a fundamental level, both radiation and matter have characteristics of particles and waves. It is also the branch of Physics that deals with the mathematical description of the motion and interaction of subatomic particles, incorporating the concepts of quantization of energy, wave-particle duality, the uncertainty principle, and the correspondence principle
554. **Quantum theory** - A theory which states that energy is not continuous, but can only be emitted or absorbed in discrete quantities called quanta. The theory that energy can only be absorbed or radiated in discrete values or quanta. All particles are subject to quantum theory
555. **Quark** - Fundamental particles, incapable of independent existence, that combines to form particles such as protons and neutrons. Types or Quarks: There are 6 different kinds or flavors of quarks. They each have different things that make them unique, like their mass (or weight) and their electric charge
556. **Quasar** - One of over a thousand known extragalactic objects, star like in appearance and having spectra with characteristically large redshifts that are thought to be the most distant and most luminous objects in the universe

R

557. **Radiation** - Transfer of heat between bodies without a change in the temperature of the intervening medium. Any release of energy from its source
558. **Radio wave** - Have the longest wavelengths in the electromagnetic spectrum. These waves can be longer than a football field or as short as a football. Electromagnetic waves that have wavelengths between 1 millimeter and 100 kilometers (or 300 GHz and 3 kHz in frequency)
559. **Radioactivity** - The spontaneous release of energy from atomic nuclei. The phenomenon, exhibited by and being a property of certain elements, of spontaneously emitting radiation resulting from changes in the nuclei of atoms of the element
560. **Radiometric dating** - A technique used to determine the age of rocks in absolute terms. -Measures sediment's based on how much isotopes remain after radioactive decay of Isotopes within sediments. -Using a special machine called a mass spectrometer-Tells you how long ago the rock was formed.
561. **RAID, Redundant Array of Independent Disks** - A redundant array of inexpensive disks. RAID is a performance-enhancing method of storing the same data in different places on multiple hard disks to achieve speed and/or data redundancy
562. **Random Access Memory (RAM)** - Is a hardware device that allows information to be stored and retrieved on a computer. RAM is usually associated with DRAM, which is a type of memory module. Because information is accessed randomly instead of sequentially like it is on a CD or hard drive, the computer can access the data much faster. However, unlike ROM or the hard drive, RAM is a volatile memory and requires power to keep the data accessible. If the computer is turned off, all data contained in RAM is lost
563. **Rational number** - A number that can be written as a fraction, its decimal terminates or repeats
564. **Real-time operating system (RTOS)** - Is an operating system designed specifically for use in real-time systems and applications. Examples include embedded applications (programmable thermostats, household appliance controllers, mobile telephones), industrial robots, industrial control, and scientific research equipment
565. **Recombinant Engineering** - Recombinant DNA technology (rDNA) is technology that is used to cut a known DNA sequence from one organism and

introduce it into another organism thereby altering the genotype (hence the phenotype) of the recipient. The process of introducing the foreign gene into another organism (or vector) is also called cloning

566. **Red giant** - Is a dying star in the last stages of stellar evolution. Our own sun will turn into a red giant star, expand and engulf the inner planets, possibly even Earth.
567. **Red Super giants** - Are supergiant stars of spectral type K-M and a luminosity class of I. They are the largest stars in the universe in terms of physical size, although they are not the most massive. Stars with more than about 10 solar masses, after burning their hydrogen become red supergiants during their helium-burning phase. These stars have very cool surface temperatures (3500-4500 K), and enormous radii. The four largest known red super giants in the Galaxy are Mu Cephei, KW Sagitarii, V354 Cephei, and KY Cygni, which all have radii about 1500 times that of the Sun (about 7 astronomical units, or 7 times as far as the Earth is from the Sun).
568. **Redshift** - Results from the relative motion of the light emitting object and the observer. If the source of light is moving away from you then the wavelength of the light is stretched out, i.e., the light is shifted towards the red. These effects, individually called the blueshift, and the redshift are together known as Doppler shifts. The shift in the wavelength is given by a simple formula
569. **Reduced Instruction Set Computer (RISC)** - Is a microprocessor that is designed to perform a smaller number of types of computer instructions so that it can operate at a higher speed (perform more millions of instructions per second, or MIPS). Since each instruction type that a computer must perform requires additional transistors and circuitry, a larger list or set of computer instructions tends to make the microprocessor more complicated and slower in operation
570. **Reflection** - The process in which a wave strikes a material and rebounds at the same angle with which it was struck
571. **Refraction**- The change in direction of a light, heat or sound from one medium to another
572. **Regression testing** - Is any type of software testing which seeks to uncover regression bugs. Regression bugs occur whenever software functionality that previously worked as desired stops working or no longer works in the same way that was previously planned. Typically, regression bugs occur as an unintended consequence of program changes. Common methods of regression testing include re-running previously run tests and checking whether previously fixed faults have reemerged

573. **Relativity** - The relative values of time, motion, mass and energy of a body in motion. The special theory of relativity was a first step for Einstein. The fuller development of his goal of relativizing physics came with his general theory of relativity. That theory was completed in its most important elements in November of 1915.
574. **Reproduction** - Is the process by which a new organism is produced. The first stage in the production of any organism is the fertilization of an ova by spermatozoa (or spores on the case of plants). Fertilization produces a single cell called a zygote which contains all the information required to build the adult organism. The progression (growth) from zygote to adult is achieved through cell division
575. **Resistance** - Opposition to current flow in a conductor
576. **Resonance** - A state where the natural frequency of a body equals an applied frequency
577. **Resonance frequency** - A resonant frequency is a natural frequency of vibration determined by the physical parameters of the vibrating object. It is easy to get an object to vibrate at its resonant frequencies, hard to get it to vibrate at other frequencies. Think of a high pitch breaking a glass
578. **Resultant Vector**- The vector resulting from the sum of two vectors
579. **Retrograde motion** - For orbital motion in the solar system, motion that is clockwise in the orbit as seen from the North Pole of the ecliptic; for an object observed on the celestial sphere, motion that is from east to west, resulting from the relative motion of the object and the Earth
580. **Ribonucleic Acid (RNA)** - The biological molecule that contains long chains of nucleotide units of a nitrogenous base, a ribose sugar and a phosphate is known as Ribonucleic acid. RNA is single stranded molecule with ribose, unlike DNA that is double stranded with deoxyribose. Another difference between a DNA and RNA molecule is presence of base uracil instead of thymine in RNA.
581. **Riemann sum** - An approximation for the area beneath a curve that is achieved by adding the areas of rectangles
582. **Robotics** - A branch of engineering that involves the conception, design, manufacture, and operation of robots. This field overlaps with electronics, computer science, artificial intelligence, mechatronics, nanotechnology and bioengineer
583. **RS-485 and RS-422** - Are serial interface standards in which data is sent in a differential pair (two wires, or twisted pair cable), which allows greater distances and higher data rates than non-differential serial schemes such as RS-

232. See: Differential Signaling. RS-485 and RS-422 can be configured for full-duplex or half-duplex bus systems

584. **Ruby** - Is a dynamic, open-source programming language with a focus on simplicity and productivity. It has an elegant syntax that is natural to read and easy to write

585. **Runtime** - In computer science, runtime or run time describes the operation of a computer program, the duration of its execution, from beginning to termination (compare compile time)

S

586. **Sampling Rate**- An A/D converter converts an analog signal into a stream of digital numbers, each representing the analog signal's amplitude at a moment in time. Each number is called a "sample." The number sample per second is called the sampling rate, measured in samples per second

587. **Schematic Diagram** – A symbolic representation of an electrical circuit

588. **Schrodinger's Cat** - Schrödinger wanted people to imagine that a cat, poison, a Geiger counter, radioactive material, and a hammer were inside of a sealed container. The amount of radioactive material was minuscule enough that it only had a 50/50 shot of being detected over the course of an hour. If the Geiger counter detected radiation, the hammer would smash the poison, killing the cat. Until someone opened the container and observed the system, it was impossible to predict if the cat's outcome. Thus, until the system collapsed into one configuration, the cat would exist in some superposition zombie state of being both alive and dead

589. **Schwarzschild radius** - The critical radius, according to the general theory of relativity, at which a body under the influence of its own gravitation becomes a black hole.

590. **Science, Technology, Engineering and Math (STEM)** - Is an educational program developed to prepare primary and secondary students for college and graduate study in the fields of science, technology, engineering, and mathematics (STEM). In addition to subject-specific learning, STEM aims to foster inquiring minds, logical reasoning, and collaboration skills

591. **Scientific method** - The way scientists learn and study the world around them; it consists of observation, hypothesis testing and prediction

592. **Scientific notation** - A notation where numbers are expressed as in the form r x 10b, where r is a real number (1 £ r < 10), and b an integer

593. Script - In computer programming, is a type of macro or batch file which contains a list of commands that can be executed without user interaction. A script language is a simple programming language with which you can write scripts. Apple Computer uses the term script to refer to programs written in its HyperCard or AppleScript language
594. Search engine - Is a computer system designed to help find information over a computer network such as the World Wide Web, inside a corporate or proprietary network or a personal computer. The search engine allows one to ask for content meeting specific criteria (typically those containing a given word or phrase) and retrieves a list of references that match those criteria. Search engines use regularly updated indexes to operate quickly and efficiently. Without further qualification, search engine usually refers to a Web search engine, which searches for information on the public Web
595. Second Law of Motion - States that if an unbalanced force acts on a body, that body will experience acceleration (or deceleration), that is, a change of speed
596. Second Law of Thermodynamics - The Second Law of Thermodynamics states that whenever energy is transformed from one form to another form, entropy increases and energy decreases. The Second Law of Thermodynamics is commonly known as the Law of Increased Entropy. While quantity remains the same (First Law), the quality of matter/energy deteriorates gradually over time. How so? Usable energy is inevitably used for productivity, growth and repair. In the process, usable energy is converted into unusable energy. Thus, usable energy is irretrievably lost in the form of unusable energy
597. Sensor - Is a device that detects and responds to some type of input from the physical environment. The specific input could be light, heat, motion, moisture, pressure, or any one of a great number of other environmental phenomena. The output is generally a signal that is converted to human-readable display at the sensor location or transmitted electronically over a network for reading or further processing
598. Serial Interface - (as distinguished from a parallel interface) is one in which data is sent in a single stream of bits, usually on a single wire-plus-ground, wire-pair, or single wireless channel (or two sets, one for each direction). Examples include USB, RS-232, I2C, and 1-Wire. By contrast, a parallel interface sends several bits at once, on separate wires. Serial interfaces have certain advantages over parallel interfaces. The most significant advantage is simpler wiring. In addition, serial interface cables can be longer than parallel interface cables, because there is much less interaction (crosstalk) among the conductors in the cable

599. **Serial peripheral interface (SPI)** - Is an interface that enables the serial (one bit at a time) exchange of data between two devices, one called a master and the other called a slave. An SPI operates in full duplex mode. This means that data can be transferred in both directions at the same time. The SPI is most often employed in systems for communication between the central processing unit (CPU) and peripheral devices. It is also possible to connect two microprocessors by means of SPI
600. **SETI** - The search for extraterrestrial intelligence (SETI) is the collective name for several activities people undertake to search for intelligent extraterrestrial life
601. **Shareware** - Is a type of software distributed based on an honor system. Most shareware is delivered free of charge, but the author usually requests that you pay a small fee if you like the program and use it regularly. By sending the small fee, you become registered with the producer so that you can receive service assistance and updates
602. **Singularity** - A point in spacetime at which a physical quantity become infinite. In scientific terms, a gravitational singularity (or spacetime singularity) is a location where the quantities that are used to measure the gravitational field become infinite in a way that does not depend on the coordinate system. In other words, it is a point in which all physical laws are indistinguishable from one another, where space and time are no longer interrelated realities, but merge indistinguishably and cease to have any independent meaning.
603. **Snowball Earth** - Refers to the contention that in the distant past the Earth froze over from pole to pole. This frozen Earth, nicknamed snowball Earth, was a setting "so severe, that the Earth's entire surface, from pole to pole, including the oceans, completely froze over. Ancient rocks suggest that ice entirely covered our planet on at least two occasions. This theory may help explain the rise of complex life that followed.
604. **Software** - Sometimes abbreviated s/w, is also called a computer program that enables a computer is also called a computer program that enables a computer to perform a specific task, as opposed to the physical components of the system (hardware). This includes application software such as a word processor, which enables a user to perform a task, and system software such as an operating system, which enables other software to run properly, by interfacing with hardware and with other software. Programs stored on non-volatile storage built from integrated circuits (e.g. ROM or PROM) are usually called firmware

605. **Solid State physics** – Is the branch of physics which covers all properties of solid materials, including electrical conduction in crystals of semiconductors and metals, superconductivity and photoconductivity
606. **Solstice** - Either of the two points on the ecliptic at which the apparent longitude of the Sun is 90° or 270°; also, the time at which the Sun is at either point. Either of the two moments in the year when the Sun's apparent path is farthest north or south from Earth's Equator. In the Northern Hemisphere the summer solstice occurs on June 20 or 21 and the winter solstice on December 21 or 22
607. **Solute** - The substance dissolved in a solvent to form a solution
608. **Solution** - A homogeneous mixture of variable composition consisting of two or more substances, one of which is usually a liquid
609. **Solvent** - A liquid which dissolves a solute to form a solution. The most widely used solvent is water since it dissolves many substances; hence, it is regarded as the 'universal solvent'
610. **Sonic boom** – Occurs when an object travels faster than the speed of sound
611. **Source Program** - Refers to a program written in a source language, which is are classified as either high-level languages or low-level languages, according to whether each notation in the source language stands for many or only one instruction in machine code. Programs in high-level languages are translated into machine code by either a compiler or an interpreter program
612. **Southern lights** - *Aurora borealis* or northern lights and *aurora australis* or southern lights are natural electric phenomena that create bright, colorful and dramatic light displays in the sky
613. **Special relativity** - A theory proposed by Albert Einstein that describes the propagation of matter and light at high speeds ($E=mc^2$). It was invented to explain the observed behavior of electric and magnetic fields. The observable effects on a body in motion. As velocity increases, time slow down, mass increases and lengths contract. According to special relativity, no wave or particle may travel at a speed greater than the speed of light c. Therefore, the usual rules from Newtonian mechanics do not apply when adding velocities that are large enough
614. **Spreadsheet** - Typically refers to a table of values arranged in rows and columns. Each value can have a predefined relationship to the other values. If you change one value, therefore, you may need to change other values as well. There are many spreadsheet applications available that enable users to create and operate the data in the table efficiently
615. **St. Petersburg paradox** - In the St. Petersburg gamble, the house offers to flip a coin until it comes up heads. The payoff doubles each time tails appears,

with this compounding stopping and payment being given at the first heads. By conventional definitions, the St. Petersburg gamble has an infinite expected value; nonetheless, most people share the intuition that they should not offer more than a few dollars to play. Explaining why people offer such small sums to play a gamble with infinite expected value remains a contentious question in economics and philosophy

616. **Stack** - In computer science, a stack is a temporary abstract data type and data structure based on the principle of Last In First Out (LIFO). Stacks are used extensively at every level of a modern computer system. For example, a modern PC uses stacks at the architecture level, which are used to run an operating system. The operating system also uses stacks, which are used to run a Java Virtual Machine, which is stack oriented, and the Java language itself has a class called "Stack", which can be used by the programmer. The stack is ubiquitous

617. **Startup Code** - Refers to a piece of assembly language code that prepares the way for software written in a high-level language. Startup code usually initializes code and data segments, safes I/O, and sets up chip selects and wait states. Most cross-compilers come with startup code that a user can modify, compile, and link with embedded programs

618. **States of matter** - Solid, Liquid, Gas and Plasma. Solids – relatively rigid, definite volume and shape. In a solid, the atoms and molecules are attached to each other. They vibrate in place but don't move around. Liquids – definite volume but able to change shape by flowing. In a liquid, the atoms and molecules are loosely bonded. They move around but stay close together. Gases – no definite volume or shape. The atoms and molecules move freely and spread apart from one another. Plasma is sometimes referred to as a fourth state of matter. While it's like a gas the electrons are free in a cloud rather than attached to individual atoms. This means that a plasma has very different properties from those of an ordinary gas. Plasmas occur naturally in flames, lightning and auroras.

619. **Stereochemistry** - Is the branch of chemistry that deals with the determination of the relative positions in space of the atoms or groups of atoms in a compound and with the effects of these positions on the properties of the compound

620. **String Theory** - Uses a model of one-dimensional strings in place of the particles of quantum physics. These strings, the size of the Planck length (i.e. 10-35 m) vibrate at specific resonant frequencies

621. **Strong nuclear force** - Is the strongest force in the universe, 1038 times stronger than gravity and 100 times stronger than the electromagnetic force.

The only catch is that it only operates on length-scales of the atomic nucleus, rapidly dropping off for longer distances

622. **Structural biology** - The study of biological phenomena at the molecular and atomic level. Structural biologists are interested in understanding related processes such as protein folding, protein dynamics, molecular modeling, drug design, and computational biology
623. **Structured Query Language (SQL)** - Is a special-purpose language for managing data in relational database management systems. It is most commonly used for its "Query" function, which searches informational databases. SQL was standardized by the American National Standards Institute (ANSI) and the International Organization for Standardization (ISO) in the 1980s.
624. **Super Conductivity** - The flow of electricity without any resistance at certain temperatures near absolute zero in certain temperatures is known as super conductivity.
625. **Supergiant** - Can have masses from 10 to 70 solar masses and brightness from 30,000 up to hundreds of thousands of times the solar luminosity. They vary greatly in radii, usually from 30 to 500, or even in excess of 1000 solar radii. Because of their extreme masses they have short lifespans of only 10 to 50 million years and are only observed in young cosmic structures such as open clusters, the arms of spiral galaxies, and in irregular galaxies
626. **Superior Conjunction** - The alignment of an inferior planet and the Sun in which the planet is at the far side of the Sun from the Earth. An inferior conjunction occurs when the planet passes approximately between Earth and Sun; if it passes exactly between them, moving across the Sun's face as seen from Earth, it is said to be in transit
627. **Supernova** - One of the most energetic explosive events known is a supernova. These occur at the end of a star's lifetime, when its nuclear fuel is exhausted, and it is no longer supported by the release of nuclear energy. If the star is particularly massive, then its core will collapse and in so doing will release a huge amount of energy. This will cause a blast wave that ejects the star's envelope into interstellar space. The result of the collapse may be, in some cases, a rapidly rotating neutron star that can be observed many years later as a radio pulsar
628. **Symmetric Function** - Looks like a mirror image of itself, typically across the X or y axis or place of origin
629. **Synchronous Dynamic Random Access Memory (SDRAM)** - Is a form of semiconductor memory can run at faster speeds than conventional DRAM and is therefore the use of SDRAM is becoming more widespread

630. **Syntax** - In computer science, refers to the spelling and grammar of a programming language
631. **System software** - Also known as system program, is a type of software which includes the operating system and all utility programs that manage computer resources at a low level. System programs include operating systems, database managers, drivers, communications and messaging protocols, basic input/output system, compilers, loaders, linkers, and debuggers, etc.
632. **System-on-Chip (SoC)** - Is an integrated circuit that includes various parts. Single monolithic systems include a processor, a bus and other elements. Integrated circuits are used in a wide range of electronic equipment like portable handheld devices. In general, System-on-Chip technology is the ability to place multiple subsystems on a single semiconductor chip. It typically uses a powerful processor and can run software such as the desktop versions of operating systems

T

633. **Tachyon** - (if they exist) have $v > c$. This means that E is imaginary! Well, what if we take the rest mass m, and take it to be imaginary? Then E is negative real, and $E^2 - p^2 = m^2 < 0$
634. **Tau Particle** - Is the heaviest of the three members of the family of fundamental particle known as leptons. The tau particle acts like an electron, the most familiar lepton, except that it has a mass 1,777 times that of an electron
635. **Taxonomic kingdom** – In biology, it is the second highest rank below domain and above pylum
636. **Taylor Series** - Is an expansion of a function into an infinite sum of terms. You can use the first few terms of a Taylor Series to get an approximate value for a function
637. **TCP/IP** - Is an acronym for Transmission Control Protocol/Internet Protocol. It's a set of communication protocols used to connect host computers on the Internet
638. **Tempest** - Was the name of a classified (secret) U.S. government project to study (probably for the purpose of both exploiting and guarding against) the susceptibility of some computer and telecommunications devices to emit electromagnetic radiation (EMR) in a manner that can be used to reconstruct intelligible data

639. **Terahertz, abbreviated THz** - Is a unit of electromagnetic (EM) wave frequency equal to one trillion hertz (1012 Hz). The terahertz is used as an indicator of the frequency of infrared (IR), visible, and ultraviolet (UV) radiation
640. **Terminator** - The boundary between the illuminated and dark areas of a celestial body. For example, during half-moon, the terminator is in the middle
641. **Tesla coil** - Is an electrical resonant transformer circuit designed by inventor Nikola Tesla around 1891 as a power supply for his "System of Electric Lighting". It is used to produce high-voltage, low-current, high frequency alternating-current electricity.
642. **The Singularity** - Is the hypothetical future creation of super intelligent machines. Superintelligence is defined as a technologically created cognitive capacity far beyond that possible for humans. Should the Singularity occur, technology will advance beyond our ability to foresee or control its outcomes and the world will be transformed beyond recognition by the application of superintelligence to humans and/or human problems, including poverty, disease and mortality
643. **Theory of General Relativity** - The theory of gravitation, developed by Einstein in 1916, extending the special theory of relativity to include acceleration and leading to the conclusion that gravitational forces are equivalent to forces caused by acceleration
644. **Theory of Special Relativity** - In 1905, Albert Einstein determined that the laws of physics are the same for all non-accelerating observers, and that the speed of light in a vacuum was independent of the motion of all observers. This was the theory of special relativity. It introduced a new framework for all of physics and proposed new concepts of space and time
645. **Thermal imaging** - Is a method of improving visibility of objects in a dark environment by detecting the objects' infrared radiation and creating an image based on that information. Near-infrared illumination, low-light imaging and are the three most used night vision technologies. Unlike the other two methods, thermal imaging works in environments without any ambient light. Like near-infrared illumination, thermal imaging can penetrate obscurants such as smoke, fog and haze. The infrared energy emitted by an object is known as its heat signature. In general, the hotter an object is, the more radiation it emits. A thermal imager (also known as a thermal camera) is essentially a heat sensor that can detect tiny differences in temperature
646. **Thermochemistry** - A branch of thermodynamics which focuses on the study of heat given off or absorbed in a chemical reaction. The laws of thermochemistry are used to describe the heat or energy flow of chemical

reactions. Understanding the values of a reaction's enthalpy will help determine whether the reaction is endothermic or exothermic

647. **Thermodynamics** - Is a branch of physics which deals with the energy and work of a system. Thermodynamics deals only with the large-scale response of a system which we can observe and measure in experiments Thermodynamics deals with the relationships between heat and other forms of energy. In particular, it describes how thermal energy is converted to and from other forms of energy and how it affects matter
648. **Third Law of Motion** - For every action, there is an equal and opposite reaction. Forces always come in pairs - equal and opposite action-reaction force pairs
649. **Time dilation** - The length of any object in a moving frame will appear foreshortened in the direction of motion or contracted. In the theory of relativity, time dilation is an actual difference of elapsed time between two events as measured by observers either moving relative to each other
650. **Torque** - The tendency of a body to rotate under an applied force
651. **Toxicology** - Is the scientific study of adverse effects that occur in living organisms due to chemicals. It involves observing and reporting symptoms, mechanisms, detection and treatments of toxic substances, in particular relation to the poisoning of humans
652. **Transfer RNA** - Abbr. tRNA Is one of a class of RNA molecules that transport amino acids to ribosomes for incorporation into a polypeptide undergoing synthesis
653. **Transformer** - An inductive electrical device for changing the voltage of alternating current. A transformer consists of two magnetically coupled coils. Alternating current in one (called the "primary") creates a changing magnetic field which induces a current in the second coil (the "secondary"). A core made of Iron or ferrite generally connects the two coils, but higher frequency devices can work without a ferrous core. Transformers have two primary functions: Voltage transformation and isolation
654. **Transistor** - A basic solid-state control device which allows or disallows current flow between two terminals, based on the voltage or current delivered to a third terminal. Usually built from silicon but can be constructed from other semiconductor materials. There are two major types: The FET (field-effect transistor) and the bipolar junction transistor (BJT). The first transistor was invented in 1947 at Bell Labs by Michael John Bardeen, Walter Brattain and William Shockley

655. **Transmitter** - A circuit that accepts signals or data in and translates them into a form that can be sent across a medium (transmitted), usually over a distance. The medium can be wireless or wired
656. **Triple point** - The temperature where the solid, liquid and gaseous phases of a substance are in simultaneous equilibrium. The triple point of water is 273.16 K
657. **Tri-State output** - Has three electrical states - One, zero, and "Hi-Z," or "open." The hi-Z state is a high-impedance state in which the output is disconnected, leaving the signal open, to be driven by another device (or to be pulled up or down by a resistor provided to prevent an undefined state). High-impedance schemes such as three-state are commonly used for a bus, in which several devices can be selected to drive the bus
658. **Turbine** - Is a machine that transforms rotational energy from a fluid that is picked up by a rotor system into usable work or energy. Turbines achieve this either through mechanical gearing or electromagnetic induction to produce electricity. Types of turbines include steam turbines, wind turbines, gas turbines or water turbines
659. **Turing Machine** – Was first described by Alan Turing is a simple abstract computational device intended to help investigate the extent and limitations of what can be computed. It is an abstract machine analogous to a computer in concept. Fed by tape. The machine reads the tape one entry at a time. It is considered the model for present day computers
660. **Turkana boy** - A skeleton that has allowed scientists to find out a lot of information about body size, body shape, and growth rates *of Homo erectus*. This skeleton is 40% complete, based on the principle that bones from one side of the body can tell what the same bone from the other side looked like even if it's missing. It is a nearly complete skeleton of a 12-year-old hominid boy who died 1.6 million years ago
661. **Turnkey** - Is a product or service that is designed, supplied, built, or installed fully complete and ready to operate. The term implies that the end user just has to turn a key and start using the product or service.

U

662. **Ultraviolet Light** - The region of the electromagnetic spectrum ranging from 9.12 nm to 350 nm is known as ultraviolet light. UV rays are emitted from the Sun which are blocked by the Earth's atmosphere
663. **Uncertainty Principle** - It is impossible to know exactly where something is and where it is going. This is a fundamental law of nature has a major effect on quantum theory. The uncertainty principle (for position and momentum) states that one cannot assign exact simultaneous values to the position and momentum of a physical system. Rather, these quantities can only be determined with some characteristic "uncertainties" that cannot become arbitrarily small simultaneously
664. **Universe** - Is commonly defined as the totality of existence, including planets, stars, galaxies, the contents of intergalactic space, and all matter and energy
665. **UNIX** - Is a computer operating system originally developed in the 1960s and 1970s by AT&T Bell Labs. Today Unix is split into various branches, developed over time by many companies and non-profit organizations, such as contributors to the GNU project
666. **URL** - Stands for Universal Resource Locator. It's a way of accessing the Internet. It is used to specify addresses on the World Wide Web. A URL is the fundamental network identification for any resource connected to the web (e.g., hypertext pages, images, and sound files). URLs have the following format protocol://hostname/other_information
667. **Uranium lead dating** - Is a radiometric dating method that uses the decay chain of uranium and lead to find the age of a rock. As uranium decays radioactively, it becomes different chemical elements until it stops at lead
668. **User Interface** - Is the aggregate of means by which people (the users) interact with a particular machine, device, computer program or other complex tool (the system). The user interface provides two critical functions: Input, allowing the users to manipulate the system, and Output, allowing the system to produce the effects of the users' manipulation. There are many types of user interfaces, such as GUI, Command Line, Web-based user interfaces, and touch interface, etc.
669. **Utility Program** - Or utility, is a type of computer programs that support using the computer, an application or a development environment. Utility programs include file management, searching for files, comparing file contents

as well as performing diagnostic routines to check performance and current health of the hardware

V

670. **Valence electrons** - Electrons in the outer electronic shell of an atom, and which are involved in chemical reactions
671. **Vaporization** - A phase change involving the conversion of a substance from the liquid state to the vapor phase. Enthalpy of vaporization is an important property of any liquid, which states that when heat or enthalpy is given to any liquid substance at a certain pressure and temperature, the liquid changes into a gaseous form. Thus, a transformation from one state to other state occurs.
672. **Vector** - A quantity that is determined by its magnitude and direction: forces and fields
673. **Vector Analysis** - A branch of mathematics that deals with quantities that have both magnitude and direction. Some physical and geometric quantities, called scalars, can be fully defined by specifying their magnitude in suitable units of measure
674. **Velocity** - A vector quantity describing the rate of change of displacement or distance with time. Velocity is a vector quantity. As such, velocity is direction aware. When evaluating the velocity of an object, one must keep track of direction. Since displacement is expressed in meters and time taken in seconds. Velocity is expressed in meters/second or m/s
675. **Venn diagram** - Is an illustration of the relationships between and among sets, groups of objects that share something in common. Usually, Venn diagrams are used to depict set intersections (denoted by an upside-down letter U). This type of diagram is used in scientific and engineering presentations, in theoretical mathematics, in computer applications, and in statistics
676. **Very large-scale integration (VLSI)** - Refers to an IC or technology with many devices on one chip. The term originated in the 1970s along with "SSI" (small-scale integration), "LSI" (large-scale), and several others, defined by the number of transistors or gates per IC
677. **Virus** - A microorganism which can only reproduce inside a living thing. any member of a unique class of infectious agents, which were originally distinguished by their smallness (hence, they were described as "filterable" because of their ability to pass through fine ceramic filters that blocked all cells, including bacteria) and their inability to replicate outside of and without

assistance of a living host cell. Because these properties are shared by certain bacteria

678. **Visual C++** - Also known as MSVC, is an Integrated Development Environment (IDE) product for the C and C++ programming languages engineered by Microsoft. It has tools for developing and debugging C++ code, especially that written for the Microsoft Windows API, the DirectX API, and the Microsoft .NET Framework

W

679. **WAN** - Is an acronym for Wide Area Network. Such a network spans over an area larger than a LAN
680. **Watt** - The SI unit of power, with symbol W. 1 W = 1 joule.sec-1
681. **Wave Particle Duality** - In physics and chemistry, wave-particle duality holds that light and matter exhibit properties of both waves and of particles. Treating Electrons as waves helps explain why only certain orbits are allowed in atoms
682. **Wavelength** - The distance between two successive points which are in phase on a wave train, e.g., the distance between two successive maxima or minima. The distance between one peak of a wave to the next corresponding peak, or between any two adjacent corresponding points, defined as the speed of a wave divided by its frequency
683. **Weak nuclear force** - Fundamental interaction that underlies some forms of radioactivity and certain interactions between subatomic particles. Some scientists believe that the weak nuclear force and the electromagnetic force are both aspects of a single force called the electroweak force
684. **Web Page** - Is an HTML/XHTML document that is included in a website. A web page is almost always accessible over the network or Internet via HTTP
685. **Weight** - The gravitational force exerted on a mass
686. **White dwarf** - White dwarf stars are the corpses of stars; what happens once they've used up all their fuel and lack the temperature and pressure to continue fusion in their core
687. **Wideband** - A classification of the information capacity or bandwidth of a communication channel. Wideband is generally understood to mean a bandwidth between 64kbits/s and 2Mbit/s

688. **Wien's Law** - Also called Wien's displacement law, relationship between the temperature of a blackbody (an ideal substance that emits and absorbs all frequencies of light) and the wavelength at which it emits the most light
689. **Wirth's Law** - States that computer software increases in complexity faster than does the ability of available hardware to run it
690. **Work** - A physical quantity being the product of a force and the distance through which that force is applied. Units, joules. The amount of energy transferred to a system
691. **Work function** - The minimum amount of energy required to eject an electron from the surface of a substance
692. **Wormhole** – An Einstein-Rosen Bridge (or wormhole) is a hypothetical topological feature of spacetime that would be, fundamentally, a "shortcut" through spacetime

X

693. **X-Chromosome** - A type of chromosome that is found in pairs in the cells of female mammals and is found with the Y chromosome in the cells of male mammals
694. **X-Intercept** - The value of x where a line or curve intersects the x-axis
695. **X-rays** - High energy electromagnetic radiation, with a frequency of 1018 - 1020 Hz

Y

696. **Y-Chromosome** - Y is the sex-determining chromosome in many species, since it is the presence or absence of Y that determines the male or female sex of offspring produced in sexual intercourse
697. **Y Chromosome Test** - Is a genealogical DNA test which is used to explore a man's patrilineal or direct father's-line ancestry. The Y chromosome, like the patrilineal surname, passes down virtually unchanged from father to son. It has also been used to determine the ancestry of humanity and along with Mitochondrial DNA has confirmed the African ancestry of humanity

698. **Y-Intercept** - The value of y where a line or curve intersects the y-axis

Z

699. **Zeckendorff decomposition** - States that every positive integer can be uniquely decomposed as a sum of non-consecutive Fibonacci numbers {Fn}{Fn}, where F1=1F1=1, F2=2F2=2 and Fn+1=Fn+Fn−1Fn+1=Fn+Fn−1
700. **Zodiac** - The Sun, the moon, and the planets travel on a set path through the sky known as the ecliptic as the Earth rotates. The list of 13 constellations they pass through are known as the stars of the Zodiac. The Zodiac constellations' names: Capricorn, Aquarius, Pisces, Aries, Taurus, Gemini, Cancer, Leo, Virgo, Libra, Scorpio, Ophiuchus, Sagittarius
701. **Zoology** - The branch of biology that deals with animals and animal life, including the study of the structure, physiology, development, and classification of animals
702. **Zygote** - A fertilized egg, the fusion of a male and female gamete

TIMELINE OF SCIENTIFIC EVENTS

This section is a chronological outline of major scientific events, discoveries and innovations from the Big Bang until the present day. By no means is this a complete list. However, it can be used as a teaching tool to expose STEM students and educators to the history of Scientific events and discoveries.

Timeline	Event
	Before the Common Era (BCE)
Designations	bya = billion years ago, mya= million year ago
14.5 bya	Birth of the Universe (in accordance with the Big Bang). The current estimate is that the universe contains 200 billion galaxies with 200 billion stars(suns):
4.5 bya	Birth of the Sun, Solar System and Earth
2 bya	Free Oxygen in the atmosphere as a result of photosynthesis
600 mya	Present atmosphere, multicellular life
550 mya	Cambrian explosion – formation of complex, hard-bodied animals
250- 65 mya	Mesozoic Era- Age of the Dinosaurs
55 mya	Earliest primates
8 mya	Common ancestor to African apes and hominids
7 mya	First hominid, perhaps *Sahelanthropus tchadensis*
4.2 mya	Frist australopithecines
2.5mya-5000 BCE	Stone Age
2.4 mya	First hominid in the Homo genus, *Homo habilis*
1.9 mya	First *Homo erectus*
1.8 mya	First migration out of Africa by *Homo erectus*
125,000	First *Homo sapiens* (Modern man) based on DNA

70,000	Homo sapiens second migration
30,000	Disappearance of Neanderthals; *Homo sapiens* is the only surviving hominid
25,000- 11,000 BCE	The Ice Age at its height 18,000 years ago
10,000 BCE	Domestication of Animals, Agriculture
12,000-10,000 BCE	End of the (Pleistocene) Ice Age
3,500-3000 BCE	Civilization in Sumer and Egypt. The Sumerians develop cuneiform writing, and the Egyptians develop hieroglyphic
2,600 BCE	Egyptians build the Pyramids
540 BCE	Pythagoras founded a school and introduced the Pythagorean Theorem
440 BCE	Democritus speculates about fundamental indivisible particles---calls them ``atoms"
387 BCE	Plato founded the Academy
300 BCE	Euclid presents his geometry in The Elements
	The Common Era (CE)
145	Ptolemy develops his theories on Astronomy which last for over 1,000 years
105	Tsai Lun invents paper
786	The House of Wisdom founded in Bagdad
820	Muhammed ibn Musa al-Kwarizmi wrote essays on Hindu arithmetic and *Al Jabr*, translated as 'the transposition,' and pronounced 'algebra.
830	Al -Jahiz proposes theory of evolution
850	Abu Yusek Yacob ibn Ishak al-Kindi commented on Aristotle and wrote numerous treatises on optics, perspective, and medicine. Thabit ibn Qurra publishes many famous works. He translates Nicomachus of Gerasa's Arithmetic and makes important discoveries in algebra and geometry. In astronomy, he is considered one of the first reformers of the Ptolemaic system, and in mechanics he is the founder of

	statics, the analysis of force and torque
855	Hunayna ibn Ishaq becomes one of the masters of the Translation Movement. His most famous translation Questions on Medicine becomes a key resource for medical scholars
900	Abu Bakr al-Razi, better known as Rhazes, distinguished smallpox from measles while writing several medical books in Arabic
976	A manuscript from non-Moslem Spain showed the first examples of the nine Hindu-Arabic numerals, 0 thru 9 in Europe.
1000	Ibn al-Haitam, or al-Hazen, in *Opticae Thesaurus*, introduced the idea that light rays emanate in straight lines in all directions from every point on a luminous surface
1025	Abu Ali al-Hussain ibn Sina (also referred to by his western name, Avicenna) is often referred to as the "Father of Modern Medicine." His primary works are The Book of Healing, which is a philosophical and scientific encyclopedia, and The Canon of Medicine, which is an overview of all aspects of medicinal knowledge. He recognizes the cure for tuberculosis, and also contributes to astronomy, alchemy, geography, psychology, theology, mathematics, and poetry.
1054	Chinese astronomers at the Sung national observatory at K'ai-feng observed the explosion of a supernova in the Crab Nebulae, visible in daylight for twenty-three days.
1126	Adelard of Bath translated Euclid's Elements and al-Kwarizmi's arithmetic and astronomical tables from Arabic into Latin
1200	Fibonacci brought the knowledge of Islamic math and numbers to Europe. University of Paris started Oxford about twenty years later
1249	Roger Bacon states formulas for gunpowder

1450	The Chinese develop wooden block movable type printing
1454	Johannes Gutenberg finishes a printing press with metal movable type
1492	Leonardo da Vinci describes a flying machine
1543	Copernicus develops the theory of a heliocentric model
1605	Francis Bacon, with the Advancement of Learning, began the publication of his philosophical works, in which he urged collaboration between the inductive and experimental methods of proof, as opposed to scholasticism's a priori method
1609	Galileo builds his first telescope
1613	Galileo Galilei uses sunspot observations to demonstrate the rotation of the Sun
1619	Kepler's introduces his laws on planetary motion
1621	Galileo discerned that the acceleration of a falling body is proportional to the time and independent of weight and density
1627	William Harvey was able to confirm his observation that the blood circulates throughout the body, which he inferred from the structure of the venal valves.
1637	Descartes, in *Discours de la Méthod pour bien conduire sa raison, et chercher la vérité dans les sciences*, held that science begins with observation, which is followed by analysis, leading to the intuition of the self-evident nature of a proposition, and synthesis, or the reconstitution of the original observation
1644	Blaise Pascal built a five-digit adding machine, driven by rising and falling weights
1660	The Royal Society is founded in London
1661	Robert Boyle, in the *Sceptical Chymist,* separated chemistry as corpuscles, from alchemy, as qualities, and gave the first precise definitions of a chemical element, a chemical reaction, chemical analysis, and made studies of acids and bases.

1662	Boyle, using a vacuum pump of his own invention, determined that the volume and pressure of a gas are inversely proportional. This is known as 'Boyle's law.' He also determined that sound doesn't travel in a vacuum.
1665	Robert Hooke describes the cells of plants Isaac Newton deduces the inverse-square gravitational force law from the ``falling'' of the Moon
1665-1879	Newton and Leibniz invent Calculus
1677	Leeuwenhoek discovers single-cell organisms
1687	Newtons work *Principia* is published
1704	Newton, in *Opticks*, presented his discoveries using light and elaborated his theory that it was composed of particles
1705	Edmund Halley publicly predicts the periodicity of Halley's comet and computes its expected path of return in 1758
1709	Gabriel Daniel Fahrenheit constructed an alcohol thermometer and, five years later, a mercury thermometer
1742	Celsius developed the centigrade temperature scale which carries his name
1747-1752	Ben Franklin's electrical experiments
1753	Carl Linné, better known as Carolus Linneas, published Species plantarum, in which he distinguished plants in terms of genera and species
1758	Linné published the 10th edition of *Systema Naturae,* where he applied the genera and species system to animals based on their external appearance
1766	Henry Cavendish discovers and studies hydrogen
1771	Joseph Priestly discovers that plants convert carbon dioxide into oxygen
1778	Carl Scheele and Antoine Lavoisier discover that air is composed mostly of Nitrogen and Oxygen
1781	Joseph Priestly creates water by igniting Hydrogen and Oxygen
1769	James Watt patented a new type of steam engine with a

	separate condensing chamber and an air pump to bring steam into the chamber. He coined the term 'horsepower' and later loaned his name to the unit of power, or work, done per unit of time
1774	Priestly discovered Sulphur dioxide, ammonia, and 'dephlogisticated air,' later named Oxygen by Lavoisier.
1781	Frederick William Herschel discovered the planet Uranus by its movement
1785	Charles Augustin de Coulomb, discovered the reciprocal attraction of the electrical fluid called positive, on the electrical fluid ordinarily called negative, is in the inverse proportion of the squares of the distance
1794	Erasmus Darwin, Charles' grandfather, proposed that "warm-blooded animals have arisen from one living filament" Eli Whitney patented the cotton gin
1799	Voltra invents the electric battery
1800	William Nicholson and Anthony Carlisle use electrolysis to separate water into Hydrogen and Oxygen
1801	Thomas Young, in "On the theory of light and colors," proposed that light striking the retina creates vibrations and the frequency of the vibrations excites a particular nerve filament, "one for each principle color." He also made the observation that if light from a single source is split into two beams, then recombined and projected on a screen, dark and light fringes appear
1801- 1817	Gauss work on differential geometry
1803	John Dalton introduces atomic ideas into chemistry and states that matter is composed of atoms of different weights
1811	Fourier, in a paper on heat diffusion, invented the formula for a trigonomic series by which any repeated physical event can be defined by its phase and its amplitude and represented as a set of simple wave forms. As this was

	incapable of expressing initial conditions in infinite bodies, he also created an integral theorem. Today these are known as Fourier series and Fourier integrals Amedeo Avogadro proposes Avogadro's law, that equal volumes of gases contain equal numbers of particles
1820s	Joseph Henry, by vastly increasing the number of wire coils around a magnet, created a powerful electromagnet
1820	Hans Oersted notices that a current in a wire can deflect a compass needle Francis Bacon, in *The New Organon*, pointed out, as an "instance of resemblance," that maps of Africa and South America show "similar isthmuses and similar promontories, and that does not happen without a reason
1822	Joseph Fourier formally introduces the use of dimensions for physical quantities in *his* {\sevenit *Theorie Analytique de la Chaleur*}
1824	Sadi Carnot scientifically analyzes the efficiency of steam engines
1826	Simon Ohm states his law of electrical resistance
1827	Robert Brown discovers the Brownian motion of pollen and dye particles in water
1831	Michael Faraday states his law of induction Robert Brown identifies the nucleus of a cell
1834	Charles Babbage designed a programmable mechanical calculating machine, or 'analytical engine,' that could carry out arithmetic operations specified on punch cards and choose the sequence of operations Benoit-Pierre Clapeyron presents a formulation of the second law of thermodynamics
1842	Christian Doppler developed the theory that the frequency of energy in the form of the form of waves changes depending on the motion of either the sender or the receiver

1843	James Joule experimentally finds the mechanical equivalent of heat Experimentally demonstrated that heat is a form of energy Samuel Morse builds the first long distance electric telegraph line
1844	Charles Goodyear discovered vulcanization of rubber by Sulphur, inorganic accelerator, and heat. Hancock in England made a parallel discovery
1846	Johann Gottfried Galle discovered the planet Neptune where Urbain Jean Joseph Le Verrier and, independently, John Couch Adams had predicted that a planet would be found. Their predictions were based on perturbations in the orbit of Uranus
1848	Lord Kelvin discovers the absolute zero point of temperature British Mathematician George Boole devised binary algebra (Boolean algebra) paving the way for the development of a binary computer almost a century later
1849	Armand Fizeau and Jean-Bernard Foucault measure the speed of light to be about 298,000 km/s
1850s	Gustav Kirchhoff and Robert Bunsen discovered that each element absorbs or emits light at specific wavelengths, producing specific spectra.
1854	Heinrich Geissler creates the first vacuum tube.
1859	Darwin publishes the Origins of Species
1860	Bunsen and Kirchhoff, in *Chemische Analyse durch Spektralbeobachtungen,* recounted their discovery of cesium and rubium and explained the Fraunhofer lines in the solar spectrum as being dark absorption lines which are created by cooler gaseous clouds absorbing energy from the Sun
1862	Alvan Clark observes Sirius B
1865	Gregor Mendel presents his experiments on the crossbreeding of pea plants and postulates dominant and

	recessive factors
1869	Mendeleev proposes the Periodic Table
1873	Maxwell develops his equations describing Electromagnetic phenomena
1876	Nikolaus Otto designs a four-stroke internal-combustion engine
1876	Alexander Graham Bell and Thomas Watson exhibit an electric telephone
1877	Thomas Edison patents the phonograph
1879	Thomas Edison patents the carbon-thread incandescent lamp Josef Stefan observes that the total radiant flux from a blackbody is proportional to the fourth power of its temperature William Crookes made headway in modern atomic theory when he used the vacuum tube made by Heinrich Geissler to discover cathode rays
1883	Tesla constructs the first polyphase AC motor Edison demonstrates a village lite by Electricity in Roselle, New Jersey
1885	Eugene Goldstein discovered positive particles (protons) by using a tube filled with hydrogen gas
1887	Tesla builds first AC Induction Motor Michelson and Edward W. Morley, using an interferometer to investigate whether the speed of light depends on the direction the light beam moves, failed to detect the motion of the Earth with respect to the aether, thereby refuting the hypothesis that the aether exists Edison patents motion picture camera
1888	Heinrich Hertz discovers radio waves Nikola Tesla patents the induction motor
1895	Roentgen discovers x-rays
1896	Tesla builds the first long-distance transmission of

	polyphase AC Power at Niagara Falls
1897	Joseph Thomson discovers the electron
1898	William Ramsay and Morris Travers discover neon, krypton, and xenon Marie Curie and Pierre Curie isolate and study radium and polonium
1899	Lorentz, in "*Théorie simplifiée des phénomènes électriques et optique dans des corps en mouvement*," treated his contraction hypothesis mathematically in terms of electrons and, except for not determining the coefficient and not generalizing the subject to any mass, the resulting transformations for space and time coordinates are the same as those in his 1904 article
1900	Planck, in "*Zur Theorie des Gesetzes der Enieverteilung im Normalspektrum*," introduced the 'quantum theory' to explain a formula, E=hf, where E is energy, f is frequency, and h is a new constant, 6.63 x 10-34 J-sec., which accounts for experimental data in black-body radiation
1901	Guglielmo Marconi transmits radio signals from Cornwall to Newfoundland
1902	Walter Sutton discovers that egg and sperm have matching chromosomes
1903	Orville Wright and Wilbur Wright fly the first motor-driven airplane
1905	Einstein presents his Theory of Special Relativity Fleming electron tube
1906	de Forest develops Audion vacuum tube amplifier
1908	Henry Ford develops the assembly line method of automobile manufacturing
1913	Niels Bohr presents his quantum model of the atom
1915	Albert Einstein completes his theory of general relativity
1922	The Russian cosmologist and mathematician Alexander Friedmann discovers the expanding universe solution to

	Einstein's general relativity field equations.
1923	Louis de Broglie suggests that electrons may have wavelike properties
1924	Hubble discovers galaxies; Wolfgang Pauli states the quantum exclusion principle
1926	Robert Goddard launches the first liquid fuel rocket Oskar Klein, attempting to explain what happened to Kaluza's fifth dimension, proposed that we do not notice it because it is "'rolled up' to a very small size [and that] what we normally think of as a point in three-dimensional space is in reality a tiny circle going around the fourth dimension"
1926	The German physicist Werner Heisenberg formulates his uncertainty principle, that the values of certain pairs of variables cannot both be known exactly (i.e. the more precisely one variable is known, the less precisely the other can be known), a central concept in quantum physics Erwin Schrodinger states his nonrelativistic quantum wave equation and formulates quantum wave mechanics, He proves that the wave and matrix formulations of quantum theory are mathematically equivalent
1927	Philo Farnsworth invents fully electronic TV (First all-electronic TV is made by RCA in 1932)
1928	Alexander Fleming discovers penicillin
1931	Kurt Gödel shows that mathematical systems are not fully self-contained
1932	The English physicist James Chadwick discovers the neutron
1935	The Indian American astrophysicist Subrahmanyan Chandrasekhar establishes the "Chandrasekhar limit" of about 1.4 solar masses, above which a star must continue to collapse into a neutron star rather than settling down into a white dwarf
1936	Electrical engineer and mathematician Claude Shannon, in

	his master's thesis, "A Symbolic Analysis of Relay and Switching Circuits," uses Boolean algebra to establish a working model for digital circuits. This paper, as well as later research by Shannon, lays the groundwork for the future telecommunications and computer industries
1943-1944	Two University of Pennsylvania professors, John Mauchly and J. Presper Eckert, build the Electronic Numerical Integrator and Calculator (ENIAC). Considered the grandfather of digital computers, it fills a 20-foot by 40-foot room and has 18,000 vacuum tubes. The military needed ENIAC for the calculation of ballistic tables.
1945	The Atomic Bomb was developed and dropped on Japan
1947	William Shockley, John Bardeen and Walter Brattain of Bell Laboratories invent the transistor.
1949	EDVAC (electronic discrete variable computer) - First computer to use Magnetic Tape. This was a breakthrough as previous computers had to be re-programmed by re-wiring them whereas EDVAC could have new programs loaded off of the tape. Proposed by John von Neumann, it was completed in 1952 at the Institute for Advance Study, Princeton, USA
1950	Alan Turing proposes the ``Turing test" criterion for an intelligent machine
1952	Watson and Crick discover the structure of the double helix
1953	Grace Hopper develops the first computer language, which eventually becomes known as COBOL. Thomas Johnson Watson Jr., son of IBM CEO Thomas Johnson Watson Sr., conceives the IBM 701 EDPM to help the United Nations keep tabs on Korea during the war
1953	Gamow began the attempts to explain the coding problem, that is, how the sequential structure of DNA could directly, physically order the sequential structure of proteins. In Gamow's scheme, several different base sequences could

	code for one amino acid
1954	The FORTRAN programming language is born.
1957	Soviet Union launch Sputnik, the first artificial satellite. The United States government formed the Advanced Research Agency, or ARPA, in response to Sputnik John Backus led the team which created 'Fortran,' the Formula Translation language for the IBM 704 computer
1958	Jack St. Clair Kilby built the first integrated circuit. Explorer 1 was the first satellite launched by the United States when it was sent into orbit on January 31, 1958
1960	Theodore Maiman makes the first laser
1961	April 12 - Russian cosmonaut Yuri Gagarin became the first human in space May 5 - Astronaut Alan Shepard became the first American in space May 25 - President Kennedy challenged the country to put a man on the moon by the end of the decade.
1962	February 20 - Astronaut John Glenn became the first American in orbit June 16 - Valentina Nikolayeva Tereshkova became the first woman in space
1964	Louis Leaky identified and named *Homo habilis* It was a pair of radio astronomers, Arno Penzias and Robert Wilson, who discovered cosmic microwave background radiation
1969	Americans land on the moon
1970	The first X-ray astronomical satellite, built by NASA, was launched and over the next three years, discovered many X-ray sources. The newly formed Intel unveils the Intel 1103, the first Dynamic Access Memory (DRAM) chip
1971	The Mariner 9 spacecraft began to map Mars Alan Shugart leads a team of IBM engineers who invent the

	"floppy disk," allowing data to be shared among computers First microprocessor, the 4004, developed by Marcian E. Hoff for Intel, was released. It contains the equivalent of 2300 transistors and was a 4-bit processor. It is capable of around 60,000 Interactions per second (0.06 MIPs), running at a clock rate of 108KHz
1972	C programming language developed at The Bell Laboratories in the USA by Dennis Ritche (one of the inventors of the UNIX operating system), its predecessor was the B programming language - also from The Bell Laboratories. It is a very popular language, especially for systems programming - as it is flexible and fast. C++, allowing for Object-Orientated Programming, was introduced in early 1980s First handheld scientific calculator released by Hewlett-Packard, the engineer's slide rule is at last obsolete. 8008 Processor released by Intel The first international connections to ARPANET are established. ARPANET later became the basis for what we now call the internet
1973	Vinton Cerf and Robert E. Kahn began development of a protocol, later called TCP/IP, which allows diverse computer networks to interconnect and communicate with each other Robert Metcalfe, a member of the research staff for Xerox, develops Ethernet for connecting multiple computers and other hardware Motorola created the world's first cell phone. Motorola engineer Martin Cooper created this revolutionary device called the DynaTAC 8000X
1974	Introduction of 8080. An 8 Bit Microprocessor from Intel
1975	The Altair 8800, widely considered the first home computer, is marketed to hobbyists by Micro Instrumentation Telemetry Systems. Bill Gates and Paul Allen propose

	writing a version of BASIC. They start the project by forming a partnership called Microsoft
1976	Z80 released by Zilog, and the basis for the computer boom in the early 1980s. It was an 8-bit microprocessor. CP/M was written for the Z80 as well as software like Wordstar and dBase II - and it formed the basis for the Sinclair Spectrum of 1982 6502, 8-bit microprocessor developed and later chosen to equip the Apple II computer. It Also fitted in the original Acorn machine, BBC Micro and Commodore 64 Sony Beta home video released Steve Jobs and Steve Wozniak start Apple computer. NASA's Viking 1 probe touched down on Mars in July 1976. It was the first time a man-made object had soft-landed on the red planet.
1977	August and September - Voyagers 1 and 2 were launched. (Voyager 2 was launched before Voyager 1, but Voyager 1 was on a faster trajectory) VHS (Video Home System) format - Matsushita
1978	Mary Leaky announced the discovery of fossilized human footprints from about 3.5 million years ago Introduction of 8086 by Intel, the first commercially successful 16-bit processor. It was too expensive to implement in early computers, so an 8-bit version was developed (the 8088), which was chosen by IBM for the first IBM PC. This ensured the success of the x86 family of processors that succeeded the 8086 since they and their clones are used in every IBM PC compatible computer
1979	The spacecraft Voyager 1 photographed Jupiter's rings. The 68000 Microprocessor launched by Motorola. Used by Apple for the Macintosh and by Atari for the ST series. Later versions of the processor include the 68020 used in the Macintosh II Language Ada introduced by Jean Ichbiah and team at

	Honeywell The video laser disc introduced
1981	April 12 - Columbia became the first Space Shuttle to be launched The first IBM personal computer code-named "Acorn," is introduced. It uses Microsoft's MS-DOS operating system. It has an Intel chip, two floppy disks and an optional color monitor. Sears & Roebuck and Computerland sell the machines, marking the first time a computer is available through outside distributors. It also popularizes the term PC
1983	June 19 - Sally Ride became the first American woman in space on Challenger's second mission August 30 - Guion Bluford became the first African American in space TCP/IP became the official protocol of ARPANET (which became the Internet). TCP/IP still forms the basis of the Internet today Sony and Philips introduce 1st CD player
1984	CD-ROM introduced: Philips and Sony combine efforts to introduce the CD-ROM (compact disc read-only memory), patented in 1970 by James T. Russell. With the advent of the CD, data storage and retrieval shift from magnetic to optical technology. The CD can store more than 300,000 pages worth of information—more than the capacity of 450 floppy disks—meaning it can hold digital text, video, and audio files. Advances in the 1990s allow users not only to read prerecorded CDs but also to download, write, and record information onto their own disks.
1987	GIF format (CompuServe), JPEG format (Joint Photographic Experts Group) VGA (Video Graphics Array) invented by IBM
1989	World Wide Web, invented by Tim Berners-Lee who saw the need for a global information exchange that would allow

	physicists to collaborate on research. The Web was a result of the integration of hypertext and the Internet
1990	The United States National Aeronautics and Space Administration (NASA) and the European Space Agency (ESA) launched the Hubble Space Telescope, or HST The Human Genome Project (HGP) started in 1990 and was completed in 2003. The project was a voyage of biological discovery led by an international group of researchers looking to comprehensively study all of the DNA (known as a genome) of a select set of organisms.
1991	The World Wide Web becomes available to the general public (see Internet)
1992	Discovery of first exoplanet. The 5,000-plus planets found so far include small, rocky worlds like Earth, gas giants many times larger than Jupiter
1993	MPEG-1 standard published GPS system introduced
1994	MPEG-2 developed and published a year later
1995	The Galileo probe began transmitting data on Jupiter
1995	The American theoretical physicist Edward Witten and others develop M-theory
1996	United States patent issued for MP3
1997	July 4 - The Mars Pathfinder arrived on Mars and later began transmitting images
1998	MPEG-4 standard announced
2001	The Wilkinson Microwave Anisotropy Probe (WMAP), measured with unprecedented accuracy the temperature of the radiation left over from the Big Bang.
2003	NASA launched the largest-diameter infrared telescope ever in space, the Spitzer Space Telescope. H.264 HD video standard
2004	Cassini-Huygens-This joint NASA/ESA spacecraft, launched in 1997, reached Saturn. Spirit and Opportunity,

	the twin Mars Exploration Rovers, landed on opposite sides of the planet.
2008	NASA's Phoenix Mars Lander landed safely and began sending images home after a 10-month, 422-million-mile journey.
2009	The NASA spacecraft Kepler was launched. Its mission is to search for planets outside our solar system, in a distant area of the Milky Way
2011	The space shuttle Atlantis became the last American space shuttle to be launched into space. Mission STS-135 and its 4-member crew brought much-needed supplies and equipment to the International Space Station (ISS)
2012	NASA's Curiosity rover successfully landed on Mars. As large as a car, it carried an array of advanced new instruments and experiments
2017	Evidence of water found on Mars
2019	First picture of a black hole
2022	NASA unveils Space Launch System (SLS), a new moon rocket 15 percent more powerful than the Saturn V rocket from the Apollo era James Webb telescope active

http://timelineindex.com/content/select/
http://www.sciencetimeline.net/1961.htm

INTERNET VIDEOS, LECTURES AND WEBSITES

Scientific Discoveries

https://youtu.be/Bpid0LBTqWg

https://youtu.be/nNXv8kZS3JE

https://youtu.be/QGQq2aB3cWE

Scientific Timelines

https://youtu.be/AJxNjkp_Zkg

https://youtu.be/9KWEYgdqM7s

https://youtu.be/deTvb0wKh-A

https://youtu.be/0yBzxC9eoog

https://youtu.be/cL7ls9GfaFo

Islamic Science

https://youtu.be/iojISmRrTq0

https://youtu.be/k4_z_LVU900

Newton

https://youtu.be/iH48Lc7wq0U

https://youtu.be/YPRV1h3CGQk

Maxwell's Equations

https://youtu.be/9Tm2c6NJH4Y

https://youtu.be/ANIkxDm8bF4

https://youtu.be/kGia0ngFq0o

Einstein

https://youtu.be/O6t5AzGEyc0

https://youtu.be/K3QdmgEv7KE

https://youtu.be/1CYqBNgqJ1A

https://youtu.be/xvZfx7iwq94

Evolution:

https://youtu.be/K7tQIB4UdiY

https://youtu.be/K3QdmgEv7KE

https://youtu.be/AYBRbCLI4zU

https://youtu.be/pzmZjO1RDOQ

Earth

Astronomy:

Solar System

https://youtu.be/mtKNH2Y2OJM
https://youtu.be/Z-qoBocsU70
https://youtu.be/QfOF0bRBFJ4
https://youtu.be/QfOF0bRBFJ4
https://youtu.be/XSVjaklLf24
https://youtu.be/htOtW0pD92Y

Overview

https://youtu.be/2xKgXY8Epvg
https://youtu.be/rkXSLBQrNSE
https://youtu.be/l7KDLHCP_Ro
https://youtu.be/NItnlKWS16s
https://youtu.be/rkXSLBQrNSE
https://youtu.be/0yPMQ2HjjX4

The Force of Gravity

https://youtu.be/YCJ6gItzXBA
https://youtu.be/rKbJjzmFCco

Hubble Deep Field

https://youtu.be/vY9H7zAofmA
https://youtu.be/Q0Brz7Vd72c
https://youtu.be/loXDVGi_lK0

Life of a Star

https://youtu.be/6-fbpMwK-eA

Sizes of Stars

https://youtu.be/HEheh1BH34Q
https://youtu.be/B4dK_083LrA

Galaxy

https://youtu.be/UE8yHySiJ4A
https://youtu.be/ttz4Sr0tZFg
Earth
https://youtu.be/0VttgxfqEtw

Constellations

https://youtu.be/7oxoAAs9TUI
https://youtu.be/G3tFtyUnnlA
https://youtu.be/Hm2MKez7atI
https://youtu.be/LMTZXXFRDa8
https://youtu.be/61GFUf1y4mY

Math

Math and Music

https://youtu.be/Ocq8OR4OydQ

BIBLIOGRAPHY

1. Abbot, Edwin, *Flatland*, Princeton University Press, New Jersey, 2008
2. Abhayananda, S, *History of Mysticism,* Atma Books, Olympia, WA,1996
3. Barrow, John, D., and Tipler, Frank, J., *The Anthropic Cosmological Principle*, Oxford University Press, 1986
4. Bartusiak, Marcia, *Archives of the Universe,* Pantheon Books, New York, 2004
5. Bell, John L., *Time and Causation in Gödel's Universe*, Transcendent Philosophy 3, 2002
6. Berg, Richard and Stock, David, *The Physics of Sound,* Pearson Prentice Hall, Upper Saddle River NJ, 2005
7. Bernard Balleine (Contributor), Joe Staines (Editor), Avril Cridlan (Editor), QPB *Dictionary of Ideas*, Quality Paperback Book Club, New York, 1996
8. Born, Max, Einstein's *Theory of Relativity*, Dover Publications, New York, 1962
9. Bova, Ben, *The Story of Light,* Sourcebooks, Inc, Naperville IL 2001
10. Brockman, John, *This Explains Everything,* HaperCollins, New York, 2013
11. Brown, Michael H., *The Search for Eve,* Harper and Row, New York, 1990
12. Browne. Robert, T., *The Mystery of Space*, BilhlioLife, 2008
13. Burke, James, *The Pinball Effect,* Little Brown and Company, New York, 1996
14. Calder, Niger, *The Key to the Universe,* Penguin Books, 1977
15. Capek, Milic, *Philosophical Impact of Contemporary Physics,* D. Van Nostrand Company, Princeton, N.J. 1961
16. Capra, Fritjof and Steindl-Rast, *Belonging to the Universe,* HarperSanFrancisco, 1991
17. Capra, Fritjof, The *Tao of Physics*, Shambhala, Boston, 2000
18. Carrington, Richard, *A Million Years of Man,* The World Publishing Company, New York, 1963
19. Casi, John, L., *The Cambridge Quintet*, Helix Books, Reading, Massachustts,1998
20. Chopra, Deepak, *Life After Death,* Harmony Books, New York, 2006
21. Close, Frank, Marten, Michael, and Sutton, Christine, *The Particle Explosion*, Oxford University Press, 1987
22. Cole, K.C., *Sympathetic Vibrations*, Bantum Books, New York, 1984

23. Combs, Allan, *The Radiance of Being,* Paragon House, St Paul, MN, 1996
24. Coveney, Peter, Highfield, Roger, *The Arrow of Time,* Fawcett Columbine, New York, 1990
25. Dalai Lama, *The Universe in a Single Nutshell,* Morgan Road Books, New York, 2005
26. Davies, Paul, *The Cosmic Blueprint,* A Touchstone Book, New York, 1989
27. Davies, Paul, *The Last Three Minutes,* Basic Books, New York, 1994
28. Dyson. Freeman, *Imagined Worlds*, Harvard University Press, Cambridge, MA, 1997
29. Einstein, Albert and Infeld, Leopold, *The Evolution of Physics,* Simon and Schuster, 1938
30. Elmore, William, C., and Heald, Mark A., *Physics of Waves,* Dover Publications, Inc., New York,1969
31. Falk, Dan, *In Search of Time,* St. Martin's Press, New York,2008
32. Ferris, Timothy, *The World Treasure of Physics, Astronomy and Mathematics,* Little Brown and Company, New York, 1991
33. Feynman, Richard, P., *QED,* Princeton University Press, Princeton, NJ, 1985
34. Gell-Mann, *The Quark and the Jaguar,* W.H. Freeman, New York, 1994
35. Goswami, Amit, *The Self-Aware Universe*, Penguin Putman Inc., New York, 1995
36. Goswami, Amit, *The Visionary Window,* Quest Books, Wheaton, IL, 2000
37. Greene, Brain, *The Fabric of the Cosmos*, Alfred A. Knopf, New York, 2004
38. Greene, Brian, *The Elegant Universe*, W.W. Norton Company, New York,1999
39. Grof, Syanislav, *Ancient Wisdom and Modern Science,* State of New York Press, 1981
40. Guth, Alan, H., *The Inflationary Universe,* Helix Books, New York, 1997
41. Hakim, Joy, *The Story of Science*, Smithsonian Books, Washington DC, 2004
42. Hawking, Stephen, *Selections from the Principles of Relativity,* Running Press, Philadelphia, PA, 2002
43. Hawking, Stephen, *The Grand Design,* Bantam Books, New York, 2010
44. Hind, Rebecca, *1000 Faces of God,* Barnes abs Noble Books, New York, 2004
45. Hofstadter, Douglas, R., *Godel, Escher, Bach,* Vintage Books, New York, 1980
46. Hooper, Dan, *Dark Cosmos,* Smithsonian Books, New York, 2006
47. Hoyle, Fred, *The Intelligent Universe*, Holt, Rinehart and Winston, New York, 1983

48. Hoyle, Fred, *The Nature of the Universe,* Harper and Brother, New York, 1950
49. James, Jamie, *The Music of the Spheres,* Grove Press, 1993
50. Jeans, James, *Science & Music,* Dover Publications, New York,1968
51. Jones, George, Thaddeus, *Music Theory,* HarperResources, New York, 1974
52. Jones, Roger, S., *Physics As Metaphor,* New American Library, New York, 1982
53. Kakalios, James, *The Amazing Story of Quantum Mechanics,* Gotham Books, New York, 2010
54. Kanigel, Robert, *The Man Who Knew Infinity,* Charles Scribner's Sons, New York, 1991
55. Kaufman, William, J,. *Black Holes and Warped Spacetime,* W. H. Freeman and Company, San Francisco, CA, 1979
56. Kean, Sam, *The Disappearing Spoon*, Little Brown and Company, New York ,2018
57. Krasuss, Lawrence, M., *Hiding in the Mirror,* Viking, New York, 2005
58. Krauss, Lawrence, *Atom,* Little Brown Book and Company, New York, 2001
59. Krauss, Lawrence, *Hiding in the Mirror*, Viking Penguin Group, New York 2005
60. Lama, Dalai, *The Universe in a Single Atom:* The Convergence of Science and Spirituality, Morgan Road Books/Random House. 2005
61. Leaderman, Leon, M. and Schramm, David, N., *From Quarks to the Cosmos,* Scientific American Library, New York, 1995
62. Lederman, Leon, *The God Particle,* Houghton Mifflin Company, 1993
63. Livio, Mario, *The Accelerating Universe,* John Wiley and Son, Inc., New York, 2000
64. Loy, Gareth, *Musimathics*: The Mathematical Foundations of Music (Volume 1), The MIT Press, Cambridge, MA, 2011
65. Macdougall, Doug, *Nature's Clock*, University of California Press, Berkeley CA
66. Maddox, John, *What Remains to be Discovered,* Martin Kessler Books, New York 1998
67. Mermin, N, David, *It's About Time,* Princeton University Press, Princeton, NJ, 2005
68. National Academy of Science, Science, *Evolution and Creation,* The National Academies Press, Washington DC 2008
69. Osserman, Robert, *Poetry of the Universe,* Anchor Books, New York, 1995

70. Panek, Richard, *The 4% Universe,* Houghton Mifflin Harcourt, Boston, MA, 2011
71. Penrose, Roger, *The Emperors New Mind,* Oxford University Press, Oxford, 1994
72. Penrose, Roger, *The Road to Reality,* Alfred A. Knopf, New York, 2005
73. Penrose, Roger, *The Shadows of the Mind,* Oxford University Press, Oxford, 1994
74. Pierce, John, R., *The Science of Musical Sound,* Scientific American Library, New York, 1983
75. Plato, *The Republic,* The Heritage Press, New York, 1944
76. Rees, Martin, *Just Six Numbers,* Basic Books, New York, 1999
77. Reeves, Hubert, *The Hour of Our Delight,* W. H. Freeman Company, New York, 1991
78. Reichenbach, Hans, *From Copernicus to Einstein,* Dover Publications, Inc., New York, 1980
79. Ricard, Matthieu and Thuan, Xuan, Trinh, *The Quantum and the Lotus,* Crown Publishers, New York, 2001
80. Rifkin, Jeremy, *Entropy,* The Viking Press, New York, 1980
81. Segre, Gino, *Faust in Copenhagen,* Viking, New York, 2007
82. Singley, Richard, *A New Perspective,* Prometheus, Germantown, MD, 2020
83. Singley, Richard, *Negritude,* Prometheus, Germantown, MD, 2022
84. Singley, Richard, *Renaissance Man,* Prometheus, Germantown, MD, 2020
85. Smoot, George, *Wrinkles in Time,* Harper Perennial, New York, 2007
86. Straub, William, O., *Kaluza-Klein Theory,* Pasadena, CA, 2008
87. Stumpf, Samuel, Enoch, *Socrates to Sartre,* McGraw-Hill, New York, 1975
88. Stwertky, Albert, *Guide to the Elements,* Oxford University press, Oxford, 1998
89. Trefil, James, S., *SpaceTimeInfinity,* Pantheon Books New York, 1985
90. Trefil, James, S., *The Moment of Creation,* Charles Scribner's Sons, New York, 1983
91. Von Baeyer, Hans, Christian, *Taming the Atom,* Random House, New York, 1992
92. Wallace, David, Foster, *Everything and More,* W.W Norton and Company, New York, 2003
93. Weinberg, Steven, *Dreams of a Final Theory,* Peantheon, Books, New York, 1992
94. Weinberg, *The First Three Minutes,* Basic Books Inc., New York, 1977

95. Whitt, Terrence, *Our Undiscovered Universe*, Adrian Publishing Corporation, Melbourne, 2007
96. Wiber, Ken, *Quantum Questions,* Shambhala, Boston, 1985
97. Wilczek, Frank and Betsy Devine, *Longing for the Harmonies,* WW Norton & Company, New York, 1988
98. Wilczek, Frank, *The Lightness of Being,* Basic Books, New York, 2008
99. Yougrau, Paul, *A World Without Time,* Basic Books, 2005

Internet

100. http://plato.stanford.edu/entries/plotinus/
101. http://www.damtp.cam.ac.uk/research/gr/public/qg_ss.html
102. http://www.seventhstring.com/resources/notefrequencies.html
103. http://plus.maths.org/content/kaluza-klein-and-their-story-fifth-dimension
104. http://stars.astro.illinois.edu/sow/spectra.html
105. http://www.astrophysicsspectator.com/topics/stars/Fusion.html

About the Author

Richard Lawson Singley has worked in the Engineering field for over three decades. Richard spent many years employed as an Electrical Engineer, but never lost his passion for exploring the intricacies of the English language and historical events. He is the author of: *In a Nutshell, A New Perspective, Renaissance Man, Mystic Chords of Memory, Unholy Crusades, The Pursuit of Happiness, Triumvirate, Negritude, Interesting Times* and several forthcoming books and essays on various subjects including, religion, science, philosophy, history and politics. He has also worked as an educator and assistant in the Montgomery County Public School (MCPS) system. He is a graduate of Drexel University with a degree in Electrical Engineering. A native of Roselle, NJ; he is the father of two sons, Bradford and Adrian, and lives in the Washington, DC area with his life-long love and wife Jeanne.

Medium: https://medium.com/@richardsingley

Email: rich.singley5234@gmail.com

LinkedIn: https://www.linkedin.com/in/richard-singley-a7218332

Acknowledgments

The author is extremely grateful to the many family members and friends for giving him the fortitude to compile this book and to pursue a writing career. The author wishes to extend his acknowledgments to the following individuals: First and foremost, to my wife Jeanne Royal Singley of 40 plus years; whose patience and endearing support is written on each page in ink only visible to me. To my two sons: Bradford and Adrian, for putting up with their dad all these years and for making me so proud to be their father. To my loving parents, Lawson and Willie Mae Singley (both deceased); whose love, wisdom and guidance I will always cherish. To my uncles, particularly my Uncle Johnny (deceased) aunts, cousins and grandparents (Grandmother and Mommy Essie both deceased), many of them loss but not forgotten, whose perception is also compiled within the confines of this book.

TO my sisters and brothers: Edward Singley (deceased), Margret Griggs (deceased), Brenda Wright (deceased), Robert Singley (deceased), Gloria Riley, Mary Crump (deceased), Linda McDaniel, David Singley and especially my older brother Amin Rasheed (Roger Singley) who was always there when I needed him and to their progeny. I would like to extend my appreciation to my Sister-in-laws: Helen Rasheed and Kitty Singley (deceased), Brother-in-law Joseph McDaniel (deceased), and to the entire Singley family. Their support was not only vital to this book but to my life. Gratitude is also extended to my godparents, Sister Ruthie and Brother Chester, for their help in raising me. My appreciation is also extended to my wife's family, Lester McDaniel and Barbara Hughes and the McDaniel and Hughes siblings.

The number of individuals that helped me during my life is far too great to acknowledge without citing some and omitting others. However, there were some that had an enduring impact that warrant special recognition:

From my college days: Steve Throne and James G. Spady (deceased), whose enlightening conversation and mentorship opened my eyes to a world hitherto unseen. Special thanks to my dearest friends, Jesse Brown, James Spigner, James Grant, Willie Dash, Bill Hart, Dr. Erasmus Feltus (deceased), Jimmy Martin and Charles Cook (deceased) who mentored me during my first job. A special thanks to Lynn Worthy, Alex Tatum (deceased), Jesse Brown, Steve Throne and Professor Joseph Kennedy for their review of my work and their faith in my abilities. To Mabel Elizabeth Singletary for her help in getting this book published. To the members of Mount Calvary Church of God in Elizabeth, NJ that prayed for me when I was sick and for providing me with a religious background.

TO my team of doctors particularly: Dr. Frekko, Mrs. Frekko (deceased) and their daughters, Dr. Bass, Dr. Lin, Dr. Juarbe, Dr. Baek, Dr. Lieberman, Dr. Anthony Roberts and Dr. Melki. To the citizens of the town of Roselle, NJ (the first town in the world to have an electrical infrastructure). My appreciation is also extended to the following individuals: Mr. Frazier who kept on me about publishing my writing, Mrs. Roberts who gave me money for running errands as a child and families like the Sumners, Cooks, Simmons, etc., that formed a neighborhood that was the laboratory for my learning.

TO my enduring friends and former coworkers, particularly my former secretary Alice Baker, former managers, John Bigelow, Alan Dunn, Randy Cline and longtime coworkers that supported me in this and other efforts: Mary Lott, Dick Alban, Laveer Jovel, Steve Becraft, Verna Wright, Jim Rader, Dave Epler, Steve Frinak, Jim Hanlon, Paul Gantz, Steve Williams, Stu Benner, Tim Bresnan, Douglas Smith (deceased), Bob McKay, Bob Peters, Dick Roberts, Scott Pettygrove, Zach Stern, Dale Mann, Dale Dahlke, Jeff Shires, Barry Constantine, Bahman Salamat, Jeff Dobek, Jim Burck, John Murphy, Bill Smith, Greg Kneer, Terry Fogle, Mark Mowen, Kevin Graul, Gary Bell, Brendan Regan, Mark and Cathy Flinchbaugh, Melvin Pierce, Pat Howley, Lee Lusby, Ralph Lee, Lenny Jacobs, John Sill, Rob Strain, Frank Hornbuckle (deceased), Wyatt Bell, Ken Grish, Jerry Carbone, Daun Green, Don Muller, Ed Kozlosky, Gordon Whitney, Rod Miller, Richard Mugg, Krista Ochlech, Ben Burt, Mike Zeher, Al Hughes, Don Eenigenburg, Wynn Aung, Carl Slutsky, Tim Malac, Stu Mullan, Tim Burns, Joe Colburn, R.T. Stokes, Mike Janes, Mark Beatson, Carmela Young and the employees at Fairchild Space and Defense, Orbital Sciences, Smith Industries, General Electric and so many others that have offered support and encouragement during this endeavor. To the many workers at Friends of the Library Montgomery County, MD.

Thanks, is also given to the following friends whose discussions helped in this endeavor. Glenn Yanis, Gene Limb, Mark Edwards, Edward Fortney, Earl Spence, Earl Jenkins, Derrick Carr, Bristol Martin, Steve Cooper, Stephan Shin, Marcus (Doc) Coleman, Vince Coleman, Lenny Chornock, Dr. Gary Greenbaum (deceased) Robert White, Mike Slater, Joe Hasshen, Marcelo Flores, Stu Keeler, Grandville Smith, Earl Meadows, Tommy Arnold, Paul Kane, Kevin Kistler, Cameron Dryden, Reuben Smith, Calton Hall, Russ (40) Deshields, Greg Estep, Trenton Holmes, Cameron Simmons, Matt Fremin, Mike Wigal, Kevin Sinai, Heath Marshall, Kevin Correa, Leroy Gettinger, Mark Alan Hill, Kenny Jenkins, Lloyd McMillian, Jimmy McMillian, Steve Bender, Don Frost, Crispian Kirk, Chalk Dawson, Rashida Anderson, Kathy Amidon, Brian Hackett and Boyd Thomas. Gratitude is also extended to my fellow coworkers, staff and students at Rocky Hill Middle School, Clarksburg, Maryland. Specifically, Daphne Williams, Marion Martin, Olivia Reyes, Marc Cosby, Kimberly Wiltshire, Allen Ambush, the Montgomery County Public School (MCPS) staff at Clarksburg, Seneca Valley, Gaithersburg, Northwest and Watkins Mill High Schools; Martin Luther King Middle School and to Dr. Karl Reid of the National Society of Black Engineers (NSBE). Special thanks to Kristin Chikes.

TO the Washington Independent Review for helping to improve my writing skills and to Chris Matthews of MSNBC, David Maraniss Associate Editor of the Washington Post and Roger Williams of New England Publishing Associates, INC. for their comments on writing and the literary business. I also want to thank my classmates and teachers at Abraham Clark High School and Drexel University for providing me with intriguing dialogue and a good educational foundation. I would like to mention some of my Drexel colleagues specifically, Alton Knight, Colvin Bert, Gwen Evans, Derese (Sharon) Fisher, Alfred Taylor, Don Hinson, John Greene, David Adams (deceased) Mike Wright and Ray Gibson (deceased). Finally, I am grateful to my homeboys "the brothers": Lynn Worthy, Dr. Albert Ford, Kevin Williamson, Alex Tatum (deceased) and Willis Tatum. Together we proved that little black boys who dare to dream can become doctors, lawyers, historians and engineers. The journey has been long, but it has also been joyous.

Let me end by saying that I know there are some that deserve personal recognition that I have failed to mention. To them I offer my sincere apology.

Notes

Notes

Notes

Notes

Made in the USA
Monee, IL
04 November 2026

67383581R00252